Cracking the

AP

PHYSICS 1 EXAM

2018 Edition

By The Staff of The Princeton Review

PrincetonReview.com

Penguin
Random
House

The Princeton Review
555 W. 18th Street
New York, NY 10011
Email: editorialsupport@review.com

Published in the United States by Penguin Random House LLC, New York, and in Canada by Random House of Canada, a division of Penguin Random House Ltd., Toronto.

Terms of Service: The Princeton Review Online Companion Tools ("Student Tools") for retail books are available for only the two most recent editions of that book. Student Tools may be activated only twice per eligible book purchased for two consecutive 12-month periods, for a total of 24 months of access. Activation of Student Tools more than twice per book is in direct violation of these Terms of Service and may result in discontinuation of access to Student Tools Services.

ISBN: 978-1-5247-1011-8
eBook ISBN: 978-1-5247-1046-0
ISSN: 2374-5401

Editor: Selena Coppock
Production Editors: Liz Rutzel, Harmony Quiroz
Production Artist: Deborah A. Silvestrini

Printed in the United States of America on partially recycled paper.

10 9 8 7 6 5 4 3 2 1

2018 Edition

Editorial
Robert Franek, Editor-in-Chief
Casey Cornelius, VP Content Development
Mary Beth Garrick, Director of Production
Selena Coppock, Managing Editor
Meave Shelton, Senior Editor
Colleen Day, Editor
Sarah Litt, Editor
Aaron Riccio, Editor
Orion McBean, Associate Editor

Random House Publishing Team
Tom Russell, VP, Publisher
Alison Stoltzfus, Publishing Director
Jake Eldred, Associate Managing Editor
Ellen Reed, Production Manager
Suzanne Lee, Designer

Acknowledgments

The Princeton Review would like to give a very special thanks to Nick Owen, Jenkang Tao, Felicia Tam, for their work on revising existing practice tests and developing the three additional practice tests in the new Premium edition of this book. In addition, The Princeton Review would like to thank top notch physics teacher TJ Smolka for his content review and input on this edition.

Contents

Register Your

1 Go to **PrincetonReview.com/cracking**

2 You'll see a welcome page where you can register your book using the following ISBN: 9781524710118.

3 After placing this free order, you'll either be asked to log in or to answer a few simple questions in order to set up a new Princeton Review account.

4 Finally, click on the "Student Tools" tab located at the top of the screen. It may take an hour or two for your registration to go through, but after that, you're good to go.

If you have noticed potential content errors, please email EditorialSupport@review.com with the full title of the book, its ISBN number (located above), and the page number of the error.

Experiencing technical issues? Please email TPRStudentTech@review.com with the following information:

- your full name
- email address used to register the book
- full book title and ISBN
- your computer OS (Mac or PC) and Internet browser (Firefox, Safari, Chrome, etc.)
- description of technical issue

Book Online!

Once you've registered, you can...

- Find any late-breaking information released about the AP Physics 1 Exam

- Take a full-length practice SAT and ACT

- Get valuable advice about the college application process, including tips for writing a great essay and where to apply for financial aid

- Sort colleges by whatever you're looking for (such as Best Theater or Dorm), learn more about your top choices, and see how they all rank according to *The Best 382 Colleges*

- Access comprehensive study guides and a variety of printable resources, including bubble sheets for the practice tests in the book as well as important equations and formulas

- Check to see if there have been any corrections or updates to this edition

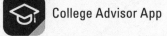

Look For These Icons Throughout The Book

Online Articles

Study Break

Proven Techniques

More Great Books

College Advisor App

Part I
Using This Book to Improve Your AP Score

- Preview: Your Knowledge, Your Expectations
- Your Guide to Using This Book
- How to Begin

PREVIEW: YOUR KNOWLEDGE, YOUR EXPECTATIONS

Your route to a high score on the AP Physics 1 Exam depends a lot on how you plan to use this book. Respond to the following questions.

1. Rate your level of confidence about your knowledge of the content tested by the AP Physics 1 Exam:

    A. Very confident—I know it all
    B. I'm pretty confident, but there are topics for which I could use help
    C. Not confident—I need quite a bit of support
    D. I'm not sure

2. If you have a goal score in mind, circle your goal score for the AP Physics 1 Exam:

    5   4   3   2   1   I'm not sure yet

3. What do you expect to learn from this book? Circle all that apply to you.

    A. A general overview of the test and what to expect
    B. Strategies for how to approach the test
    C. The content tested by this exam
    D. I'm not sure yet

YOUR GUIDE TO USING THIS BOOK

This book is organized to provide as much—or as little—support as you need, so you can use this book in whatever way will be most helpful for improving your score on the AP Physics 1 Exam.

*   The remainder of **Part I** will provide guidance on how to use this book and help you determine your strengths and weaknesses.

*   **Part II** of this book contains your first practice test, answers and explanations, and a scoring guide. (Bubble sheets can be found in the very back of the book for easy tear-out.) This is where you should begin your test preparation in order to realistically determine
    o   your starting point right now
    o   which question types you're ready for and which you might need to practice
    o   which content topics you are familiar with and which you will want to carefully review
    Once you have nailed down your strengths and weaknesses with regard to this exam, you can focus your preparation and be efficient with your time.

- **Part III** of this book will
  - o provide information about the structure, scoring, and content of the AP Physics 1 Exam
  - o help you to make a study plan
  - o point you toward additional resources

- **Part IV** of this book will explore the following strategies:
  - o how to attack multiple-choice questions
  - o how to approach free-response questions
  - o how to manage your time to maximize the number of points available to you

- **Part V** of this book covers the content you need to know for the AP Physics 1 Exam.

- **Part VI** of this book contains Practice Test 2, its answers and explanations, and a scoring guide. (Bubble sheets can be found in the very back of the book for easy tear-out.) If you skipped Practice Test 1, we recommend that you do both (with at least a day or two between them) so that you can compare your progress between the two. Additionally, this will help to identify any external issues: If you get a certain type of question wrong both times, you probably need to review it. If you only got it wrong once, you may have run out of time or been distracted by something. In either case, this will allow you to focus on the factors that caused the discrepancy in scores and to be as prepared as possible on the day of the test.

You may choose to use some parts of this book over others, or you may work through the entire book. This will depend on your needs and how much time you have.

HOW TO BEGIN

1. **Take a Test**

   Before you can decide how to use this book, you need to take a practice test. Doing so will give you insight into your strengths and weaknesses, and the test will also help you make an effective study plan. If you're feeling test-phobic, remind yourself that a practice test is a tool for diagnosing yourself—it's not how well you do that matters but how you use information gleaned from your performance to guide your preparation.

   So, before you read further, take AP Physics 1 Practice Test 1 starting at page 9 of this book. Be sure to do so in one sitting, following the instructions that appear before the test.

2. **Check Your Answers**

Using the answer key on page 30, count how many multiple-choice questions you got right and how many you missed. Don't worry about the explanations for now, and don't worry about why you missed questions. We'll get to that soon.

3. **Reflect on the Test**

After you take your first test, respond to the following questions:

- How much time did you spend on the multiple-choice questions?

- How much time did you spend on each free-response question?

- How many multiple-choice questions did you miss?

- Do you feel you had the conceptual understanding to address the subject matter of the free-response questions?

4. **Read Part III of this Book and Complete the Self-Evaluation**

As discussed in the Guide section, Part III will provide information on how the test is structured and scored. It will also set out areas of content that are tested.

As you read Part III, re-evaluate your answers to the questions in section 3. At the end of Part III, you will revisit and refine the answers. You will then be able to make a study plan, based on your needs and the time available, that will allow you to use this book most effectively.

5. **Engage with Parts IV and V as Needed**

Notice the word **engage**. You'll get more out of this book if you use it intentionally than if you read it passively, hoping for an improved score through osmosis.

Strategy chapters will help you think about your approach to the question types on this exam. Part IV opens with a reminder to think about how you approach questions now and then closes with a reflection section asking you to think about how/whether you will change your approach in the future.

Part V is designed to provide a review of the content tested on the AP Physics 1 Exam, including the level of detail you need to know and how the content is tested. You will have the opportunity to assess your mastery of the content of each chapter through test-appropriate questions and a reflection section.

6. **Take Test 2 and Assess Your Performance**

Once you feel you have developed the strategies you need and gained the knowledge you lacked, you should take Test 2, which begins at page 363 of this book. You should do so in one sitting, following the instructions at the beginning of the test.

When you are done, check your answers to the multiple-choice sections. Approach a teacher to read your essays and provide feedback.

Once you have taken the test, reflect on what areas you still need to improve upon, and revisit the respective chapters. Through this reflective and engaging approach, you will continue to improve.

7. **Keep Working**

After you have revisited certain chapters, continue the process of testing, reflection, and engaging with the next practice test in this book. Consider what additional work you need to do and how you will change your strategic approach to adapt to different parts of the test.

As we will discuss in Part III, there are other resources available to you, including a wealth of information on AP Central. You should continue to explore areas that can stand to improve right up to the day of the test.

Part II
Practice Test 1

Practice Test 1

AP® Physics 1 Exam

SECTION I: Multiple-Choice Questions

DO NOT OPEN THIS BOOKLET UNTIL YOU ARE TOLD TO DO SO.

At a Glance

Total Time
90 minutes
Number of Questions
50
Percent of Total Grade
50%
Writing Instrument
Pen required

Instructions

Section I of this examination contains 50 multiple-choice questions. Fill in only the ovals for numbers 1 through 50 on your answer sheet.

CALCULATORS MAY BE USED ON BOTH SECTIONS OF THE AP PHYSICS 1 EXAM.

Indicate all of your answers to the multiple-choice questions on the answer sheet. No credit will be given for anything written in this exam booklet, but you may use the booklet for notes or scratch work. Please note that there are two types of multiple-choice questions: single-select and multi-select questions. After you have decided which of the suggested answers is best, completely fill in the corresponding oval(s) on the answer sheet. For single-select, you must give only one answer; for multi-select you must give BOTH answers in order to earn credit. If you change an answer, be sure that the previous mark is erased completely. Here is a sample question and answer.

Sample Question   Sample Answer

Chicago is a
(A) state
(B) city
(C) country
(D) continent

Use your time effectively, working as quickly as you can without losing accuracy. Do not spend too much time on any one question. Go on to other questions and come back to the ones you have not answered if you have time. It is not expected that everyone will know the answers to all the multiple-choice questions.

About Guessing

Many candidates wonder whether or not to guess the answers to questions about which they are not certain. Multiple-choice scores are based on the number of questions answered correctly. Points are not deducted for incorrect answers, and no points are awarded for unanswered questions. Because points are not deducted for incorrect answers, you are encouraged to answer all multiple-choice questions. On any questions you do not know the answer to, you should eliminate as many choices as you can, and then select the best answer among the remaining choices.

GO ON TO THE NEXT PAGE.

ADVANCED PLACEMENT PHYSICS 1 TABLE OF INFORMATION

CONSTANTS AND CONVERSION FACTORS	
Proton mass, $m_p = 1.67 \times 10^{-27}$ kg	Electron charge magnitude, $e = 1.60 \times 10^{-19}$ C
Neutron mass, $m_n = 1.67 \times 10^{-27}$ kg	Coulomb's law constant, $k = 1/4\pi\varepsilon_0 = 9.0 \times 10^9$ N·m^2/C^2
Electron mass, $m_e = 9.11 \times 10^{-31}$ kg	Universal gravitational constant, $G = 6.67 \times 10^{-11}$ m^3/kg·s^2
Speed of light, $c = 3.00 \times 10^8$ m/s	Acceleration due to gravity at Earth's surface, $g = 9.8$ m/s^2

UNIT SYMBOLS	meter,	m	kelvin,	K	watt,	W	degree Celsius,	°C
	kilogram,	kg	hertz,	Hz	coulomb,	C		
	second,	s	newton,	N	volt,	V		
	ampere,	A	joule,	J	ohm,	Ω		

PREFIXES		
Factor	Prefix	Symbol
10^{12}	tera	T
10^{9}	giga	G
10^{6}	mega	M
10^{3}	kilo	k
10^{-2}	centi	c
10^{-3}	milli	m
10^{-6}	micro	μ
10^{-9}	nano	n
10^{-12}	pico	p

VALUES OF TRIGONOMETRIC FUNCTIONS FOR COMMON ANGLES							
θ	$0°$	$30°$	$37°$	$45°$	$53°$	$60°$	$90°$
$\sin\theta$	0	1/2	3/5	$\sqrt{2}/2$	4/5	$\sqrt{3}/2$	1
$\cos\theta$	1	$\sqrt{3}/2$	4/5	$\sqrt{2}/2$	3/5	1/2	0
$\tan\theta$	0	$\sqrt{3}/3$	3/4	1	4/3	$\sqrt{3}$	∞

The following conventions are used in this exam.
   I.   The frame of reference of any problem is assumed to be inertial unless otherwise stated.
   II. Assume air resistance is negligible unless otherwise stated.
   III. In all situations, positive work is defined as work done <u>on</u> a system.
   IV. The direction of current is conventional current: the direction in which positive charge would drift.
   V.   Assume all batteries and meters are ideal unless otherwise stated.

GO ON TO THE NEXT PAGE.

ADVANCED PLACEMENT PHYSICS 1 EQUATIONS, EFFECTIVE 2015

MECHANICS

$v_x = v_{x0} + a_x t$

$x = x_0 + v_{x0}t + \frac{1}{2}a_x t^2$

$v_x^2 = v_{x0}^2 + 2a_x(x - x_0)$

$\vec{a} = \dfrac{\sum \vec{F}}{m} = \dfrac{\vec{F}_{net}}{m}$

$\left|\vec{F}_f\right| \le \mu \left|\vec{F}_n\right|$

$a_c = \dfrac{v^2}{r}$

$\vec{p} = m\vec{v}$

$\Delta \vec{p} = \vec{F}\,\Delta t$

$K = \frac{1}{2}mv^2$

$\Delta E = W = F_{\parallel}d = Fd\cos\theta$

$P = \dfrac{\Delta E}{\Delta t}$

$\theta = \theta_0 + \omega_0 t + \frac{1}{2}\alpha t^2$

$\omega = \omega_0 + \alpha t$

$x = A\cos(2\pi ft)$

$\vec{\alpha} = \dfrac{\sum \vec{\tau}}{I} = \dfrac{\vec{\tau}_{net}}{I}$

$\tau = r_{\perp}F = rF\sin\theta$

$L = I\omega$

$\Delta L = \tau\,\Delta t$

$K = \frac{1}{2}I\omega^2$

$\left|\vec{F}_s\right| = k|\vec{x}|$

$U_s = \frac{1}{2}kx^2$

$\rho = \dfrac{m}{V}$

a = acceleration
A = amplitude
d = distance
E = energy
f = frequency
F = force
I = rotational inertia
K = kinetic energy
k = spring constant
L = angular momentum
ℓ = length
m = mass
P = power
p = momentum
r = radius or separation
T = period
t = time
U = potential energy
V = volume
v = speed
W = work done on a system
x = position
y = height
α = angular acceleration
μ = coefficient of friction
θ = angle
ρ = density
τ = torque
ω = angular speed

$\Delta U_g = mg\,\Delta y$

$T = \dfrac{2\pi}{\omega} = \dfrac{1}{f}$

$T_s = 2\pi\sqrt{\dfrac{m}{k}}$

$T_p = 2\pi\sqrt{\dfrac{\ell}{g}}$

$\left|\vec{F}_g\right| = G\dfrac{m_1 m_2}{r^2}$

$\vec{g} = \dfrac{\vec{F}_g}{m}$

$U_G = -\dfrac{Gm_1 m_2}{r}$

ELECTRICITY

$\left|\vec{F}_E\right| = k\left|\dfrac{q_1 q_2}{r^2}\right|$

$I = \dfrac{\Delta q}{\Delta t}$

$R = \dfrac{\rho\ell}{A}$

$I = \dfrac{\Delta V}{R}$

$P = I\,\Delta V$

$R_s = \sum_i R_i$

$\dfrac{1}{R_p} = \sum_i \dfrac{1}{R_i}$

A = area
F = force
I = current
ℓ = length
P = power
q = charge
R = resistance
r = separation
t = time
V = electric potential
ρ = resistivity

WAVES

$\lambda = \dfrac{v}{f}$

f = frequency
v = speed
λ = wavelength

GEOMETRY AND TRIGONOMETRY

Rectangle
$A = bh$

Triangle
$A = \frac{1}{2}bh$

Circle
$A = \pi r^2$
$C = 2\pi r$

Rectangular solid
$V = \ell wh$

Cylinder
$V = \pi r^2 \ell$
$S = 2\pi r\ell + 2\pi r^2$

Sphere
$V = \frac{4}{3}\pi r^3$
$S = 4\pi r^2$

A = area
C = circumference
V = volume
S = surface area
b = base
h = height
ℓ = length
w = width
r = radius

Right triangle
$c^2 = a^2 + b^2$

$\sin\theta = \dfrac{a}{c}$

$\cos\theta = \dfrac{b}{c}$

$\tan\theta = \dfrac{a}{b}$

GO ON TO THE NEXT PAGE.

THIS PAGE IS LEFT INTENTIONALLY BLANK.

GO ON TO THE NEXT PAGE.

AP PHYSICS 1

SECTION I

Note: To simplify calculations, you may use $g = 10$ m/s^2 in all problems.

Directions: Each of the questions or incomplete statements is followed by four suggested answers or completions. Select the one that is best in each case and then fill in the corresponding circle on the answer sheet.

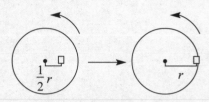

Top view

1. If a ball is rolling down an inclined plane without slipping, which force is responsible for causing its rotation?

   (A) Normal force
   (B) Gravity
   (C) Kinetic friction
   (D) Static friction

3. An object is resting on a platform that rotates at a constant speed. At first, it is a distance of half the platform's radius from the center. If the object is moved to the edge of the platform, what happens to the centripetal force that it experiences? Assume the platform continues rotating at the same speed.

   (A) Increases by a factor of 4
   (B) Increases by a factor of 2
   (C) Decreases by a factor of 2
   (D) Decreases by a factor of 4

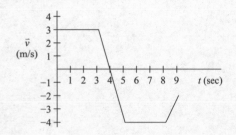

2. The graph above shows the velocity of an object as a function of time. What is the net displacement of the object over the time shown?

   (A) −30.5 m
   (B) −6.5 m
   (C) 6.5 m
   (D) 30.5 m

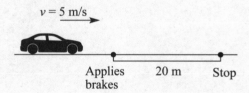

4. A car of mass 1000 kg is traveling at a speed of 5 m/s. The driver applies the breaks, generating a constant friction force, and skids for a distance of 20 m before coming to a complete stop. Given this information, what is the coefficient of friction between the car's tires and the ground?

   (A) 0.25
   (B) 0.2
   (C) 0.125
   (D) 0.0625

GO ON TO THE NEXT PAGE.

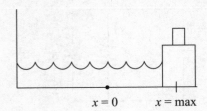

x = 0   x = max

Pre-Collision | Post-Collision

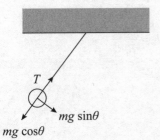

5. A spring-block system is oscillating without friction on a horizontal surface. If a second block of equal mass were placed on top of the original block at a time when the spring is at maximum compression, which of the following quantities would NOT be affected?

(A) Frequency
(B) Maximum speed
(C) Amplitude
(D) All of the above quantities would be affected.

7. Two balls collide as shown above. Given the final direction of the second ball's motion after the collision, which of the following is a possible direction for the first ball to move after the collision?

(A) ↖
(B) ←
(C) →
(D) ↘

6. A certain theme park ride involves people standing against the walls of a cylindrical room that rotates at a rapid pace, making them stick to the walls without needing support from the ground. Once the ride achieves its maximum speed, the floor drops out from under the riders, but the circular motion holds them in place. Which of the following factors could make this ride dangerous for some riders but not others?

(A) The mass of the individuals
(B) The coefficient of friction of their clothing in contact with the walls
(C) Both of the above
(D) None of the above

8. As a pendulum swings back and forth, it is affected by two forces: gravity and tension in the string. Splitting gravity into component vectors, as shown above, produces $mg\sin\theta$ (the restoring force) and $mg\cos\theta$. Which of the following correctly describes the relationship between the magnitudes of tension in the string an $mg\cos\theta$?

(A) Tension $> mg\cos\theta$
(B) Tension $= mg\cos\theta$
(C) Tension $< mg\cos\theta$
(D) The relationship depends on the position of the ball.

GO ON TO THE NEXT PAGE.

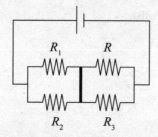

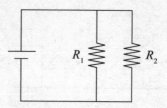

9. In order to ensure that no current will pass through the cross path (bold in the image above), what must the resistance of R be in terms of R_1, R_2, and R_3?

(A) $R = \dfrac{R_1 + R_3}{R_2}$

(B) $R = \dfrac{R_2}{R_1 + R_3}$

(C) $R = \dfrac{R_1 R_3}{R_2}$

(D) $R = \dfrac{R_2}{R_1 R_3}$

10. The circuit above has two resistors in parallel. The first, R_1, will have its resistance steadily increased. The second, R_2, will have a constant resistance of $R_2 = 4\ \Omega$. Which of the following graphs correctly depicts the total resistance of the circuit, R_T, as a function of R_1?

(A)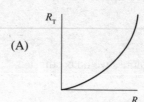

(B)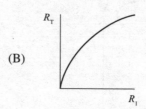

(C)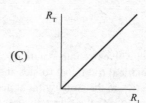

(D)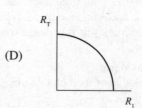

GO ON TO THE NEXT PAGE.

11. A piano tuner needs to double the frequency of note that a particular string is playing. Should he/she tighten or loosen the string, and by what factor?

    (A) Tighten, 4
    (B) Tighten, 2
    (C) Loosen, 2
    (D) Loosen, 4

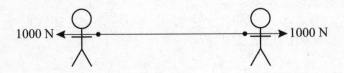

12. If two people pull with a force of 1000 N each on opposite ends of a rope and neither person moves, what is the magnitude of tension in the rope?

    (A) 0 N
    (B) 500 N
    (C) 1000 N
    (D) 2000 N

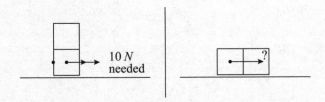

13. Two identical blocks are stacked on top of each other and placed on a table. To overcome the force of static friction, a force of 10 N is required. If the blocks were placed side by side and pushed as shown in the figure above, how much force would be required to move them?

    (A) $\dfrac{10\sqrt{2}}{2\,\text{N}}$

    (B) 10 N

    (C) $10\sqrt{2}$ N

    (D) 20 N

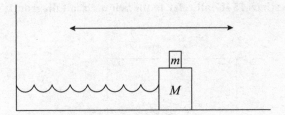

14. A block of known mass M is connected to a horizontal spring that is sliding along a flat, frictionless surface. There is an additional block of known mass m resting on top of the first block. Which of the following quantities would NOT be needed to determine if the top block will slide off the bottom block?

    (A) The maximum coefficient of static friction between the blocks
    (B) The amplitude of the system's motion
    (C) The spring constant
    (D) The average speed of the blocks

GO ON TO THE NEXT PAGE.

Questions 15–17 all refer to the below circuit diagram.

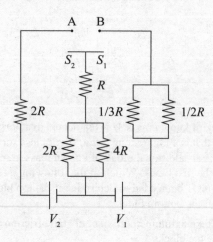

15. If switch S_1 is connected to point B but switch S_2 is left unconnected, what is the current through the resistor labeled R ?

    (A) $\dfrac{47V_1}{6R}$

    (B) $\dfrac{38V_1}{15R}$

    (C) $\dfrac{15V_1}{38R}$

    (D) $\dfrac{6V_1}{47R}$

16. If switch S_2 is connected to point A but switch S_1 is left unconnected, what is the current through the resistor labeled R ?

    (A) $\dfrac{9V_2}{R}$

    (B) $\dfrac{13V_2}{3R}$

    (C) $\dfrac{3V_2}{13R}$

    (D) $\dfrac{V_2}{9R}$

17. If both switches S_1 and S_2 are left in the unconnected positions, what is the current through the resistor labeled R ?

    (A) 0

    (B) $\dfrac{15(V_1 + V_2)}{68R}$

    (C) $\dfrac{15(V_1 + V_2)}{103R}$

    (D) $\dfrac{15(V_2 - V_1)}{68R}$

18. When a person runs to the right, as shown above, which of the following could be the direction of the force from the ground on his/her foot?

    (A)

    (B)

    (C)

    (D) None of these are possible.

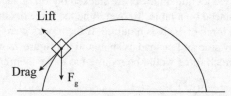

19. In real projectile motion, an object experiences three forces: gravity, drag, and lift. These are depicted in the picture above. Given this information, how would lift affect the speed of a projectile?

    (A) It would increase speed.
    (B) It would decrease speed.
    (C) Its effect would vary throughout the flight of the object.
    (D) It would have no effect.

GO ON TO THE NEXT PAGE.

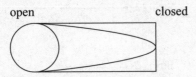

open   closed

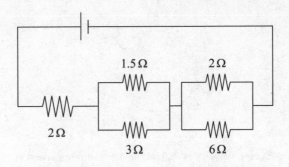

20. A standing wave is formed in a tube, as shown above. If this wave has a frequency of f, what would be the frequency of the next harmonic that can be formed in this tube?

(A) $\frac{1}{2}f$

(B) $2f$

(C) $3f$

(D) The above wave shows the highest possible harmonic frequency for this system.

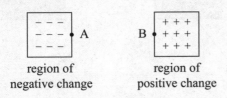

region of   region of
negative change   positive change

21. Which of the following correctly describes an electron moving from point A to point B in the situation above? Assume the two regions of charge are identical in magnitude and only different in sign.

(A) The electron moves with increasing speed and increasing acceleration and loses potential energy.

(B) The electron moves with increasing speed and constant acceleration and loses potential energy.

(C) The electron moves with increasing speed and increasing acceleration and gains potential energy.

(D) The electron moves with decreasing speed and decreasing acceleration and gains potential energy.

22. Which of the following circuits would have an equivalent resistance equal to that of the circuit depicted above?

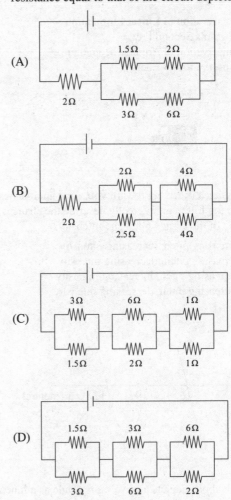

GO ON TO THE NEXT PAGE.

String 1 •————————•
                2L

String 2 •————•
              L

23. Escape velocity is defined as the minimum speed at which an object must be launched to "break free" from a massive body's gravitational pull. Which of the following principles could be used to derive this speed for a given planet?

(A) Conservation of Linear Momentum
(B) Newton's Second Law
(C) Conservation of Angular Momentum
(D) Conservation of Energy

26. Both of the above strings have their ends locked in place. The first string, S_1, is twice as long as the second string, S_2. If sound waves are going to be sent through both, what is the correct ratio of the fundamental frequency of S_1 to the fundamental frequency of S_2?

(A) 2:1
(B) $\sqrt{2}$:1
(C) 1:$\sqrt{2}$
(D) 1:2

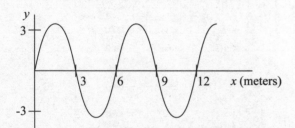

24. An ambulance is driving toward you. As it approaches, which of the following correctly describe the changes in the sound of the siren's pitch and intensity?

(A) Increasing pitch, increasing intensity
(B) Increasing pitch, decreasing intensity
(C) Decreasing pitch, increasing intensity
(D) Decreasing pitch, decreasing intensity

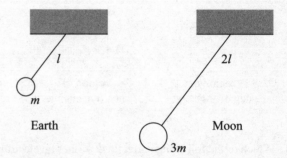

Earth   Moon

27. A pendulum with a ball of mass m hanging from a string of length l is set in motion on Earth, and the system is found to have a frequency of f. If the length of the string were doubled, the hanging mass tripled, and the system moved to the moon, what would be the new frequency? NOTE: Acceleration due to gravity of the Moon is approximately $\frac{1}{6}$ of Earth's.

(A) $\frac{1}{12}f$

(B) $\sqrt{\frac{1}{12}}f$

(C) $\sqrt{12}f$

(D) $12f$

25. The graph above depicts a wave's amplitude as a function of distance. If the wave has a speed of 600 m/s, which off the following is the best approximation of the wave's frequency?

(A) 50 Hz
(B) 100 Hz
(C) 200 Hz
(D) Cannot be determined without additional information

GO ON TO THE NEXT PAGE.

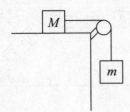

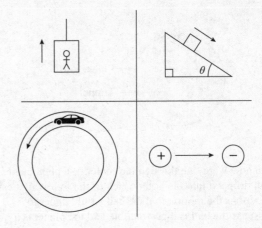

28. A block of mass M is at rest on a table. It is connected by a string and pulley system to a block of mass m hanging off the edge of the table. Assume the hanging mass is heavy enough to make the resting block move. If the acceleration of the system and the masses of the blocks are known, which of the following could NOT be calculated?

    (A) Net force on each block
    (B) Tension in the string
    (C) Coefficient of kinetic friction between the table and the block of mass M
    (D) The speed of the block of mass M when it reaches the edge of the table

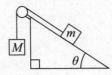

29. A block of mass m is connected by a string which runs over a frictionless pulley to a heavier block of mass M. The smaller block rests on an inclined plane of angle θ, and the larger block hangs over the edge, as shown above. In order to prevent the blocks from moving, the coefficient of static friction must be

    (A) $\dfrac{mg\sin\theta}{Mg - mg\cos\theta}$

    (B) $\dfrac{Mg - mg\sin\theta}{Mg\cos\theta}$

    (C) $\dfrac{Mg - mg\sin\theta}{mg\cos\theta}$

    (D) $\dfrac{Mg - mg\cos\theta}{mg\sin\theta}$

30. Which of the following forces does not do work in its given situation?

    (A) Normal force as a person goes up in an elevator
    (B) Frictional force as a box slides down a ramp
    (C) Centripetal force as a car drives around a circular track
    (D) Electrical force as a positive charge moves toward a negative charge

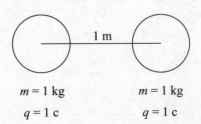

$m = 1$ kg $m = 1$ kg
$q = 1$ c $q = 1$ c

31. Two identical spheres of mass 1 kg are placed 1 m apart from each other. Each sphere pulls on the other with a gravitational force, F_g. If each sphere also holds 1 C of positive charge, then the magnitude of the resulting repulsive electric force is

    (A) $1.82 \times 10^{40} F_g$
    (B) $1.35 \times 10^{20} F_g$
    (C) $7.42 \times 10^{-21} F_g$
    (D) $5.50 \times 10^{-41} F_g$

GO ON TO THE NEXT PAGE.

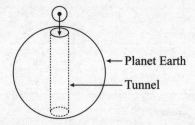

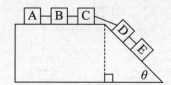

32. If a hole were dug through the center of a planet and a ball dropped into the hole, which of the following best describes the motion that the ball would undergo? Assume the ball is indestructible and the planet is a perfect sphere.

   (A) It would continuously gain speed and eventually escape the gravitational pull of the planet.
   (B) It would fall to the center of the planet and get stuck there because gravity is always pulling things toward the center of the planet.
   (C) It would fall to the other end of the hole, come to a momentary stop, fall back to the starting location, and then repeat this back-and-forth motion indefinitely.
   (D) None of the above is correct.

34. Five boxes are linked together, as shown above. If both the flat and slanted portions of the surface are frictionless, what will be the acceleration of the box marked B ?

   (A) $\dfrac{1}{5} g\sin\theta$

   (B) $\dfrac{2}{5} g\sin\theta$

   (C) $\dfrac{5}{7} g\sin\theta$

   (D) $\dfrac{2}{5} g\cos\theta$

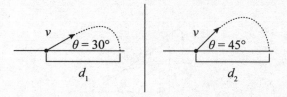

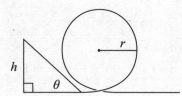

33. An empty mine car of mass m starts at rest at the top of a hill of height h above the ground, then rolls down the hill and into a semicircular banked turn. Ignoring rolling friction so that the only forces acting on the mine car are the normal force from the track and gravity, what is the magnitude of centripetal force on the car as it rounds the banked curve?

   (A) mgh
   (B) $2mgh/r$
   (C) mgh/r
   (D) $mgh/(2r)$

35. If a ball is kicked at an angle of 30 degrees such that it has an initial velocity v, it will travel some distance, d_1, before falling back to the ground. Another ball is kicked at an angle of 45 degrees so that it also has an initial velocity of v, and it travels a distance, d_2, before falling back to the ground. How much farther will the second ball travel before striking the ground?

   (A) $\dfrac{v^2}{10} (2 - \sqrt{2})$

   (B) $\dfrac{v^2}{10} (2 - \sqrt{3})$

   (C) $\dfrac{v^2}{20} (2 - \sqrt{2})$

   (D) $\dfrac{v^2}{20} (2 - \sqrt{3})$

GO ON TO THE NEXT PAGE.

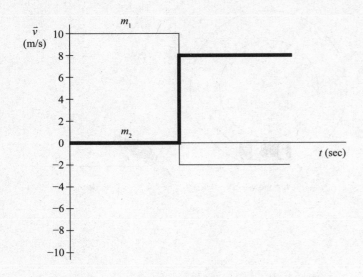

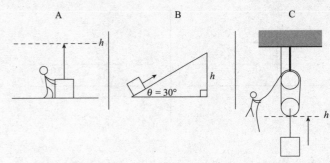

Questions 36–37 refer to the graph above.

36. Two objects of masses m_1 and m_2 undergo a collision. The graph above shows their velocities with respect to time both before and after the collision. If $m_1 = 10$ kg, then m_2 must be

   (A) 5 kg
   (B) 10 kg
   (C) 15 kg
   (D) 20 kg

37. If the two objects have masses of $m_1 = 4$ kg and $m_2 = 6$ kg, what type of collision does the graph represent?

   (A) Perfectly elastic
   (B) Perfectly inelastic
   (C) Neither of the above
   (D) Cannot be determined

38. The diagrams above show a box of mass m being lifted from the ground up to a height of h via three different methods. In situation A, the box is simply lifted by a person. In B, it is pushed up a ramp with an incline angle of 30 degrees. In C, it is lifted by a pulley system. Assuming ideal conditions (no friction) for all of these situations, which of the following correctly ranks the amount of work required to lift the box in each case?

   (A) A > B > C
   (B) A > B = C
   (C) C > B > A
   (D) A = B = C

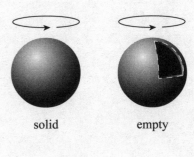

solid   empty

$$m_1 = m_2$$

$$\tau_1 = \tau_2$$

39. Two spheres of equal size and equal mass are rotated with an equal amount of torque. One of the spheres is solid with its mass evenly distributed throughout its volume, and the other is hollow with all of its mass concentrated at the edges. Which sphere would rotate faster?

   (A) Solid sphere
   (B) Hollow sphere
   (C) They would rotate at equal rates.
   (D) Additional information is required to determine the relative rates of rotation.

GO ON TO THE NEXT PAGE.

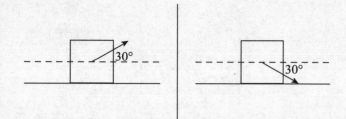

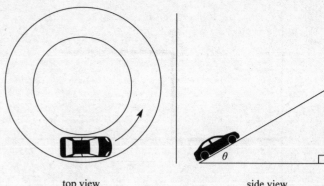

top view   side view

40. Pulling a block of mass *m* to the right by a connected string at an angle of 30 degrees above the horizontal (as shown in the left picture) with a force equal to the block's weight produces a friction force *f*. If the same block were to be pulled at an angle of 30 degrees beneath the horizontal (as shown in the right picture), what would be the friction force? Assume that the applied force is enough to make the block move in both cases.

(A) 3*f*

(B) 2*f*

(C) *f*/2

(D) *f*/3

41. A car is driving in a circle of radius *r* at a constant speed *v*. At what angle must the road be banked in order to prevent sliding even if the road has no friction with the tires?

(A) $\tan^{-1}\dfrac{v^2}{rg}$

(B) $\tan^{-1}\dfrac{rg}{v^2}$

(C) $\cot^{-1}\dfrac{v^2}{rg}$

(D) $\cot^{-1}\dfrac{rg}{v^2}$

GO ON TO THE NEXT PAGE.

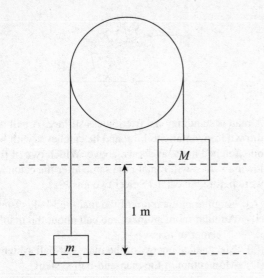

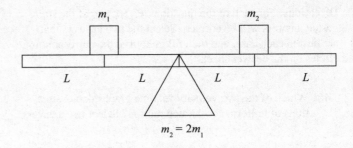

$m_2 = 2m_1$

42. Assuming the pulley above is both frictionless and mass-less, how long will it take for the two blocks to reach an equal height if block M of mass 15 kg starts 1 m above block m of mass 5 kg?

(A) $\sqrt{\dfrac{2}{5}}$ s

(B) $\sqrt{\dfrac{1}{5}}$ s

(C) $\sqrt{\dfrac{1}{20}}$ s

(D) $\dfrac{1}{5}$ s

44. As it is, the system above is not balanced. Which of the following changes would NOT balance the system so that there is 0 net torque? Assume the plank has no mass of its own.

(A) Adding a mass equal to m_2 on the far left side and a mass equal to m_1 and on the far right side
(B) Stacking both masses directly on top of the fulcrum
(C) Moving the fulcrum a distance L/3 to the right
(D) Moving both masses a distance L/3 to the left

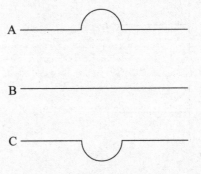

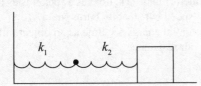

43. The spring-block system above has a block of mass m oscillating in simple harmonic motion. However, instead of one spring, there are two springs connected to each other. The first spring has a spring constant k_1, and the second has a spring constant k_2. What would be the effective spring constant of the system above if $k_2 = 3k_1$?

(A) $4k_1$
(B) $3k_1$
(C) $4k_1/3$
(D) $3k_1/4$

45. Three identical balls are rolled from left to right across the three tracks above with the same initial speed. Assuming the tracks all have negligible friction and the balls have enough initial speed to reach the ends of each track, which set correctly orders the average speed of the balls on the three tracks?

(A) $A = B = C$
(B) $B > C > A$
(C) $C > B > A$
(D) Cannot be determined

GO ON TO THE NEXT PAGE.

Directions: For each of the questions 46-50, <u>two</u> of the suggested answers will be correct. Select the two answers that are best in each case, and then fill in both of the corresponding circles on the answer sheet.

46. Which of the two velocity vs. time graphs depict situations of uniform accelerated motion? Select two answers.

(A)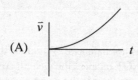

(B)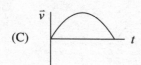

(C)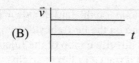

(D)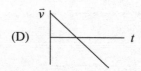

47. A man is standing on a frictionless surface. A ball is thrown horizontally to him, and he catches it with his outstretched hand, as shown above. Which two of the following values will remain the same after the catch as they were before the catch? Select two answers.

(A) Angular momentum of the man-and-ball system
(B) Angular momentum of the ball about the man's center of mass
(C) Mechanical energy of the man-and-ball system
(D) Momentum of the man-and-ball system

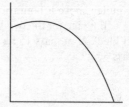

48. The graph above could be a representation of which two of the following situations? Select two answers.

(A) Vertical displacement vs. time as an object falls with no air resistance
(B) Kinetic energy vs. time as an object falls with no air resistance
(C) Kinetic energy vs. time as a rocket flies at constant speed but steadily burns fuel
(D) Potential energy vs. time as an object falls with no air resistance

GO ON TO THE NEXT PAGE.

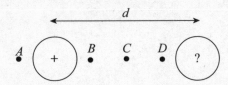

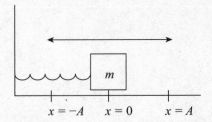

49. Two spheres are separated by a distance d. The first sphere has a known positive charge. The second sphere has a charge of a magnitude greater than the magnitude of the charge on the first sphere but unknown sign. Given this information, at which two of the above locations could a positive test charge potentially experience 0 net electrical force? Select two answers.

    (A) A
    (B) B
    (C) C
    (D) D

50. A spring-block system with a block of known mass m is oscillating under ideal conditions. Which two of the following pieces of additional sets of information would allow you to calculate the amplitude of the block's motion? Select two answers.

    (A) The speed of the block at maximum displacement from equilibrium and the maximum potential energy of the system
    (B) The spring constant and the period
    (C) The period and the maximum kinetic energy of the system
    (D) The maximum kinetic energy of the system and the spring constant

END OF SECTION I

DO NOT CONTINUE UNTIL INSTRUCTED TO DO SO.

AP PHYSICS 1

SECTION II

Free-Response Questions

Time—90 minutes

Percent of total grade—50

<u>General Instructions</u>

Use a separate piece of paper to answer these questions. Show your work. Be sure to write CLEARLY and LEGIBLY. If you make an error, you may save time by crossing it out rather than trying to erase it.

GO ON TO THE NEXT PAGE.

AP PHYSICS 1

SECTION II

Directions: Questions 1, 2, and 3 are short free-response questions that require about 13 minutes to answer and are worth 8 points. Questions 4 and 5 are long free-response questions that require about 25 minutes each to answer and are worth 13 points each. Show your work for each part in the space provided after that part.

1. A car of known mass m_1 will collide with a second car of known mass m_2. The collision will be head on, and both cars will only move linearly both before and after the collision. In a clear, coherent, paragraph-length response, explain a method for determining whether the collision is perfectly elastic, perfectly inelastic, or neither. If the collision is perfectly inelastic, include at least one possible cause of energy loss.

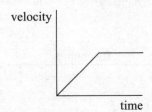

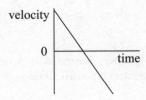

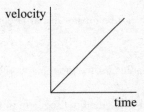

2. The three graphs above depict the velocity of different objects as functions of time. Given that information, draw potential displacement vs. time and acceleration vs. time graphs for each of them. Additionally, give an example of a situation that could produce each of the three graphs above.

GO ON TO THE NEXT PAGE.

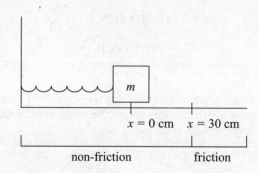

3. In the diagram above, a spring-block system is oscillating on a flat horizontal surface. Part of the surface is frictionless, and part of the surface is not frictionless. The block starts at rest at position $x = 0$ cm. The block is then pushed to the left, compressing the spring, until it reaches the position $x = -50$ cm. After being pushed in, the block is released and allowed to move naturally.

   (a)   Draw a graph of the system's position as a function of time starting from the moment it is released and explain your reasoning behind the graph you draw. Be sure to label important values on the graph.

   (b)   Given that the mass $m = 2$ kg and the spring constant $k = 100$ N/m, what is the magnitude of work done by the non-frictionless surface?

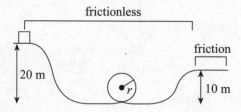

4. A roller coaster cart rides along the track shown above. The initial drop, the loop, and the ascending ramp are all frictionless. The flat portion of the track after the ascending ramp is not frictionless.

   (a)   How fast would the cart be moving just before it enters the loop?

   (b)   Would the normal force on the cart be greater just after entering the loop or at the peak of the loop? Explain why using relevant equations.

   (c)   What is the greatest possible radius for the loop that would allow the cart to still make it through?

   (d)   If the coefficient of static friction for the final segment of the track is 0.2, how long does the segment need to be in order to allow the cart to come to a complete stop due to friction alone?

GO ON TO THE NEXT PAGE.

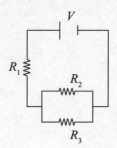

5. The above circuit contains a battery of voltage V and three resistors with resistances R_1, R_2, and R_3, respectively. As part of an experiment, a student has been given two measuring devices: a voltmeter and an ammeter. The first can be used to measure the changes in voltage of a circuit. The second can be used to measure the current flowing through a particular segment of wire. For answering the questions below, a voltmeter and ammeter look like ———(V)——— and ———(A)———, respectively, when drawn in a circuit diagram.

    (a)   In terms of the known variables, what is the voltage lost in passing through the first resistor?

    (b)   Draw a diagram showing how you would integrate the voltmeter to measure the voltage lost in the resistor labeled R_1. Explain the reasoning behind your decision.

    (c)   Draw a diagram showing how you would integrate the ammeter to measure the current passing through the resistor labeled R_1. Explain the reasoning behind your decision.

    (d)   What would be the ideal resistances for each device to have? Explain why each would be ideal for that device.

STOP

Practice Test 1:
Answers and
Explanations

ANSWER KEY AP PHYSICS 1 TEST 1

1.	D	26.	D
2.	B	27.	B
3.	B	28.	D
4.	D	29.	D
5.	C	30.	C
6.	B	31.	B
7.	D	32.	C
8.	D	33.	B
9.	C	34.	B
10.	B	35.	D
11.	A	36.	C
12.	C	37.	A
13.	B	38.	D
14.	D	39.	A
15.	C	40.	A
16.	C	41.	A
17.	A	42.	B
18.	A	43.	D
19.	D	44.	A
20.	C	45.	C
21.	B	46.	B, D
22.	D	47.	A, D
23.	D	48.	A, D
24.	A	49.	A, B
25.	B	50.	C, D

SECTION I

1. **D**   As a ball rolls down the hill, the four forces affecting it are as shown below.

     Gravity affects the center of mass of an object, so it cannot cause rotation. Normal force also cannot cause rotation in this case because of its direction. Finally, the friction force must be static because the question specifies that there is no slipping.

2. **B**   In a velocity vs. time graph, the displacement of an object is the area between the curve and the x-axis. In this case, the area can be broken into multiple rectangles and triangles.

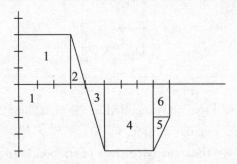

     In order of the number labels, each shape has an area of 9, 1.5, 2, 12, 1, and 2, respectively. The last four, however, are under the x-axis, so they must be treated as negative values. Therefore, displacement is $9 + 1.5 - 2 - 12 - 1 - 2 = -6.5$ m.

3. **B**   The formula for centripetal force is $F_c = mv^2/r$, which initially seems to indicate the force is inversely proportional to the radius. However, in the case of circular motion, an object's linear speed is $v = \omega r$. Substituting this value into the equation gives $F_c = m(\omega r)^2/r = m\omega^2 r$. So it turns out that the force is directly proportional to r, which means doubling the radius will double the force on the object.

4. **D**   First, you know $F_f = \mu F_N = \mu(mg)$. Second, you know that $W = \Delta KE = KE_f - KE_0 = -KE_0$ (since $KE_f = 0$ in this case) $= -\frac{1}{2}mv_0^2$. Plugging in numbers, you get $W = -\frac{1}{2}(1000 \text{ kg})(5 \text{ m/s})^2 = -12,500$ J. You also know that $W = Fd\cos\theta = [\mu(mg)]d\cos\theta$. Solving for μ, you get $\mu = W/[(mg)d\cos\theta]$. Plugging in the numbers then gives $\mu = -12,500 \text{ J}/[(1000 \text{ kg})(10 \text{ m/s}^2)(20 \text{ m})(\cos 180°) = 0.0625$.

5. **C**   The frequency of spring-block system is $f = \frac{1}{2\pi}\sqrt{\dfrac{k}{m}}$, so it would be affected by the change in mass. Furthermore, when a spring is at maximum compression or extension, all of its energy is potential

energy, which is given by $PE = \frac{1}{2}kx^2$. In adding this block, none of the relevant values are changed, so the spring will still extend to the same length, which means the amplitude is unchanged. Finally, maximum speed will be limited by the maximum KE of the system (which will be unchanged since the maximum PE was unchanged). $KE = \frac{1}{2}mv^2$, so the increased mass would have to be balanced by a decrease in speed to leave the KE unaltered.

6. **B** The diagram below shows the forces involved.

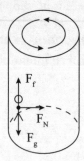

In this case, the person will remain suspended in the air as long $F_f = F_g$. Furthermore, because this is an example of uniform circular motion, we know $F_N = F_C = mv^2/r$. So we can rewrite the first equation as $\mu(mv^2/r) = mg$. Thus, the coefficient of friction is an important factor, but the mass of the person is not since it exists on both sides of the equation and will cancel out.

7. **D** In any collision, the total momentum of the system must be conserved. Prior to the collision, we see that all of the momentum horizontal. Therefore, the net momentum after the collision must also be purely horizontal. If one ball is moving up after the collision, the other must be moving down to cancel out that vertical momentum.

8. **D** Whenever the ball is in motion, it will be experiencing circular motion, which means there must a centripetal force. Centripetal force is a net force toward the center, so that means tension must be the greater force whenever the ball is moving. However, the ball is not always moving. At the two extreme edges of its motion, the ball is motionless for an instant as it changes directions. At those times, there is no net force, meaning the two given forces are equal.

9. **C** In a circuit with multiple paths, the current distributes itself inversely proportional to the resistance along each path. In order for there to be no current passing through the indicated segment of wire, those proportions must be the same before and after the cross path. Thus, $\frac{R_1}{R_2} = \frac{R}{R_3}$, which means $R = \frac{R_1 R_3}{R_2}$.

10. **B** For resistors in parallel, you know $\frac{1}{R_T} = \frac{1}{R_1} + \frac{1}{R_2} = \frac{R_2 + R_1}{R_1 R_2}$, which means $R_T = \frac{R_1 R_2}{R_1 + R_2}$. You know $R_2 = 4\ \Omega$, so plug that in to get $R_T = \frac{4R_1}{R_1 + 4}$. The graph of this equation looks like (B). It will always be increasing, but it will increase more and more slowly.

11. **A** The frequency of a wave traveling through a string is $f = \sqrt{\dfrac{T}{\dfrac{m}{1}}}$. Because tension is in the numerator, tightening will increase the frequency. This eliminates (C) and (D). Furthermore, because frequency depends on the square root of tension, the tension must be multiplied by 4 to double the frequency.

12. **C** Just focus on one end of the rope. Nothing is moving, so the net force must be 0. If a person pulls with a force of 1000 N and nothing moves, then the resulting tension must also be 1000 N. Additionally, tension is a constant magnitude throughout a string. The direction of the tension can change (as in the case of a pulley system), but the magnitude will remain the same.

13. **B** The force of static friction will be $F_f = \mu F_N = \mu(mg)$. Changing the arrangement of the blocks does not change any of these three quantities, so the force will remain the same. Thus, 10 N will again be required to move them.

14. **D** In this case, the blocks will remain stuck together as long as $F_F = F_{Spring}$ at the point of maximum displacement. In this case, that becomes $\mu F_N = kA$. And $F_N = (m + M)g$, so the final equation for this situation is $\mu[(m + M)g] = kA$. The only one of the listed values we don't need is the average speed.

15. **C** In this case, only the right hand loop would be connected. First, find the equivalent resistance for the right-hand loop. The total resistance of parallel resistors on the far right, $R_{eq,R}$, can be combined by doing $\frac{1}{R_{eq,R}} = \frac{1}{R_1} + \frac{1}{R_2}$. In this case, that gives $\frac{1}{R_{eq,R}} = \frac{1}{\frac{1}{2}R} + \frac{1}{\frac{1}{3}R} = \frac{5}{R}$, so $R_{eq,R} = \frac{R}{5}$.

Next, find the total resistance for the parallel resistors in the center, $R_{eq,C}$, in the same way. In this case, that gives $\frac{1}{R_{eq,C}} = \frac{1}{2R} + \frac{1}{4R} = \frac{3}{4R}$, so $R_{eq,C} = \frac{4R}{3}$.

Finally, these two parallel sets are in series with each other and with the resistor R, so the equivalent resistance for the whole loop, R_{eq}, would be $R_{eq} = R + \frac{R}{5} + \frac{4R}{3} = \frac{38R}{15}$. From Ohm's Law, we know $I = V/R$, so this gives $I = \frac{V_1}{\frac{38R}{15}} = \frac{15V_1}{38R}$.

16. **C** This is similar to the previous question except you need to include the far left resistor instead of the parallel set on the far right. Thus, $R_{eq} = 2R + \dfrac{4R}{3} + R = \dfrac{13R}{3}$. Using Ohm's Law then gives $I = V/R = \dfrac{V_2}{\dfrac{13R}{3}} = \dfrac{3V_2}{13R}$.

17. **A** If neither switch is connected, then no loop can be made, which means no current will flow.

18. **A** The forces on the person's foot are shown in the diagram below:

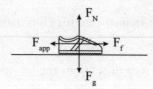

Remember that friction always opposes motion or intended motion. Imagine the ground had no friction (like a sheet of ice). In that case, the person's foot would simply slide to the left. Thus, friction must be opposing the intended motion and point to the right.

Of these forces, both the normal force and the friction force are coming from the ground, so the sum of those vectors is the correct answer.

19. **D** Because the lift force is perpendicular to the velocity of the object, lift can have no effect on the magnitude of velocity. The magnitude of velocity is also known as speed, so (D) is correct.

20. **C** Remember that in a system with one closed end and one open end there will be a node at one end and an antinode at the other. From the drawing, you can see that the diagram shows a wave with the greatest possible wavelength in this system. If that is the greatest possible wavelength, then it has the smallest possible, or fundamental, frequency. This type of system cannot make even harmonics, so the next harmonic would be the third, which will be $3f$.

21. **B** In the diagram, the electron, which has a negative charge, will move from a region of concentrated negative charge to a location of concentrated positive charge. Moving from negative to positive is something a negative charge will do naturally, and natural motion always causes a loss of potential energy. This eliminates (C) and (D).

Acceleration in this case will be constant. At the beginning of its movement, the electron will be primarily pushed away by the negative charges and only gently pulled on by the positive charges. As it travels, the balance of these two forces will trade off, meaning that as the electron approaches point B, it will be primarily pulled by the positive charge and only gently pushed by the negative charge. The total force on the electron, however, will be unchanged throughout, so the acceleration will be constant.

22. **D** First, find the total resistance of the original circuit. Remember that parallel resistors require you to use the reciprocal sum, while series resistors simply add. For example, the first set of parallel resistors would be combined by doing $\frac{1}{R} = \frac{1}{1.5} + \frac{1}{3} = 1$, which means $R = 1\ \Omega$. For the second set of parallel resistors, you would get $\frac{1}{R} = \frac{1}{2} + \frac{1}{6} = \frac{4}{6}$, which means $R = 1.5\ \Omega$. Adding those two values to the lone series resistor gives $R_{Total} = 2 + 1 + 1.5 = 4.5\ \Omega$.

   The only answer that matches this value is (D). The first and last set of parallel resistors match the two sets in the original, which means the last parallel set must match the $2\ \Omega$ resistor in the original. Solving for that middle parallel set's combined resistance gives $\frac{1}{R} = \frac{1}{3} + \frac{1}{6} = \frac{1}{2}$, which means $R = 2\ \Omega$.

23. **D** To escape the pull of gravity means you no longer have potential energy (since potential energy comes from the planet's pull on you). Furthermore, launching an object with just enough speed to achieve this would mean it ends with 0 remaining speed. This means the final energy will be 0, which (by conservation of energy) implies the initial energy is 0. Thus, $KE = -PE$, which gives $\frac{1}{2}mv^2 = -mgh = -m(GM/r^2)h$. In this case, r and h would be the same thing, so this can be solved for v, giving the escape velocity.

24. **A** The intensity of a wave is proportional to $1/r^2$, so it will increase as the ambulance approaches. This eliminates (B) and (D). The change in pitch will be determined by the Doppler effect. Pitch is simply another term for frequency, and the Doppler effect says that if the source of a wave and the object detecting the wave are growing closer, then the frequency will increase. Thus, (A) is correct.

25. **B** The distance from one crest to another is the wavelength. In this case, that distance is 6 m. Furthermore, you know that $v = f\lambda$ for any wave. Solving for f gives $f = v/\lambda$. Plugging in the known values gives $f = (600)/(6) = 100$ Hz.

26. **D** A wave passing along a string with both ends held in place will have a fundamental frequency of $f = v/(2l)$, where v is the speed of the wave and l is the length of the string. Thus, fundamental frequency is inversely proportional to the length of the string, so the first string, which is twice as long as the second, will have a fundamental frequency that is half of the second string's. Expressed as a ratio, that is 1:2.

27. **B** The frequency of a pendulum is $f = \frac{1}{2\pi}\sqrt{\frac{g}{l}}$. Therefore, the change in mass would have no effect on the system, and the others would change the equation to $f = \frac{1}{2\pi}\sqrt{\frac{\frac{1}{6}g}{2l}}$. Thus, the original frequency f will become $\sqrt{\frac{1}{12}}f$.

28. **D** The net force on each block can be found by using Newton's Second Law, $F_{Net} = ma$. The tension in the string can be found by focusing on the hanging mass. You know the net force on it will be $F_{Net} = ma = F_g - T = (mg) - T$, which means $T = (mg) - (ma)$. Finally, the coefficient of kinetic friction can be found by the looking at the top block. For that block, the net force in the horizontal direction will be the same as the overall net force since the two vertical forces (normal and gravity) will cancel out. Thus, you get $F_{Net} = ma = T - F_f = T - \mu F_N = T - \mu mg$. The only value in the answers you cannot calculate is the speed of the block when it reaches the edge. In order to compute this value, you would need to know how far from the edge the block is when it begins moving.

29. **D** If nothing is moving, then we know that the net force will be 0. Looking first at the forces perpendicular to the plane, we get $F_N = F_g \sin\theta = mg\sin\theta$. Next, using Newton's Second Law and defining "up the ramp" as positive, we can say $F_{Net} = T - F_g\cos\theta - F_f = 0$. Solving for tension and plugging in all the variables gives us $T = mg\cos\theta + \mu mg\sin\theta$.

Next, looking at the hanging block, you can again use Newton's Second Law to determine that $F_{Net} = F_g - T = Mg - (mg\cos\theta + \mu mg\sin\theta) = 0$. Thus, solving for μ gives us $\mu = (Mg - mg\cos\theta)/(mg\sin\theta)$.

30. **C** The basic formula for work is $W = Fd\cos\theta$, where θ is the angle between the force and the direction of motion. Centripetal force, by definition, is always perpendicular to motion. Therefore, the θ will always be 90, and $\cos 90 = 0$, which means this force cannot do work.

31. **B** The force of gravity between two objects is given by $F_g = GmM/r^2$. This comes out to $(6.67 \times 10^{-11})(1)(1)/(1)^2 = 6.67 \times 10^{-11}$. Electric force is given by $F_e = kqQ/r^2$. This comes out to $(8.99 \times 10^9)(1)(1)/(1)^2 = 8.99 \times 10^9$. Therefore, in terms of F_g, $F_e = [(8.99 \times 10^9)/(6.67 \times 10^{-11})]F_g = (1.35 \times 10^{20})F_g$.

32. **C** Consider the situation from the perspective of conservation of energy. When it is initially dropped on one end of the planet, the ball has some amount of potential energy and no kinetic energy. When it reaches the center of the Earth, it will have no more potential energy, so all of that energy is now kinetic. Due to momentum, the ball will continue in that direction until the energy has again been converted entirely to potential energy. This would happen just as it reaches the other end of the hole. At that point, it would fall again and the process would repeat infinitely. This is exactly the same motion as an ideal spring system.

33. **B** At the top of the ramp, the car will have potential energy $PE = mgh$. At the bottom, this same amount of energy has been converted into kinetic energy $KE = \frac{1}{2}mv^2 = mgh$. Multiplying both sides by 2 produces $mv^2 = 2mgh$. Finally, divide both sides by r to get $mv^2/r = 2mgh/r$. The first term is the general formula for centripetal force, so the second term must also be equivalent to that force in this situation.

34. **B** First, find the net force on the system. Each of the top three blocks will contribute nothing to the net force. The forces on each of the two blocks on the ramp will be the following:

The normal force and the perpendicular component of gravity will cancel out. This leaves just the parallel component of gravity to be a net force. Using Newton's Second Law, you get $F_{Net} = F_{g'parallel}$ = $2mg\sin\theta = 5ma$. The 2 comes in because there are two blocks on the ramp and each one will contribute the net force discussed earlier. The 5 is necessary because the blocks cannot accelerate individually. Either they all move or none do. Solving for a then gives $a = \frac{2}{5}g\sin\theta$.

35. **D** First, find the time the ball would be in the air. You do this by focusing only on the vertical movement of the ball. The equation to use is $v_f = v_0 + at$. Solving for t, you get $t = \frac{v_f - v_0}{a}$. Using up as the positive direction and plugging in the values from the problem gives $t = \frac{-v\sin\theta - v\sin\theta}{-10}$. For 45 degrees, this gives $t = \frac{\sqrt{2}v}{10}$. For 30 degrees, it gives $t = \frac{v}{10}$.

Next, solve for the horizontal displacement by using $d = v_0 + at$. In this case, our horizontal acceleration is 0, so it simplifies to $d = v_0 t = v\cos\theta t$. For the 45 degree case, this gives $d = \frac{v\sqrt{2}}{2}\frac{\sqrt{2}v}{10} = \frac{2v^2}{20}$. For 30 degrees, this gives $d = \frac{v\sqrt{3}}{2}\frac{v}{10} = \frac{\sqrt{3}v^2}{20}$. Thus, the difference is $\frac{2v^2}{20} - \frac{\sqrt{3}v^2}{20} = \frac{v^2}{20}\left(2 - \sqrt{3}\right)$.

36. **C** For any collision, momentum must be conserved. That means $m_1 v_{1,0} + m_2 v_{2,0} = m_1 v_{1,f} + m_2 v_{2,f}$. Plugging in known values, that gives $(10)(10) + m_2(0) = (10)(-2) + m_2(8)$. Solving for m_2 gives $m_2 = 120/8 = 15$ kg.

37. **A** In a perfectly inelastic collision, the two objects stick together. That is not true in the graph because the objects have different velocities post-collision.

For a perfectly elastic collision, kinetic energy must be conserved. Check this using $\frac{1}{2}m_1 v_{1,0}^2 + \frac{1}{2}m_2 v_{2,0}^2 = \frac{1}{2}m_1 v_{1,f}^2 + \frac{1}{2}m_2 v_{2,f}^2$. Plugging in the values gives $\frac{1}{2}(4)(10)^2 + \frac{1}{2}(6)(0)^2 = \frac{1}{2}(4)(-2)^2 + \frac{1}{2}(6)(8)^2$. Calculating each side gives $200 + 0 = 8 + 192$, which is true. Therefore, kinetic energy is conserved, which makes this a perfectly elastic collision.

38. **D** In all of these cases, mechanical energy must be conserved. Thus, the energy required will be mgh for all of them.

39. **A** You know $T_{Net} = I\alpha$. The torque applied to each is equal, so the sphere with a smaller moment of inertia (I) will rotate more quickly. Moment of inertia is smaller for an object with its mass concentrated closer to the center, so the solid sphere will have a smaller moment of inertia and thus rotate faster.

40. **A** You know $F_f = \mu F_N$. In both cases, μ will be the same, so you need only concern yourself with F_N. There are three vertical forces in this problem: gravity, the normal force, and the vertical component of the applied force. In the first case, that means $F_N = F_g - mg\sin\theta = mg - mg\sin30 = mg - mg/2 = mg/2$.

In the second case, $F_N = F_g + mg\sin\theta = mg + mg\sin30° = mg + mg/2 = 3mg/2$. Therefore, the force in this case will be 3 times what it was in the first case.

41. **A** First, the force diagram for this situation would look like the following picture:

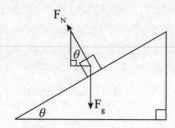

Furthermore, because this is uniform circular motion, you know there must be a net centripetal force toward the center, and you know it must be such that $F_C = mv^2/r$. If the net force is completely inward, then the net vertical force must be 0, which means $0 = F_N\cos\theta - mg$. Solving for F_N gives $F_N = mg/\cos\theta$.

Next, the only horizontal force is the horizontal component of F_N. This means $mv^2/r = F_N\sin\theta = (mg/\cos\theta)\sin\theta = mg\tan\theta$. Solving for θ then gives $\theta = \tan^{-1}[v^2/(rg)]$.

42. **B** Using Newton's Second Law and taking up as the positive direction, you get $F_{Net} = ma = T - mg$ for the first block and $F_{Net} = M(-a) = T - Mg$ for the second block. Subtracting the second equation from the first gives $ma + Ma = Mg - mg$. Solving for a then gives $a = \dfrac{g(M-m)}{m+M} = \dfrac{10(15-5)}{5+15} = \dfrac{100}{20} = 5$ m/s^2.

From there, the problem is simply a situation of uniform accelerated motion. The blocks need to move a distance of $d = 0.5$ m each, the starting speed is 0 m/s, and the acceleration is 5 m/s^2. Using $d = v_0 t + \dfrac{1}{2}at^2$ gives $(0.5) = (0)t + \dfrac{1}{2}(5)t^2$, so $t = \sqrt{\dfrac{1}{5}}$ s.

43. **D** Consider a situation in which the block is displaced some distance x from the equilibrium position.

The first spring would be displaced some amount x_1, and the second spring would be displaced some amount x_2. Then you can say $x = x_1 + x_2$. Knowing that, $F = -kx$ for all springs, you can say $x_1 = -F/k_1$

and $x_2 = -F/k_2$. Substituting those into the first equation gives $x = -\dfrac{F}{k_1} + \left(-\dfrac{F}{k_2}\right) = -F\left(\dfrac{1}{k_1} + \dfrac{1}{k_2}\right)$.

Solving for F then gives $F = -\dfrac{k_1 k_2}{k_1 + k_2}$. Therefore, the effective spring constant is $k_{eff} = \dfrac{k_1 k_2}{k_1 + k_2}$.

Plugging in the given values for this problem, you get $k_{eff} = \dfrac{k_1 3k_1}{k_1 + 3k_1} = \dfrac{3k_1^2}{4k_1} = \dfrac{3k_1}{4}$.

44. **A** In this situation, the left-hand block will provide counter-clockwise torque, and the right-hand block will provide clockwise torque. Therefore, the two must be equal in magnitude for the system to be balanced. We know that the formula for torque is $\tau = Fr\sin\theta$. In this problem, the θ will always be 90 degrees, and $\sin 90° = 1$, so that term will be neglected for the rest of the explanation. Furthermore, the only forces involved in this problem are the forces of gravity on the blocks, and you know $F_g = mg$.

For (A), the net counter-clockwise torque (left side of the system) would be $\tau = (m_2 g)(2L) + (m_1 g)(L)$ $= (2m_1 g)(2L) + (m_1 g)(L) = 5m_1 gL$. The clockwise torque (right side of the system) would be $\tau = (m_2 g)(L) + (m_1 g)(2L) = (2m_1 g)(L) + (m_1 g)(2L) = 4m_1 gL$. Thus, it would not be balanced. Choice (B) would make both torques 0, so that would be balanced. Choices (C) and (D) both result in the fulcrum being twice as far from m_1 as it is from m_2, which would counteract the difference in their weights.

45. **C** The correct answer is (C). For the left and right sides of the tracks, all three tracks are identical. Looking only at the middle segment of each track, (B) would hold steady throughout. Choice (A) would decrease the speed of the ball as it climbed the hill, then increase the speed back to the original amount as it descended the hill. That means it spends the entire segment at a speed less than the initial speed. Choice (C) would be the opposite of (A). The balls gains speed as it descends, and then loses speed as it climbs back up, eventually ending at the original speed. Therefore, it spends the entire middle segment at a speed greater than the initial amount.

46. **B, D** Uniform accelerated motion means the acceleration must be a constant, which looks like a flat line on a graph. If you are given a velocity vs. time graph, the acceleration is the slope of the graph at any given point. Thus, the correct answers are the two graphs with consistent slopes, (B) and (D).

47. **A, D** First, the momentum of a system will always be conserved, regardless of the type of collision that takes place. Thus, (D) must be true. You can eliminate (C) because that is only true for perfectly elastic collisions, and this collision is not. Much like linear momentum, angular momentum is always conserved for an entire system. Therefore, (A) is correct, but (B) is not because it does not encompass the entire system.

48. **A, D** Choice (B) can eliminated because KE would increase in this situation due to the object's increasing speed. Choice (C) is also incorrect. In this situation, KE would decrease, but $KE = \dfrac{1}{2}mv^2$, meaning it would be a linear decrease. Thus, (A) and (D) are correct.

49. **A, B** First, assume the second charge is positive. If this is the case, both charges will repel the positive test charge. For two repelling forces to cancel out, the test charge must be between the two spheres. Additionally, because $F_e = kq_1q_2/r^2$, the test charge must be closer to the left-hand sphere to counteract the greater magnitude of charge on the right-hand sphere. Thus, location B would be acceptable for this situation.

Next, assume the second charge is negative. If this is the case, placing the positive test charge anywhere between the two sphere would result in a net force to the right since the electric force from each sphere would be in that direction. That leaves only location A as an acceptable answer in this situation.

50. **C, D** Recall that the potential energy of a spring-block system is given by $PE = \frac{1}{2}kx^2$. Using conservation of energy, you can then say $\frac{1}{2}mv_{max}^2 = \frac{1}{2}kA^2$, where the left side represents a time when all the energy is kinetic (at equilibrium) and the right side represents when all the energy is potential (at either maximum displacement). Therefore, one of the correct answers is (D). The maximum kinetic energy would give a value for the left side of the equation, and a value for k would leave A as the only variable, which you could then calculate.

Next, recall that the period of a spring system is given by $T = 2\pi\sqrt{\dfrac{m}{k}}$. Because m is already known, knowing T would allow you to solve for k. This means (C) is also correct. You could calculate k from the previous equation and then use the maximum kinetic energy in the same way you did in the first paragraph to solve for A.

SECTION II

1.   The defining characteristic of a perfectly elastic collision is that no kinetic energy is lost. Therefore, you must record the speeds of each car before and after the collision. Then, the equation $\frac{1}{2}m_1v_{1,0}^2 + \frac{1}{2}m_2v_{2,0}^2 = \frac{1}{2}m_1v_{1,f}^2 + \frac{1}{2}m_2v_{2,f}^2$ can be used to check whether or not it is elastic.

The defining feature of a perfectly inelastic collision is that the objects stick together after the collision. This can be simply observed. If the two cars remain intact post-collision, then the collision is perfectly inelastic.

If neither of the above is true, then it is neither perfectly elastic nor perfectly inelastic.

Finally, a few possible sources of energy loss would be the heat generated as metal is warped, the sound made by the crash, and even the light of any sparks generated.

2.   The other two graphs for the first situation would look like this:

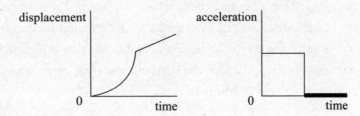

These graphs could be made from a car or plane accelerating up to a certain speed and then maintaining it.

The next set of graphs should look like this:

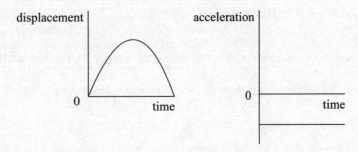

These could be made from any case of ideal projectile motion.

The final set would look like this:

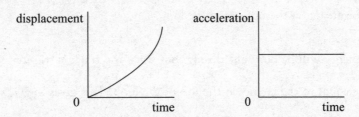

These could be made from an object undergoing uniform accelerated motion.

3.   (a)   Your graph should look like this:

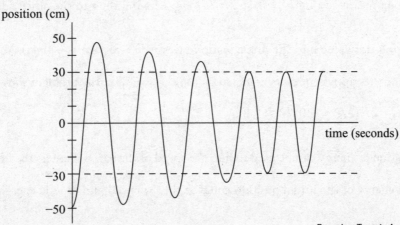

First, it must start at position −50 as the problem indicates. Second, it must go up to a point that is nearly 50 but not quite. This is because the frictional surface will remove some of the energy from the system as the block slides across it. Each time the block swings back to the frictional surface, it will drain a little more of the energy, making the amplitude slowly taper off until it eventually is only on the frictionless surface. Because the frictionless surface extends to 30 cm, it will oscillate between that point and −30 cm indefinitely.

(b)   The energy of a spring system can be found by $PE = \frac{1}{2} kA^2$, where A is the amplitude of the system, so the initial energy of my system will be $PE = \frac{1}{2}(100)(0.5)^2 = 100/8 = 12.5$ J. After the oscillations have tapered off, it will have a new energy of $PE = \frac{1}{2}(100)(0.3)^2 = 4.5$ J. Therefore, the frictional surface has caused 8 J of energy to be lost.

4.   (a)   Using conservation of energy, we set the potential energy at the top of the ramp equal to the kinetic energy at the bottom to get $mgh = \frac{1}{2} mv^2$. Solving for v gives $v = \sqrt{2gh} = \sqrt{(2)(10)(20)} = \sqrt{400}$ = 20 m/s.

(b)   Throughout the entire loop, there are only two forces acting on the cart: F_g and F_N. The loop can be considered as a case of uniform circular motion, so the net force will always be directed toward the center and will be equal to mv^2/r. At the bottom of the loop, that means $mv^2/r = F_N - F_g$, so $F_N = mv^2 + F_g$. At the top of the loop, $mv^2/r = F_N + F_g$, so $F_N = mv^2 - F_g$. Thus, the normal force must always be greater at the bottom of the loop.

(c)   Mechanical energy will be constant throughout since the track is frictionless, so set the energy at the beginning equal to the energy at the top of the loop. This gives $mgh = mg(2r) + \frac{1}{2} mv^2$, where r is the radius of the circle and v is the speed at the top of the loop.

Next, recall that in order for the car to complete the loop, it must be going fast enough to create some normal force at the top. In general, this equation would look like $mv^2/r = F_N + F_g$. Maximizing the loop means setting $F_N = 0$, so $mv^2/r = mg$. Moving the r to the other side gives $mv^2 = mgr$.

Substituting this value into the first equation gives $mgh = mg(2r) + \frac{1}{2}(mgr)$. You can cancel out m and g since they appear in every term, so $h = 2r + \frac{1}{2}r = 2.5r$. The problem shows $h = 20$, so solving for r gives $r = h/(2.5) = (20)/(2.5) = 8$ m.

(d)   Conservation of energy can be used to find the speed of the car as it enters the final segment. Set the potential energy of the initial position equal to the potential and kinetic energies as the cart enters the final stretch. This gives $mgh_1 = mgh_2 + \frac{1}{2}mv^2$. First, realize m can be dropped since it appears in every term. Solving for v then gives $v = \sqrt{2g(h_1 - h_2)} = \sqrt{(2)(10)(20-10)} = \sqrt{200} = 10\sqrt{2}$ m/s.

In order for the friction to bring the cart to a stop, it must do work to the cart. We know $W = \Delta KE$, so $Fd\cos\theta = \frac{1}{2}mv_f^2 - \frac{1}{2}mv_0^2$. The first KE term can be dropped since the final speed will be 0. We can substitute $F_f = \mu F_N = \mu mg$ for F term since friction is the force doing the work. Additionally, $\theta = 180°$ because the frictional force will act in a direction opposite the cart's motion. All of this gives $(\mu mg)d\cos180° = -\frac{1}{2}mv_0^2$. Canceling the m on each side and solving for d then gives $d = -\frac{1}{2}v_0^2/(\mu g\cos180°) = -\dfrac{\frac{1}{2}(10\sqrt{2})^2}{(0.2)(10)(-1)} = 50$ m.

5. (a) First, find the equivalent resistance for the whole circuit. First, the total resistance of the two parallel resistors can be found by using $\dfrac{1}{R_T} = \dfrac{1}{R_2} + \dfrac{1}{R_3} = \dfrac{R_2 + R_3}{R_2 R_3}$, which means $R_T = \dfrac{R_2 R_3}{R_2 + R_3}$. Then this would be in series with the first resistor, so the equivalent resistance of the whole circuit, R_{eq}, would be $R_{eq} = \dfrac{R_1 + R_2 R_3}{R_2 + R_3} = \dfrac{R_1 R_2 + R_1 R_3 + R_2 R_3}{R_2 + R_3}$.

Next, we know from Ohm's Law that $V = IR$, so you can say $I = V/R = \dfrac{V}{\dfrac{R_1 R_2 + R_1 R_3 + R_2 R_3}{R_2 + R_3}} = \dfrac{V(R_2 + R_3)}{R_1 R_2 + R_1 R_3 + R_2 R_3}$. This will be the total current of the circuit. Because there are no branches before R_1, this will also be the current flowing through that resistor. So the voltage drop of that resistor, V_1, can be found by again using Ohm's Law for that location rather than the whole resistor. This gives $V_1 = I_1 R_1 = \dfrac{V(R_2 + R_3)}{R_1 R_2 + R_1 R_3 + R_2 R_3}R_1$.

(b) The diagram should look like this:

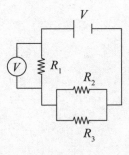

In order to measure the voltage loss of a particular resistor, the voltmeter must be arranged in parallel with that resistor. This is because parallel elements always have equal voltage drops. Thus, whatever voltage drop the device measures will be the same as the voltage drop in the resistor being measured.

(c) The diagram should look like this:

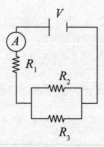

In order to measure the current passing through a particular resistor, the ammeter must be arranged in series with that resistor. This is because elements in series always have equal currents. Thus, whatever current the device measures will be the same as the current flowing through the resistor being measured.

(d) For the voltmeter, infinite resistance would be ideal. As in all measurements, you don't want to disturb the system in any way as you take the measurement. Otherwise, your readings would not be accurate for the original system. Thus, in order to leave the circuit as it was, we need to maintain the flow of current that exists before we started our measurement. Having an infinite resistance would ensure no current flows into the voltmeter, leaving it all on its original path.

For the ammeter, 0 resistance would be ideal. Again, we want to have minimal disturbance on the system. If the ammeter did have resistance, then it would be a source of voltage loss that did not previously exist. In essence, it would be an additional resistor that would need to considered, forcing us to recalculate everything from the ground up.

Part III
About the AP Physics 1 Exam

- The Structure of the AP Physics 1 Exam
- A Quick Word About Equations
- How AP Exams Are Used
- Other Resources
- Designing Your Study Plan

THE STRUCTURE OF THE AP PHYSICS 1 EXAM

The AP Physics 1 Exam consists of two sections: a multiple-choice section and a free-response section. The multiple-choice section consists of two question types. Single-select questions are each followed by four possible responses, only one of which is correct. Multi-select questions are a new addition to the AP Physics 1 Exam, and require two of the listed answer choices to be selected to answer the question correctly. There are five multi-select questions that always appear at the end of the multiple-choice section.

Section	Timing	Scoring	Question Type	Number of Questions
I: Multiple Choice	90 minutes	50% of exam score	Single-select (discrete questions and questions in sets with one correct answer)	45
			Multi-select (discrete questions with two correct answers)	5
				Total—50

Section	Timing	Scoring	Question Type	Number of Questions
II: Free Response	90 minutes	50% of exam score	Experimental Design	1
			Qualitative/ Quantitative Translation	1
			Short Answer	3
				Total—5

The free-response section consists of five multi-part questions, which require you to write out your solutions, showing your work. The total amount of time for this section is 90 minutes. Unlike the multiple-choice section, which is scored by a computer, the free-response section is graded by high school and college teachers. They have guidelines for awarding partial credit, so you may still receive partial points should you not correctly respond to every part of the question. You are allowed to use a calculator on the entire AP Physics 1 Exam—including both the multiple-choice and free-response sections. Scientific or graphing calculators may be used, provided that they don't have any unapproved features or capabilities (a list of approved graphing calculators is available on the College Board's website). In addition, a table of equations commonly used in physics will be provided to you at the exam site. This can be found online and we've included it with both practice tests.

Grades on the AP Physics 1 Exam are reported as a number: either 1, 2, 3, 4, or 5. Here is a description of each of these five numerical scores plus data on how students scored on the May 2016 test:

Score (Meaning)	Percentage of Test Takers	Equivalent Grade in a first-year college course
5 (extremely qualified)	4.6%	A
4 (well qualified)	14%	A–, B+, B
3 (qualified)	21.2%	B–, C+, C
2 (possibly qualified)	30.2%	C-
1 (no recommendation)	30%	D

Colleges are generally looking for a 4 or 5, but some may grant credit for a 3. How well do you have to do to earn each grade? Each test is curved so scores vary from year to year, but as we see above, in May 2016 around 60% of test takers earned scores of 1 or 2, so you'll want to study hard and prepare for this very difficult exam. We'll get to that soon, we promise!

So, what is on the exam and how do you prepare? Take a look at the following list of the major topics covered on the AP Physics 1 Exam.

AP Physics 1: Algebra-Based Course Content

You may be using this book as a supplementary text as you take an AP Physics 1 course at your high school or you may be using it on your own. The College Board is very detailed in what they require your AP teacher to cover in his or her AP Physics 1 course. They explain that you should be familiar with the following topics:

- Kinematics
- Dynamics: Newton's laws
- Circular motion and universal law of gravitation
- Simple harmonic motion: simple pendulum and mass-spring systems
- Impulse, linear momentum, and conservation of linear momentum: collisions
- Work, energy, and conservation of energy
- Rotational motion: torque, rotational kinematics and energy, rotational dynamics, and conservation of angular momentum
- Electrostatics: electric charge and electric force
- DC circuits: resistors only
- Mechanical waves and sound

Want to know which colleges are best for you? Check out The Princeton Review's College Advisor app to build your ideal college list and find your perfect college fit! Available for free in the iOS App Store and Google Play Store.

The College Board also maps out the areas that you should study as "Big Ideas" and those are as follows:

- Objects and systems have properties such as mass and charge. Systems may have internal structure.
- Fields existing in space can be used to explain interactions.
- The interactions of an object with other objects can be described by forces.
- Interactions between systems can result in changes in those systems.
- Changes that occur as a result of interactions are constrained by conservation laws.
- Waves can transfer energy and momentum from one location to another without the permanent transfer of mass and serve as a mathematical model for the description of other phenomena.

Naturally, it's important to be familiar with the topics—to understand the fundamentals of the theory, to define fundamental quantities, and to recognize and be able to use the equations. Then, you must practice at applying what you've learned to answering questions like those you will see on the exam. This book is designed to review all of the content areas covered on the exam. Also, each chapter is followed by practice multiple-choice and free-response questions, and perhaps even more important, answers and explanations for every example and question in this book. You'll learn as much—if not more—from actively reading the solutions as you will from reading the text and examples.

A QUICK WORD ABOUT EQUATIONS

As you know, you will be given an Equation Sheet on the day of your exam and that Equation Sheet is yours for the duration of the test. You will see many equations in a different forms and it can be hard to recall what is equivalent to what, so we have made a handy chart for you. Here is a tally of a few equations that are equivalent to other equations and where you can find them in this book and on your Equation Sheet.

Equation(s)	Page	Location on AP Physics 1 Equations Sheet
Big Five #2, #3, #5	90	First three equations in top left.
Newton's Second Law	111	Fourth equation below the first three.
Force of Friction	121	Fifth equation from top.
Power	149	$P = \Delta E/\Delta t$ is the equation on the sheet. Note specifically that the change in energy is the work done, so these two equations are equivalent.
Gravitational Force	176	Third equation from the bottom of the second column. Don't be thrown by the use of absolute value bars; they mean the same thing.

HOW AP EXAMS ARE USED

Different colleges use AP Exams in different ways, so it is important that you go to a particular college's website to determine how it uses AP Exams. The three categories below represent the main ways in which AP Exam scores can be used.

- **College Credit.** Some colleges will award you college credit if you score well on an AP Exam. These credits count toward your graduation requirements, meaning that you can take fewer courses while in college. Given the cost of college, this could be quite a benefit, indeed.
- **Satisfy Requirements.** Some colleges will allow you to "place out" of certain requirements if you do well on an AP Exam, even if they do not give you actual college credits. For example, you might not need to take an introductory-level course, or perhaps you might not need to take a class in a certain discipline at all.
- **Admissions Plus.** Even if your AP Exam will not result in college credit or even allow you to place out of certain courses, most colleges will respect your decision to push yourself by taking an AP course or even an AP Exam outside of a course. A high score on an AP Exam shows mastery of more difficult content than is taught in many high school courses, and colleges may take that into account during the admissions process.

Are You Preparing for College?
Check out all of the useful books from The Princeton Review, including *The Best 381 Colleges, Cracking the SAT, Cracking the ACT*, and more!

OTHER RESOURCES

There are many resources available to help you improve your score on the AP Physics 1 Exam, not the least of which are your teachers. If you are taking an AP class, you may be able to get extra attention from your teacher, such as obtaining feedback on your essays. If you are not in an AP course, reach out to a teacher who teaches AP Physics 1 and ask if the teacher will review your essays or otherwise help you with content.

Another wonderful resource is **AP Central**, the official site of the AP Exams. The scope of the information at this site is quite broad and includes

- a course description, which includes details on what content is covered and sample questions
- sample questions from the AP Physics 1 Exam
- free-response question prompts and multiple-choice questions from previous years

The AP Central home page address is: **http://apcentral.collegeboard.com**.

Finally, The Princeton Review offers tutoring for the AP Physics 1 Exam. Our expert instructors can help you refine your strategic approach and add to your content knowledge. For more information, call 1-800-2REVIEW.

Train As You Will Test
The AP Physics 1 Exam is grueling and you need to build up your endurance, so be sure to take practice tests throughout the school year.

DESIGNING YOUR STUDY PLAN

In Part I, you identified some areas for potential improvement. Let's now delve further into your performance on Test 1, with the goal of developing a study plan appropriate to your needs and time commitment.

Read the answers and explanations associated with the multiple-choice questions (starting at page 35). After you have done so, respond to the following questions:

- Review the topic list on pages 51–52. Next to each topic, indicate your rank of the topic as follows: "1" means "I need a lot of work on this," "2" means "I need to beef up my knowledge," and "3" means "I know this topic well."

- How many days/weeks/months away is your exam?

- What time of day is your best, most focused study time?

- How much time per day/week/month will you devote to preparing for your exam?

- When will you do this preparation? (Be as specific as possible: Mondays and Wednesdays from 3 to 4 P.M., for example)

- Based on the answers above, will you focus on strategy (Part IV) or content (Part V) or both?

- What are your overall goals in using this book?

Study Breaks Are Important
Don't burn yourself out before test day. Remember to take breaks every so often—go for a walk, listen to a favorite album, get some fresh air.

Part IV
Test-Taking Strategies for the AP Physics 1 Exam

PREVIEW

Review your Test 1 results and then respond to the following questions:
- How many multiple-choice questions did you miss even though you knew the answer?
- On how many multiple-choice questions did you guess blindly?
- How many multiple-choice questions did you miss after eliminating some answers and guessing based on the remaining answers?
- Did you find any of the free-response questions easier or harder than the others—and, if so, why?

HOW TO USE THE CHAPTERS IN THIS PART

For the following Strategy chapters, think about what you are doing now before you read the chapters. As you read and engage in the directed practice, be sure to think critically about the ways you can modify your approach.

A QUICK NOTE ABOUT NAMING CONVENTIONS

Each classroom uses different conventions for naming things. Here in this book we are using using the AP conventions set by the College Board, but please note that your AP Physics 1 course teacher may or may not decide to norm to the test the same way.

For example, v_0 and v_i are usually the same thing but not always the same thing. Some teachers and books may treat these as the same, though. These mean velocity at time zero and velocity initial, respectively.

Take a look at this example:

> A car starts from rest and accelerates to 20 m/s in 5 seconds. The car holds this velocity for 10 seconds. the driver then speeds up to 28 m/s in 5 more seconds. What is the magnitude of the second acceleration?

$v_0 = 0$ miles per hour but v_i in this instance will be 20 m/s because the problem is looking at the second acceleration. Be careful not to confuse v_0 and v_i. This may seem silly, but it's a common error so we wanted to address it right up front. Now, onto the good stuff!

Chapter 1
How to Approach
Multiple-Choice
Questions

CRACKING THE MULTIPLE-CHOICE SECTION

All the multiple-choice questions will have a similar format: Each will be followed by four answer choices. At times, it may seem that there could be more than one possible correct answer. With the exception of the five multi-select questions, there is only one! Answers resulting from common mistakes are often included in the four answer choices to trap you.

The Two-Pass System

The AP Physics 1 Exam covers a broad range of topics. There's no way, even with our extensive review, that you will know everything about every topic in algebra-based physics. So, what should you do?

Adopt a two-pass system. The two-pass system entails going through the test and answering the easy questions first. Save the more time-consuming questions for later. (Don't worry—you'll have time to do them later!) First, read the question and decide if it a "now" or "later" question. If you decide this is a "now" question, answer it in the test booklet. If it is a "later" question, come back to it. Once you have finished all the "now" questions on a double page, transfer the answers to your bubble sheet. Flip the page and repeat the process.

Once you've finished all the "now" questions, move on to the "later" questions. Start with the easier questions first. These are the ones that require calculations or that require you to eliminate the answer choices (in essence, the correct answer does not jump out at you immediately). Transfer your answers to your bubble sheet as soon as you answer these "later" questions.

Watch Out for Those Bubbles!

Because you're skipping problems, you need to keep careful track of the bubbles on your answer sheet. One way to accomplish this is by answering all the questions on a page and then transferring your choices to the answer sheet. If you prefer to enter them one by one, make sure you double-check the number beside the ovals before filling them in. We'd hate to see you lose points because you forgot to skip a bubble!

Process of Elimination (POE)

On most tests, you need to know your material backward and forward in order to get the right answer. In other words, if you don't know the answer beforehand, you probably won't answer the question correctly. This is particularly true of fill-in-the-blank and essay questions. We're taught to think that the only way to get a question right is by knowing the answer. However, that's not the case on Section I of the AP Physics 1 Exam. You can get a perfect score on this portion of the test without knowing a single right answer—provided you know all the wrong answers!

What are we talking about? This is perhaps the most important technique to use on the multiple-choice section of the exam. Let's take a look at an example that probably looks quite familiar:

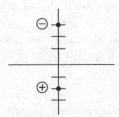

26. In the diagram above, a positive charge and negative charge are placed at $y = -2$ and $y = 3$, respectively. If the negative charge has a greater magnitude, then the only place of 0 net electric field would be

   (A) along the positive x-axis
   (B) along the negative x-axis
   (C) along the positive y-axis
   (D) along the negative y-axis

Now, if this were a fill-in-the-blank-style question, you might be in a heap of trouble. But let's take a look at what we've got and how we can eliminate answer choices in order to reach the correct answer.

Remember that the electric field for any given particle will point toward a negative charge and away from a positive charge. Anywhere between the two charges here (regardless of x-coordinate) would have an electric field that pointed up. The negative charge would pull it up, and the positive one would push it up. This eliminates (A) and (B). Finally, the negative charge is greater magnitude. Therefore, the fields could only be in balance somewhere closer to the positive charge. This eliminates (C). So the correct answer must be (D).

We think we've illustrated our point: Process of Elimination is the best way to approach the multiple-choice questions. Even when you don't know the answer right off the bat, you'll surely know that two or three of the answer choices are not correct. What then?

Aggressive Guessing

As mentioned earlier, you are scored only on the number of questions you get right, so we know guessing can't hurt you. But can it help you? It sure can. Let's say you guess on four questions; odds are you'll get one right. So you've already increased your score by one point. Now, let's add POE into the equation. If you can eliminate as many as two answer choices from each question, your chances of getting them right increase, and so does your overall score. Remember, don't leave any bubbles blank on test day!

General Advice

Answering 50 multiple-choice questions in 90 minutes can be challenging. Make sure to pace yourself accordingly and remember that you do not need to answer every question correctly to do well. Exploit the multiple-choice structure of this section. There are three wrong answers and only one correct one (and your odds are even better with the multi-select questions!), so even if you don't know exactly which one is the right answer, you can eliminate some that you know for sure are wrong. Then you can make an educated guess from among the answers that are left and greatly increase your odds of getting that question correct.

Problems with graphs and diagrams are usually the fastest to solve, and problems with an explanation for each answer usually take the longest to work through. Do not spend too much time on any one problem or you may not get to easier problems further into the test.

These practice exams are written to give you an idea of the format of the test, the difficulty of the questions, and to allow you to practice pacing yourself. Take them under the same conditions that you will encounter during the real exam.

Reflect

Respond to the following questions:

- How long will you spend on multiple-choice questions?

- How will you change your approach to multiple-choice questions?

- What is your multiple-choice guessing strategy?

- Will you seek further help, outside of this book (such as a teacher, tutor, or AP Central), on how to approach the questions that you will see on the AP Physics 1 Exam?

Chapter 2
How to Approach
Free-Response
Questions

FREE-RESPONSE SECTION

On the free-response section, be sure to show the graders what you're thinking. Write clearly—that is *very* important—and show your steps. If you make a mistake in one part and carry an incorrect result to a later part of the question, you can still earn valuable points if your method is correct. However, the graders cannot give you credit for work they can't follow or can't read. And, where appropriate, be sure to include units in your final answers.

Free-Response Terms Defined

You will be asked to both **describe** and **explain** natural phenomena. Both terms require the ability to demonstrate an understanding of physics principles by providing an accurate and coherent description or explanation. You will also be asked to **justify** a previously given answer. A justification is an argument supported by evidence. Evidence may consist of statements of physical principles, equations, calculations, data, graphs, and diagrams as appropriate. The argument, or equation used to support justifications and explanations, may in some cases refer to fundamental ideas or relations in physics, such as Newton's laws, conservation of energy, or Coulomb's Law. In other cases, the justification or explanation may take the form of analyzing the behavior of an equation for large or small values of a variable in the equation.

Calculate means that a student is expected to show work leading to a final answer, which may be algebraic but more often is numerical. **Derive** is more specific and indicates that you need to begin your solutions with one or more fundamental equations, such as those given on the AP Physics 1 Exam equation sheet. The final answer, usually algebraic, is then obtained through the appropriate use of mathematics. **What is** and **determine** are indicators that work need not necessarily be explicitly shown to obtain full credit. Showing work leading to answers, however, is a good idea, as you can earn partial credit in the case of an incorrect answer. Strict rules regarding significant digits are usually not applied to the scoring of numerical answers. However, in some cases, answers containing too many digits may be penalized. In general, two to four significant digits are acceptable. Exceptions to these guidelines usually occur when rounding makes a difference in obtaining a reasonable answer.

The words **sketch** and **plot** relate to producing graphs. "Sketch" means to draw a graph that illustrates key trends in a particular relationship, such as slope, curvature intercept(s), or asymptote(s). Numerical scaling or specific data are not required in a sketch. "Plot" means to draw the data points given in the problem on the grid provided, either using the given scale or indicating the scale and units when none are provided.

Exam questions that require the drawing of free-body diagrams or force diagrams will direct the students to **draw and label the forces (not components) that act on the [object]**, where [object] is replaced by a reference specific to the question, such as "the car when it reaches the top of the hill." Any components that are included in the diagram will be scored in the same way as incorrect or extraneous forces. In addition, in any subsequent part asking for a solution that would typically make use of the diagram, the following will be included: "If you need to draw anything other than what you have shown in part [x] to assist in your solutions, use the space below. Do NOT add anything to the figure in part [x]." This will give you the opportunity to construct a working diagram showing any components that are appropriate to the solution of the problem. This second diagram will *not* be scored.

Some questions will require students to **design** an experiment or **outline** a procedure that investigates a specific phenomenon or would answer a guiding question. You are expected to provide an orderly sequence of statements that specifies the necessary steps in the investigation needed to reasonably answer the questions or investigate the phenomenon.

CRACKING THE FREE-RESPONSE SECTION

Section II is worth 50 percent of your grade on the AP Physics 1 Exam. This section is composed of five free-response questions. You're given a total of 90 minutes for this section. It is recommended that you spend the first 60 minutes on the three short answer questions, and the next 30 minutes on the experimental design and qualitative/quantitative translation questions.

Clearly Explain and Justify Your Answers

Remember that your answers to the free-response questions are graded by readers and not by computers. Communication is a very important part of AP Physics 1. Compose your answers in precise sentences. Just getting the correct numerical answer is not enough. You should be able to explain your reasoning behind the technique that you selected and communicate your answer in the context of the problem. Even if the question does not explicitly say so, always explain and justify every step of your answer, including the final answer. Do not expect the graders to read between the lines. Explain everything as though somebody with no knowledge of physics is going to read it. Be sure to present your solution in a systematic manner using solid logic and appropriate language. And remember, although you won't earn points for neatness, the graders can't give you a grade if they can't read and understand your solution!

Use Only the Space You Need

Do not try to fill up the space provided for each question. The space given is usually more than enough. The people who design the tests realize that some students write in big letters and some students make mistakes and need extra space for corrections. So if you have a complete solution, don't worry about the extra space. Writing more will not earn you extra credit. In fact, many students tend to go overboard and shoot themselves in the foot by making a mistake after they've already written the right answer.

Read the Whole Question!

Some questions might have several subparts. Try to answer them all, and don't give up on the question if one part is giving you trouble. For example, if the answer to part (b) depends on the answer to part (a), but you think you got the answer to part (a) wrong, you should still go ahead and do part (b) using your answer to part (a) as required. Chances are that the grader will not mark you wrong twice, unless it is obvious from your answer that you should have discovered your mistake.

Use Common Sense

Always use your common sense in answering questions. For example, on one free-response question that asked students to compute the weight of a newborn baby on the Moon, some students answered 70 pounds. It should have been immediately obvious that the answer was probably off by a decimal point. A 70-pound baby would be a giant! This is an important mistake that should be easy to fix. Some mistakes may not be so obvious from the answer. However, the grader will consider simple, easily recognizable errors to be very important.

Don't Forget Units!
In your free-response questions, you'll be doing calculations right there by hand. Once you get to your final answer, DO NOT FORGET to include units or dimensions in it.

Think Like a Grader

When answering questions, try to think about what kind of answer the grader is expecting. Look at past free-response questions and grading rubrics on the College Board website. These examples will give you some idea of how the answers should be phrased. The graders are told to keep in mind that there are two aspects to the scoring of free-response answers: showing a comprehensive knowledge of physics and communicating that knowledge. Again, responses should be written as clearly as possible in complete sentences. You don't need to show all the steps of a calculation, but you must explain how you got your answer and why you chose the technique you used.

Think Before You Write

Abraham Lincoln once said that if he had eight hours to chop down a tree, he would spend six of them sharpening his axe. Be sure to spend some time thinking about what the question is, what answers are being asked for, what answers might make sense, and what your intuition is before starting to write. These questions aren't meant to trick you, so all the information you need is given. If you think you don't have the right information, you may have misunderstood the question. In some calculations, it is easy to get confused, so think about whether your answers make sense in terms of what the question is asking. If you have some idea of what the answer should look like before starting to write, then you will avoid getting sidetracked and wasting time on dead-ends.

REFLECT

Respond to the following questions:

- How much time will you spend on the short free-response questions? What about the long free-response questions?

- What will you do before you begin writing your free-response answers?

- Will you seek further help, outside of this book (such as a teacher, tutor, or AP Central), on how to approach the questions that you will see on the AP Physics 1 Exam?

Part V
Content Review for the AP Physics 1 Exam

Chapter 3
Vectors

Vectors will show up all over the place in our study of physics. Some physical quantities that are represented as vectors are displacement, velocity, acceleration, force, momentum, and electric and magnetic fields. Since vectors play such a recurring role, it's important to become comfortable working with them; the purpose of this chapter is to provide you with a mastery of the fundamental vector algebra we'll use in subsequent chapters. For now, we'll restrict our study to two-dimensional vectors (that is, ones that lie flat in a plane).

DEFINITION

A **vector** is a quantity that involves both magnitude and direction. A quantity that does not involve direction is a **scalar**. For example, the quantity *55 miles per hour* is a scalar, while the quantity *55 miles per hour to the north* is a vector. Other examples of scalars include mass, work, energy, power, temperature, and electric charge.

Vectors can be denoted in several ways, including:

$$\mathbf{A}, \; \mathit{A}, \; \overline{A}, \; \vec{A}$$

In textbooks, you'll usually see one of the first two, but when it's handwritten, you'll see one of the last two.

Displacement (which is net distance traveled including direction) is an example of a vector:

$$\underbrace{\mathbf{A}}_{\text{displacement}} = \underbrace{4 \text{ miles}}_{\text{magnitude}} \; \underbrace{\text{to the north}}_{\text{direction}} \qquad \mathbf{B} = \underbrace{3 \text{ miles}}_{\text{magnitude}} \; \underbrace{\text{to the east}}_{\text{direction}}$$

Vectors obey the Commutative Law for Addition, which states:

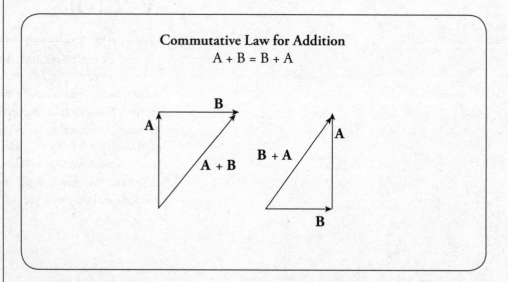

Commutative Law for Addition
A + B = B + A

The vector sum **A** + **B** means the vector **A** followed by **B**. The vector sum of **B** + **A** means the vector **B** followed by **A**, and the result is an identical vector to **A** + **B**. Vectors are always added tail to end to find their sum, so **A** + **B** or **B** + **A**—both are examples of tail to end.

Two vectors are equal when they have the same magnitude and the same direction.

Example 1 Add the following two vectors:

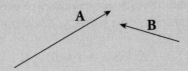

Solution. Place the tail of **B** at the tip of **A** and connect them:

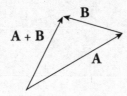

SCALAR MULTIPLICATION

A vector can be multiplied by a scalar (that is, by a number), and the result is a vector. If the original vector is **A** and the scalar is k, then the scalar multiple k**A** is as follows:

> **Also Worth Noting!**
> Scalar multiplication indicates a change in magnitude by the numerical multiple, and direction if there is a negative sign only (negative sign indicates opposite direction). An example of this is that of a car travelling east at a velocity of 100 km per hour. When multiplied by $k = -2$, the car's new velocity is 200 km per hour WEST.

Scalar Multiplication

magnitude of $k\,\mathbf{A} = |k| \times$ (magnitude of **A**)

direction of $k\mathbf{A} = \begin{cases} \text{the same as } \mathbf{A} \text{ if } k \text{ is positive} \\ \text{the opposite of } \mathbf{A} \text{ if } k \text{ is negative} \end{cases}$

Example 2 Sketch the scalar multiples 2**A**, $\frac{1}{2}$**A**, –**A**, and –3**A** of the vector **A**:

Solution.

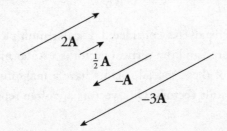

VECTOR SUBTRACTION

To subtract one vector from another, for example, to get **A – B**, simply form the vector **–B**, which is the scalar multiple (–1)**B**, and add it to **A**:

$$A - B = A + (-B)$$

Example 3 For the two vectors **A** and **B**, find the vector **A – B**.

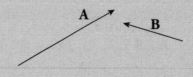

Tail To End!
By adding the negative of B, we are allowing the process to follow the tail to end convention that we discussed earlier.

Solution. Flip **B** around (thereby forming **–B**) and add that vector to **A**:

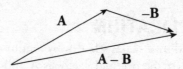

A Note About Direction
Make sure to pay attention to direction if you are not using a coordinate system. If you set a vector point to the right as positive, then you must set a vector pointing to the left as negative.

It is important to know that vector subtraction is **not** commutative; you must perform the subtraction in the order stated in the problem.

TWO-DIMENSIONAL VECTORS

Two-dimensional vectors are vectors that lie flat in a plane and can be written as the sum of a horizontal vector and a vertical vector. For example, in the following diagram, the vector **A** is equal to the horizontal vector **B** plus the vertical vector **C**:

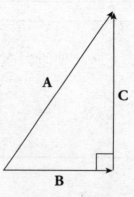

The horizontal vector is always considered a scalar multiple of what's called the **horizontal basis vector**, **i**, and the vertical vector is a scalar multiple of the **vertical basis vector**, **j**. Both of these special vectors have a magnitude of 1, and for this reason, they're called **unit vectors**. Unit vectors are often represented by placing a hat (caret) over the vector; for example, the unit **vectors i** and **j** are sometimes denoted **î** and **ĵ**.

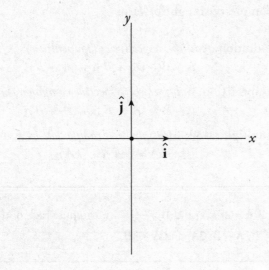

Coordinate System
Think of *i* as your *x*-coordinate system and *j* as your *y*-coordinate system. *i* is just a unit vector that points in the positive *x* direction, and *j* is just a unit vector that points in the positive *y* direction.

For instance, the vector **A** in the figure below is the sum of the horizontal vector **B** = 3**î** and the vertical vector **C** = 4**ĵ**.

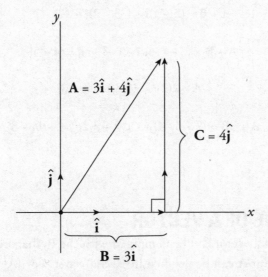

The vectors **B** and **C** are called the **vector components** of **A**, and the scalar multiples of **î** and **ĵ** which give **A**—in this case, 3 and 4—are called the **scalar components** of **A**. So vector **A** can be written as the sum $A_x \hat{\imath} + A_y \hat{\jmath}$, where A_x and A_y are the scalar components of **A**. The component A_x is called the **horizontal** scalar component of **A**, and A_y is called the **vertical** scalar component of **A**. In general, any vector in a plane can be described in this manner.

VECTOR OPERATIONS USING COMPONENTS

The use of components makes the vector operations of addition, subtraction, and scalar multiplication pretty straightforward:

Vector addition: *Add the respective components.*
$$\mathbf{A} + \mathbf{B} = (A_x + B_x)\hat{\imath} + (A_y + B_y)\hat{\jmath}$$

Vector subtraction: *Subtract the respective components.*
$$\mathbf{A} - \mathbf{B} = (A_x - B_x)\hat{\imath} + (A_y - B_y)\hat{\jmath}$$

Scalar multiplication: *Multiply each component by k.*
$$k\mathbf{A} = (kA_x)\hat{\imath} + (kA_y)\hat{\jmath}$$

Example 4 If $\mathbf{A} = 2\hat{\imath} - 3\hat{\jmath}$ and $\mathbf{B} = -4\hat{\imath} + 2\hat{\jmath}$, compute each of the following vectors: $\mathbf{A} + \mathbf{B}$, $\mathbf{A} - \mathbf{B}$, $2\mathbf{A}$, and $\mathbf{A} + 3\mathbf{B}$.

Solution. It's very helpful that the given vectors $\mathbf{A}$ and $\mathbf{B}$ are written explicitly in terms of the standard basis vectors $\hat{\imath}$ and $\hat{\jmath}$:

$$\mathbf{A} + \mathbf{B} = (2 - 4)\hat{\imath} + (-3 + 2)\hat{\jmath} = -2\hat{\imath} - \hat{\jmath}$$

$$\mathbf{A} - \mathbf{B} = [2 - (-4)]\hat{\imath} + (-3 - 2)\hat{\jmath} = 6\hat{\imath} - 5\hat{\jmath}$$

$$2\mathbf{A} = 2(2)\hat{\imath} + 2(-3)\hat{\jmath} = 4\hat{\imath} - 6\hat{\jmath}$$

$$\mathbf{A} + 3\mathbf{B} = [2 + 3(-4)]\hat{\imath} + [-3 + 3(2)]\hat{\jmath} = -10\hat{\imath} + 3\hat{\jmath}$$

MAGNITUDE OF A VECTOR

The magnitude of a vector can be computed with the Pythagorean Theorem. The magnitude of vector $\mathbf{A}$ can be denoted in several ways: A or $|\mathbf{A}|$ or $\|\mathbf{A}\|$. In terms of its components, the magnitude of $\mathbf{A} = A_x\hat{\imath} + A_y\hat{\jmath}$ is given by the equation

This is merely an interpretation of the Pythagorean Theorem. Make sure to brush up on geometry and trigonometry.

$$A = \sqrt{\left(A_x\right)^2 + \left(A_y\right)^2}$$

which is the formula for the length of the hypotenuse of a right triangle with sides of lengths A_x and A_y.

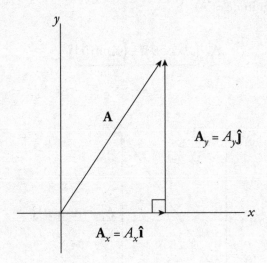

DIRECTION OF A VECTOR

The direction of a vector can be specified by the angle it makes with the positive x-axis. You can sketch the vector and use its components (and an inverse trig function) to determine the angle. For example, if θ denotes the angle that the vector $\mathbf{A} = 3\hat{\mathbf{i}} + 4\hat{\mathbf{j}}$ makes with the $+x$-axis, then $\tan \theta = 4/3$, so $\theta = \tan^{-1}(4/3) = 53.1°$.

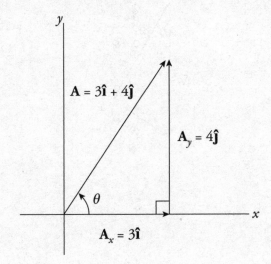

In general, the axis that θ is made to is known as the adjacent axis. The adjacent component is always going to get the cos θ. For example, if **A** makes the angle θ with the +x-axis, then its x- and y-components are $A \cos \theta$ and $A \sin \theta$, respectively (where A is the magnitude of **A**).

$$\mathbf{A} = \underbrace{(A\cos\theta)\hat{\mathbf{i}}}_{A_x} + \underbrace{(A\sin\theta)\hat{\mathbf{j}}}_{A_y}$$

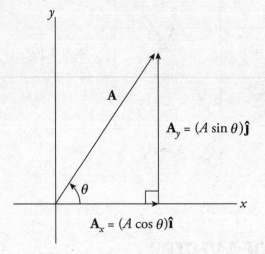

In general, any vector in the plane can be written in terms of two perpendicular component vectors. For example, vector **W** (shown below) is the sum of two component vectors whose magnitudes are $W \cos \theta$ and $W \sin \theta$:

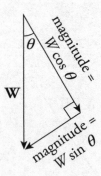

Chapter 3 Review Questions

Answers and Explanations can be found in Chapter 13.

Section I: Multiple-Choice

1. Two vectors, A and B, have the same magnitude, m, but vector A points north whereas vector B points east. What is the sum, A + B ?

   (A) m, northeast

   (B) $m\sqrt{2}$, northeast

   (C) $m\sqrt{2}$, northwest

   (D) $2m$, northwest

2. If $\mathbf{F}_1 = -20\hat{\jmath}$, $\mathbf{F}_2 = -10\hat{\imath}$, and $\mathbf{F}_3 = 5\hat{\imath} + 10\hat{\jmath}$, what is the sum $\mathbf{F}_1 + \mathbf{F}_2 + \mathbf{F}_3$?

   (A) $-15\hat{\imath} + 10\hat{\jmath}$

   (B) $-5\hat{\imath} - 10\hat{\jmath}$

   (C) $5\hat{\imath}$

   (D) $5\hat{\imath} - 10\hat{\jmath}$

3. Both the x- and y-components of a vector are doubled. Which of the following describes what happens to the resulting vector?

   (A) Magnitude increases by $\sqrt{2}$
   (B) Magnitude increases by $\sqrt{2}$, and the direction changes
   (C) Magnitude increases by 2
   (D) Magnitude increases by 2, and the direction changes

4. If vectors $v_0 = 15$ m/s north and $v_f = 5$ m/s south, what is $v_f - v_0$?

   (A) 10 m/s north

   (B) 10 m/s south

   (C) 20 m/s north

   (D) 20 m/s south

5. The magnitude of vector **A** is 10. Which of the following could be the components of A ?

   (A) $A_x = 5, A_y = 5$
   (B) $A_x = 6, A_y = 8$
   (C) $A_x = 7, A_y = 9$
   (D) $A_x = 10, A_y = 10$

6. If the vector $\mathbf{A} = \hat{\imath} - 2\hat{\jmath}$ and the vector $\mathbf{B} = 4\hat{\imath} - 5\hat{\jmath}$, what angle does $\mathbf{A} + \mathbf{B}$ form with the x-axis?

   (A) $\theta = \tan^{-1}\frac{3}{5}$

   (B) $\theta = \tan^{-1}\frac{7}{5}$

   (C) $\theta = \sin^{-1}\frac{3}{5}$

   (D) $\theta = \sin^{-1}\frac{5}{7}$

7. An object travels along the vector $d_1 = 4$ m $\hat{\imath} + 5$ m $\hat{\jmath}$ and then along the vector $d_2 = 2$ m $\hat{\imath} - 3$ m $\hat{\jmath}$. How far is the 5j m object from where it started?

   (A) 6.3 m
   (B) 8 m
   (C) 10 m
   (D) 14 m

8. If $\mathbf{A} = \nearrow$ and $\mathbf{B} = \nearrow$, which of the following best represents the direction of $\mathbf{A} - \mathbf{B}$?

   (A) $\longrightarrow$

   (B) $\longleftarrow$

   (C) $\nearrow$

   (D) $\searrow$

9. The x-component of vector **A** is –42, and the angle it makes with the positive x-direction is 130°. What is the y-component of vector **A** ?

   (A) –65.3
   (B) –50.1
   (C) 50.1
   (D) 65.3

10. If two non-zero vectors are added together, and the resultant vector is zero, what must be true of the two vectors?

    (A) They have equal magnitude and are pointed in the same direction.
    (B) They have equal magnitude and are pointed in opposite directions.
    (C) They have different magnitudes and are pointed in opposite directions.
    (D) It is not possible for the sum of two non-zero vectors to be zero.

Section II: Free-Response

1. Let the vectors, **A, B,** and **C** be defined by: $\mathbf{A} = 3\hat{\imath} + 6\hat{\jmath}$, $\mathbf{B} = -\hat{\imath} + 4\hat{\jmath}$, and $\mathbf{C} = 5\hat{\imath} - 2\hat{\jmath}$.

   (a) What is the magnitude of vector **A**?

   (b) Sketch the vector subtraction problem **B** – **C** and find the components of the resultant vector.

   (c) Find the components of **A** + 2**B**.

   (d) Express **A** – **B** – **C** as a magnitude and angle relative to the horizontal.

2. An ant walks 20 cm due north, 30 cm due east, and then 14 cm northeast.

   (a) Assuming that each portion of the ant's journey is a vector, sketch the ant's path.

   (b) How far has the ant travelled from its original position?

   (c) If the ant does not want to travel farther than 80 cm from its original position, how march farther north could it walk?

3. Consider an airplane taking off at an angle of 10° relative to the horizontal. After 2 s, the airplane has travelled a total of 140 m through the air.

   (a) Express the position, **p**, of the airplane compared with the point of liftoff as a vector using components and the $\hat{\imath}$ and $\hat{\jmath}$ basis vectors.

   (b) If the airplane had been taking off in a headwind, the position of the airplane would instead have been **p'** = **p** + **w**, in which the impact of the wind, **w**, is the vector **w** = –30$\hat{\imath}$. Draw a sketch to show the relationship between the vectors **p'**, **p**, and **w**.

   (c) What angle would the plane make with the horizontal as a result of the headwind after 2 s?

4. As a boat travels in a river, its velocity, **v**, is determined by the current, **c**, and its relative velocity compared to the water, **r**. For instance, at one point in the boat's journey, these vectors are related in the following way:

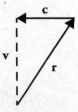

   (a) Write a vector equation that shows how **v** is related to **c** and **r**.

   (b) If the angle between **v** and **c** is 90°, as shown, write an equation for the magnitude of **v** as a function of **c** and **r**.

   (c) If the current is flowing due east at 5 m/s and the boat wants to travel due south at 10 m/s, what should the captain set its relative velocity to be (magnitude and direction)?

Summary

o   Vectors are quantities that have both magnitude and direction. Many important physical quantities such as forces and velocities are vector quantities.

o   Vectors can be represented graphically with an arrow, numerically with a magnitude and direction, or numerically with components.

o   Vectors can be added (or subtracted) graphically by drawing the first vector and then starting the tail of the second vector at the end of the first (remembering to flip the direction of the second vector for subtraction).

o   Vectors can be added (or subtracted) numerically by adding (or subtracting) individual components.

o   Multiplying a vector by a scalar can change the length of the vector or flip the direction by 180° if the scalar is negative.

o   If the magnitude, A, and angle relative to the horizontal, θ, are known, the x- and y-components can be calculated using:

$$A_x = A \cos \theta$$
$$A_y = A \sin \theta$$

o   If the components, A_x and A_y are known, the magnitude, A, and angle relative to the horizontal, θ, can be calculated using:

$$A = \sqrt{(A_x)^2 + (A_y)^2}$$
$$\theta = \tan^{-1} \frac{A_y}{A_x}$$

Chapter 4
Kinematics

And yet it moves.

—Galileo Galilei

Galileo, the "father of modern physics," stated a distinction between the cause of motion and the description of motion. Kinematics is that modern description of motion that answers questions such as:

How far does this object travel?

How fast and in what direction does it move?

At what rate does its speed change?

Kinematics are the mathematical tools for describing motion in terms of displacement, velocity, and acceleration.

POSITION

An object's position is its location in a certain space. Since it is difficult to describe an object's location, in mathematics, we typically use a coordinate system to show where an object is located. Using a coordinate system with an origin also helps to determine positive and negative positions. Typically we set the object in question as the origin and relate its surroundings to the object.

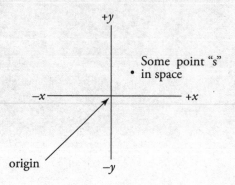

DISPLACEMENT

Pay Attention To What The Question Is Asking
Asking for displacement and total distance will give different answers for a problem.

Displacement is an object's change in position. It's the vector that points from the object's initial position to its final position, regardless of the path actually taken. Since displacement means *change in position*, it is generically denoted Δs, where Δ denotes *change in* and s means <u>s</u>patial location. The displacement is only the distance from an object's final position and initial position. The net distance takes into account the total path taken (meaning if an object went backward and then forward, those distances are included in the total distance). Since a distance is being measured, the SI unit for displacement is the meter $[\Delta s] = m$.

> **Example 1** A rock is thrown straight upward from the edge of a 30 m cliff, rising 10 m, then falling all the way down to the base of the cliff. Find the rock's displacement.

Solution. Displacement only refers to the object's initial position and final position, not the details of its journey. Since the rock started on the edge of the cliff and ended up on the ground 30 m below, its displacement is 30 m, downward. Its distance traveled is 50 m; 10 m on the way up and 40 m on the way down.

> **Example 2** An infant crawls 5 m east, then 3 m north, then 1 m east. Find the magnitude of the infant's displacement.

Solution. Although the infant crawled a *total* distance of 5 + 3 + 1 = 9 m, this is not the displacement, which is merely the *net* distance traveled.

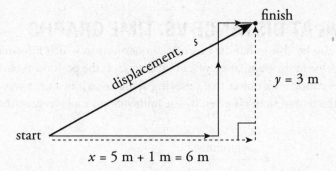

Using the Pythagorean Theorem, we can calculate that the magnitude of the displacement is

$$\Delta s = \sqrt{\left(\Delta x\right)^2 + \left(\Delta\left(y\right)\right)^2} = \sqrt{\left(6\,\text{m}\right)^2 + \left(3\,\text{m}\right)^2} = \sqrt{45\,\text{m}^2} = 6.7\,\text{m}$$

Example 3 In a track-and-field event, an athlete runs exactly once around an oval track, a total distance of 500 m. Find the runner's displacement for the race.

Solution. If the runner returns to the same position from which she left, then her displacement is zero.

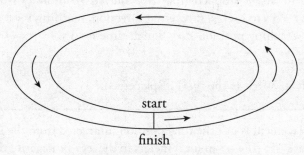

The *total* distance covered is 500 m, but the net distance—the displacement—is 0.

A Note About Notation

Δs is a more general term that works in any direction in space. The term x or "$\Delta x = x_f - x_i$" has a specific meaning that is defined in the x direction. However, to be consistent with AP notation and to avoid confusion between spacial location and speed, from this point on we will use x in our development of the concepts of speed, velocity, and acceleration. x is just a reference and can refer to any direction in terms of the object's motion (for example straight down or straight up).

LOOKING AT DISTANCE VS. TIME GRAPHS

You should also be able to handle kinematics questions in which information is given graphically. One of the popular graphs in kinematics is the position-versus-time graph. For example, consider an object that's moving along an axis in such a way that its position x as a function of time t is given by the following position-versus-time graph:

Geometry!

You will have to brush up on your coordinate geometry: slope and basic area of geometric shapes.

$$slope = \frac{rise}{run}$$

$$m = \frac{(y_2 - y_1)}{(x_2 - x_1)}$$

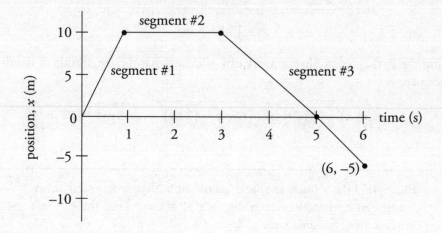

What does this graph tell us? It says that at time $t = 0$, the object was at position $x = 0$. Then, in the first second, its position changed from $x = 0$ to $x = 10$ m. Then, at time $t = 1$ s to 3 s, it stopped (that is, it stayed 10 m away from wherever it started). From $t = 3$ s to $t = 6$ s, it reversed direction, reaching $x = 0$ at time $t = 5$ s, and continued, reaching position $x = -5$ m at time $t = 6$ s.

Example 4 What is the total displacement?

Solution. Displacement is just the final position subtracted from the initial position. In this case, that's $-5 - 0 = -5$ m (or 5 meters to the left, or negative direction).

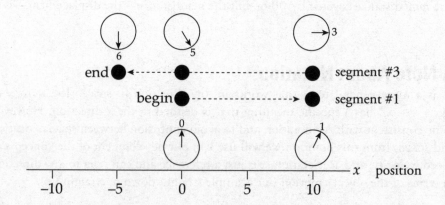

What is the total distance traveled?

| 10 m | + | 15 m | = | 25 meters |
| Segment #1 | | Segment #3 | | |

SPEED AND VELOCITY

When we're in a moving car, the speedometer tells us how fast we're going; it gives us our speed. But what does it mean to have a speed of say, 10 m/s? It means that we're covering a distance of 10 meters every second. By definition, **average speed** is the ratio of the total distance traveled to the time required to cover that distance:

$$\text{average speed} = \frac{\text{total distance}}{\text{time}}$$

The car's speedometer doesn't care what direction the car is moving. You could be driving north, south, east, west, whatever; the speedometer would make no distinction. *55 miles per hour, north* and *55 miles per hour, east* register the same on the speedometer: 55 miles per hour. Remember that speed is scalar, so it does not take into account the direction.

However, we will also need to include *direction* in our descriptions of motion. We just learned about displacement, which takes both distance (net distance) and direction into account. The single concept that embodies both speed and direction is called **velocity**, and the definition of average velocity is:

$$\text{average velocity} = \frac{\text{displacement}}{\text{time}}$$

$$\bar{\mathbf{v}} = \frac{\Delta \mathbf{x}}{\Delta t}$$

$$\text{units} = \text{meter/seconds}$$

Note that the bar over the **v** means *average*. Because $\Delta \mathbf{x}$ is a vector, $\bar{\mathbf{v}}$ is also a vector, and because Δt is a *positive* scalar, the direction of $\bar{\mathbf{v}}$ is the same as the direction of $\Delta \mathbf{x}$. The magnitude of the velocity vector is called the object's **speed**, and is expressed in units of meters per second (m/s).

Note the distinction between speed and velocity. *Velocity is speed plus direction.* An important note: The magnitude of the velocity is the speed. The magnitude of the average velocity is not called the average speed (as we will see in these next two examples).

> **Example 5** Assume that the runner in Example 3 completes the race in 1 minute and 18 seconds. Find her average speed and the magnitude of her average velocity.

Solution. *Average speed is total distance divided by elapsed time.* Since the length of the track is 500 m, the runner's average speed was (500 m)/(78 s) = 6.4 m/s. However, since her displacement was zero, her average velocity was zero also: $\bar{v} = \Delta x/\Delta t = (0 \text{ m})/(78 \text{ s}) = 0$ m/s.

> **Example 6** Is it possible to move with constant speed but not constant velocity? Is it possible to move with constant velocity but not constant speed?

Solution. The answer to the first question is *yes*. For example, if you set your car's cruise control at 55 miles per hour but turn the steering wheel to follow a curved section of road, then the direction of your velocity changes (which means your velocity is not constant), even though your speed doesn't change.

The answer to the second question is *no*. Velocity means speed and direction; if the velocity is constant, then that means both speed and direction are constant. If speed were to change, then the velocity vector's magnitude would change.

ACCELERATION

When you step on the gas pedal in your car, the car's speed increases; step on the brake and the car's speed decreases. Turn the wheel, and the car's direction of motion changes. In all of these cases, the velocity changes. To describe this change in velocity, we need a new term: **acceleration**. In the same way that velocity measures the rate-of-change of an object's position, acceleration measures the rate-of-change of an object's velocity. An object's average acceleration is defined as follows:

$$\text{average acceleration} = \frac{\text{change in velocity}}{\text{time}}$$

$$\bar{a} = \frac{\Delta v}{\Delta t}$$

$$\text{units} = (\text{meters/second})^2 = \text{m/s}^2$$

Note that an object can accelerate even if its speed doesn't change. (Again, it's a matter of not allowing the everyday usage of the word *accelerate* to interfere with its technical, physics usage). This is because acceleration depends on Δv, and the velocity vector v changes if (1) speed changes, or (2) direction changes, or (3) both speed and direction change. For instance, a car traveling around a circular racetrack is constantly accelerating even if the car's *speed* is constant, because the direction of the car's velocity vector is constantly changing.

Example 7 A car is traveling in a straight line along a highway at a constant speed of 80 miles per hour for 10 seconds. Find its acceleration.

Solution. Since the car is traveling at a constant velocity, its acceleration is zero. If there's no change in velocity, then there's no acceleration.

Example 8 A car is traveling along a straight highway at a speed of 20 m/s. The driver steps on the gas pedal and, 3 seconds later, the car's speed is 32 m/s. Find its average acceleration.

Solution. Assuming that the direction of the velocity doesn't change, it's simply a matter of dividing the change in velocity, $v_f - v_i$, 32 m/s – 20 m/s = 12 m/s, by the time interval during which the change occurred: $\bar{a} = \Delta v/\Delta t = (12 \text{ m/s})/(3 \text{ s}) = 4 \text{ m/s}^2$.

Example 9 Spotting a police car ahead, the driver of the car in Example 8 slows from 32 m/s to 20 m/s in 2 seconds. Find the car's average acceleration.

Solution. Dividing the change in velocity, 20 m/s – 32 m/s = –12 m/s, by the time interval during which the change occurred, 2 s, give us $\bar{a} = \Delta v/\Delta t = (-12 \text{ m/s})/(2 \text{ s}) = -6 \text{ m/s}^2$. The negative sign here means that the direction of the acceleration is opposite the direction of the velocity, which describes slowing down.

Let's next consider an object moving along a straight axis in such a way that its velocity, v, as a function of time, t, is given by the following velocity-versus-time graph:

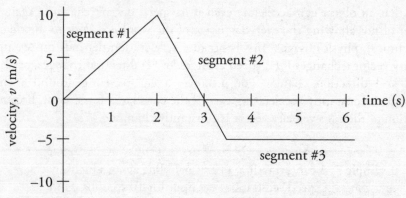

What does this graph tell us? It says that, at time $t = 0$, the object's velocity was $v = 0$. Over the first two seconds, its velocity increased steadily to 10 m/s. At time $t = 2$ s, the velocity then began to decrease (eventually becoming $v = 0$, at time $t = 3$ s). The velocity then became negative after $t = 3$ s, reaching $v = -5$ m/s at time $t = 3.5$ s. From $t = 3.5$ s on, the velocity remained a steady -5 m/s.

Segment #1: What can we ask about this motion? First, the fact that the velocity changed from $t = 0$ to $t = 2$ s tells us that the object accelerated. The acceleration during this time was

$$a = \frac{\Delta v}{\Delta t} = \frac{(10 - 0) \text{ m/s}}{(2 - 0) \text{ s}} = 5 \text{ m/s}^2$$

Note, however, that the ratio that defines the acceleration, $\Delta v / \Delta t$, also defines the slope of the v versus t graph. Therefore,

> *The slope of a velocity-versus-time graph gives the acceleration.*

Segment #2: What was the acceleration from time $t = 2$ s to time $t = 3.5$ s? The slope of the line segment joining the point $(t, v) = (2 \text{ s}, 10 \text{ m/s})$ to the point $(t, v) = (3.5 \text{ s}, -5 \text{ m/s})$ is

$$a = \frac{\Delta v}{\Delta t} = \frac{(-5 - 10) \text{ m/s}}{(3.5 - 2) \text{ s}} = -10 \text{ m/s}^2$$

Between time $t = 2$ s to time $t = 3.5$ s, the object experienced a negative acceleration. Between time $t = 2$ s to time $t = 3$ s, the object's velocity was slowed down till it reached 0 m/s. Between time $t = 3$ s to time $t = 3.5$ s, the object began increasing its velocity in the opposite direction.

When the slope of a velocity versus time graph is zero, the acceleration is zero.

Segment #3: After time $t = 3.5$ s, the slope of the graph is zero meaning the object experienced zero acceleration. This, however, does not mean the object did not move since its velocity was constant and in the negative direction.

Let's look at Example 9 on page 85 again: How far did the object travel during a particular time interval? For example, let's figure out the displacement of the object from time $t = 4$ s to time $t = 6$ s. During this time interval, the velocity was a constant -5 m/s, so the displacement was $\Delta x = v\Delta t = (-5 \text{ m/s})(2 \text{ s}) = -10$ m.

Geometrically, we've determined the area between the graph and the horizontal axis. After all, the area of a rectangle is *base × height* and, for the shaded rectangle shown below, the *base* is Δt, and the *height* is v. So, *base × height* equals $\Delta t \times v$, which is displacement.

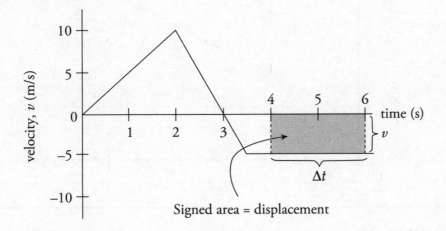

Signed area = displacement

We say *signed area* because regions below the horizontal axis are negative quantities (since the object's velocity is negative, its displacement is negative). Thus, by counting areas above the horizontal axis as positive and areas below the horizontal axis as negative, we can make the following claim:

> Given a velocity-versus-time graph, the area between the graph and the t-axis equals the object's displacement.

What is the object's displacement from time $t = 0$ to $t = 3$ s? Using the fact that displacement is the area bounded by the velocity graph, we figure out the area of the triangle shown below:

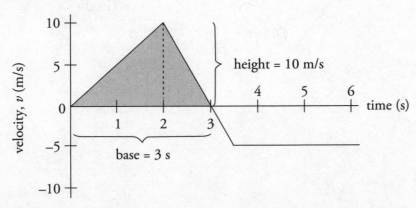

Since the area of a triangle is $\left(\dfrac{1}{2}\right) \times$ base $\times$ height, we find that $\Delta x = \dfrac{1}{2}(3\text{ s})(10\text{ m/s}) = 15\text{ m}$.

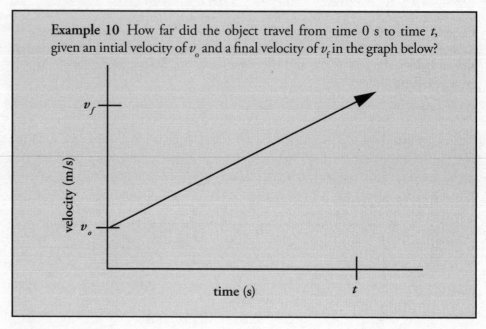

Example 10 How far did the object travel from time 0 s to time t, given an intial velocity of v_o and a final velocity of v_f in the graph below?

Solution.

This Solution Looks Familiar
And this is how Big Five #1 equation is derived.

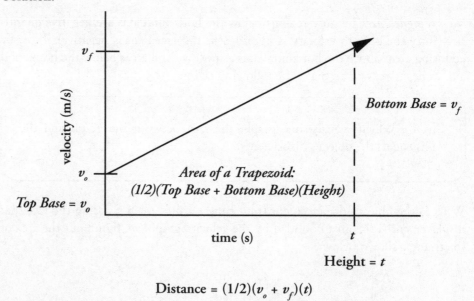

$$\text{Distance} = (1/2)(v_o + v_f)(t)$$

Let's consider the relationship between acceleration and velocity in a velocity versus time graph.

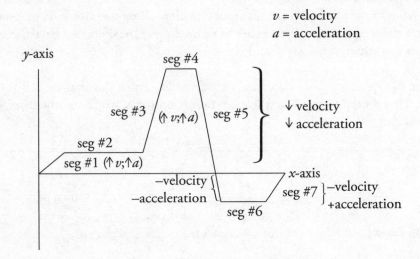

If we take an object's original direction of motion to be positive, then an increase in speed corresponds to positive acceleration. This is indicative of Segment #1 and Segment #3.

A decrease in velocity corresponds to negative acceleration (as indicated in top part of Segment #5).

However, if an object's original direction is negative, then an increase in speed corresponds to negative acceleration, indicated by the bottom part of Segment #5. Whatever is below the *x*-axis shows us that you are speeding up backward.

When the velocity and acceleration are in opposite directions, the object slows down, as is indicated in Segment #7. Note that Segments #2, #4, and #6 indicate no acceleration.

UNIFORMLY ACCELERATED MOTION AND THE BIG FIVE

The simplest type of motion to analyze is motion in which the acceleration is *constant*.

Another restriction that will make our analysis easier is to consider only motion that takes place along a straight line. In these cases, there are only two possible directions of motion. One is positive, and the opposite direction is negative. Most of the quantities we've been dealing with—displacement, velocity, and acceleration—are vectors, which means that they include both a magnitude and a direction. With straight-line motion, direction can be specified simply by attaching a + or − sign to the magnitude of the quantity. Therefore, although we will often abandon the use of bold letters to denote the vector quantities of displacement, velocity, and acceleration, the fact that these quantities include direction will still be indicated by a positive or negative sign.

Let's review the quantities we've seen so far. The fundamental quantities are position (x), velocity (v), and acceleration (a). Acceleration is a change in velocity, from

In the real world, truly uniform acceleration hardly occurs due to multiple factors. However for the purposes of the AP Physics 1 Exam, everything is treated in ideal situations unless otherwise noted.

an initial velocity (v_i or v_0) to a final velocity (v_f or simply v—with no subscript). And, finally, the motion takes place during some elapsed time interval, Δt. Also, if we agree to start our clocks at time $t_i = 0$, then $\Delta t = t_f - t_i = t - 0 = t$, so we can just write t instead of Δt in the first four equations. This simplification in notation makes these equations a little easier to write down. Therefore, we have five kinematics quantities: Δx, v_0, v, a, and Δt.

These five quantities are related by a group of five equations that we call the *Big Five*. They work in cases where acceleration is uniform, which are the cases we're considering.

When doing problems, writing down what is given and what you are solving for can help determine which Big Five equation to use.

		Variable that's missing
Big Five #1:	$\Delta x = \dfrac{1}{2}(v_0 + v)t$	a
Big Five #2:	$v = v_0 + at$	x
Big Five #3:	$x = x_0 + v_0 t + \dfrac{1}{2}at^2$	v
Big Five #4:	$x = x_0 + vt - \dfrac{1}{2}at^2$	v_0
Big Five #5:	$v^2 = v_0^2 + 2a(x - x_0)$	t

Big Five #1 is the definition of velocity (this is the area under a velocity versus time graph, which will be covered a little further in this chapter). Big Five #2 is the definition of acceleration (this is the slope at any given moment of a velocity versus time graph).

Equations #1, #2, and #5 are the important three that can be used to solve any problem (equations #3 and 4 are derivations). However, it is advisable to memorize all five equations to speed up problem-solving during the test.

Big Five #1 and #2 are simply the definitions of $\bar{v}$ and $\bar{a}$ written in forms that don't involve fractions. The other Big Five equations can be derived from these two definitions and the equation $\bar{v} = \dfrac{1}{2}(v_0 + v)$, using a bit of algebra.

Example 11 An object with an initial velocity of 4 m/s moves along a straight axis under constant acceleration. Three seconds later, its velocity is 14 m/s. How far did it travel during this time?

Solution. We're given v_0, Δt, and v, and we're asked for x. So a is missing; it isn't given and it isn't asked for, and we use Big Five #1:

$$x = \bar{v}t = \frac{1}{2}(v_0 + v)t = \frac{1}{2}(4 \text{ m/s} + 14 \text{ m/s})(3 \text{ s}) = 27 \text{ m}$$

Structured Example Method to Solve Problems

Given:

v(initial) = 4 m/s
v(final) = 14 m/s
$t = 3$ s

Missing:

acceleration (a)

Unknown (What we are solving for):

distance (x)

Equation:

Big Five #1

It's okay to leave off the units in the middle of the calculation as long as you remember to include them in your final answer. **Leaving units off your final answer will cost you points on the AP Exam.**

Example 12 A car that's initially traveling at 10 m/s accelerates uniformly for 4 seconds at a rate of 2 m/s², in a straight line. How far does the car travel during this time?

Solution. We're given v_0, t, and a, and we're asked for x. So, v is missing; it isn't given, and it isn't asked for. Therefore, we use Big Five #3:

$$x = x_0 + v_0 \Delta t + \frac{1}{2} a \left(\Delta t \right)^2 = \left(10 \text{ m/s} \right) \left(4 \text{ s} \right) + \frac{1}{2} \left(2 \text{ m/s}^2 \right) \left(4 \text{ s} \right)^2 = 56 \text{ m}$$

Example 13 A rock is dropped off a cliff that's 80 m high. If it strikes the ground with an impact velocity of 40 m/s, what acceleration did it experience during its descent?

Solution. If something is dropped, then that means it has no initial velocity: $v_0 = 0$. So, we're given v_0, Δx, and v, and we're asked for a. Since t is missing, we use Big Five #5:

$$v^2 = v_0^2 + 2a(x - x_0) \Rightarrow v^2 = 2a(x - x_0) \quad \left(\text{since } v_0 = 0 \right)$$

$$a = \frac{v^2}{2(x - x_0)} = \frac{\left(40 \text{ m/s} \right)^2}{2 \left(80 \text{ m} \right)} = 10 \text{ m/s}^2$$

Initial Velocity?
Some problems do not give initial velocity in numerical terms. If you see the statement "starting from rest" or "dropped," we can assume our initial velocity is zero.

Note that since a has the same sign as $(x - x_0)$, the acceleration vector points in the same direction as the displacement vector. This makes sense here, since the object moves downward, and the acceleration it experiences is due to gravity, which also points downward.

ADDITIONAL KINEMATIC GRAPHICAL ASPECTS

Not all graphs have nice straight lines as shown so far. Straight line segments represent constant slopes and therefore constant velocities, or accelerations, depending on the type of graph. What happens as an object changes its velocity in a position versus time graph? The "lines" become "curves." Let's look at a typical question that might be asked about such a curve.

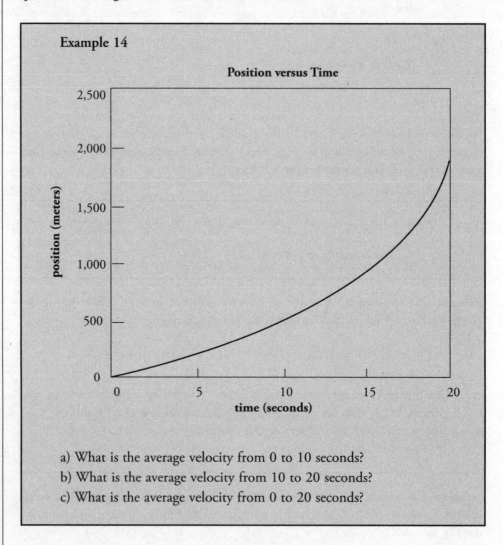

Example 14

a) What is the average velocity from 0 to 10 seconds?
b) What is the average velocity from 10 to 20 seconds?
c) What is the average velocity from 0 to 20 seconds?

Solution. This is familiar territory. To find the average velocity, use

a) $v_{avg} = \dfrac{\Delta x}{\Delta t} \Rightarrow \dfrac{(500 - 0)}{(10 - 0)} \Rightarrow 50 \text{ m/s}$

b) $v_{avg} = \dfrac{\Delta x}{\Delta t} \Rightarrow \dfrac{(2,000 - 500)}{(20 - 10)} \Rightarrow 150 \text{ m/s}$

c) $v_{avg} = \dfrac{\Delta x}{\Delta t} \Rightarrow \dfrac{(2,000 - 0)}{(20 - 0)} \Rightarrow 100 \text{ m/s}$

The instantaneous velocity is the velocity at a given moment in time. When you drive in a car and look down at the speedometer, you see the magnitude of your instantaneous velocity at that time.

To find the instantaneous velocity at 10 seconds, we will need to approximate. The velocity from 9–10 seconds is close to the velocity at 10 seconds, but it is still a bit too slow. The velocity from 10–11 seconds is also close to the velocity at 10 seconds, but it is still a bit too fast. You can find the middle ground between these two ideas, or the slope of the line that connects the point before and the point after 10 seconds. This is very close to the instantaneous velocity. A true tangent line touches the curve at only one point, but this line is close enough for our purposes.

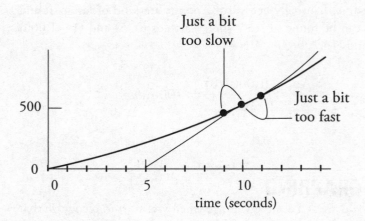

What Is a Tangent?
A line that touches a curved line at only one point.

Example 15 Using the graph below, what is the instantaneous velocity at 10 seconds?

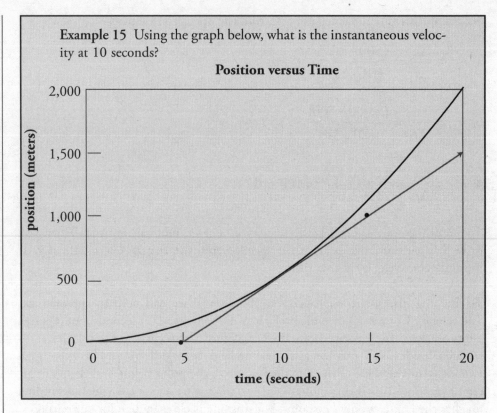

Solution. Draw a tangent line. Find the slope of the tangent by picking any two points on the tangent line. It usually helps if the points are kind of far apart and it also helps if points can be found at "easy spots" such as (0, 5) and (15, 1,000) and not (6.27, 113) and (14.7, 983)

$$v_{ins} = v_{tan} = \frac{\Delta x}{\Delta t} \Rightarrow \frac{(1,000 - 0)}{(15 - 5)} \Rightarrow 100 \text{ m/s}$$

QUALITATIVE GRAPHING

Beyond all the math, you are at a clear advantage when you start to recognize that position-versus-time and velocity-versus-time graphs have a few basic shapes, and that all the graphs you will see will be some form of these basic shapes. Having a feel for these building blocks will go a long way toward understanding kinematics graphs in physics.

Either of the following two graphs represents something that is not moving.

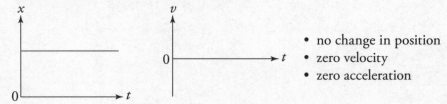

- no change in position
- zero velocity
- zero acceleration

Either of the following two graphs represents an object moving at a constant velocity in the positive direction.

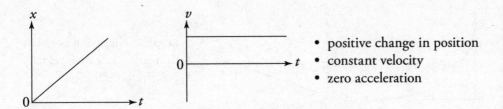

- positive change in position
- constant velocity
- zero acceleration

Either of the following two graphs represents an object moving at a constant velocity in the negative direction.

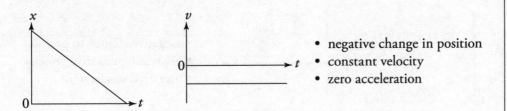

- negative change in position
- constant velocity
- zero acceleration

Either of the following two graphs represents an object speeding up in the positive direction.

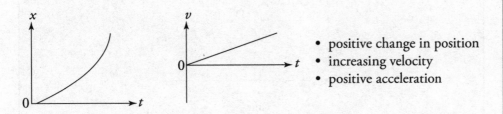

- positive change in position
- increasing velocity
- positive acceleration

Either of the following two graphs represents an object slowing down in the positive direction.

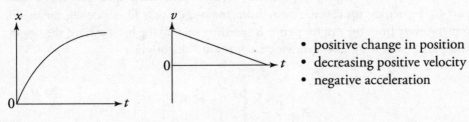

- positive change in position
- decreasing positive velocity
- negative acceleration

Either of the following two graphs represents an object slowing down in the negative direction.

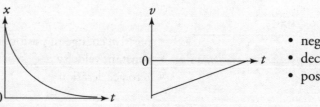

- negative change in position
- decreasing negative velocity
- positive acceleration

Either of the following two graphs represents an object speeding up in the negative direction.

- negative change in position
- increasing negative velocity
- negative acceleration

Example 16 Below is a position-versus-time graph. Describe in words the motion of the object and sketch the corresponding velocity-versus-time graph.

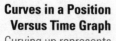

Curves in a Position Versus Time Graph
Curving up represents a positive acceleration. Curving down represents a negative acceleration.

Solution. Part A is a constant speed moving away from the origin, part B is at rest, part C is speeding up moving away from the origin, part D is slowing down still moving away from the origin, part E is speeding up moving back toward the origin, and part F is slowing down moving back toward the origin.

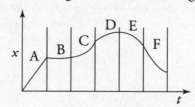

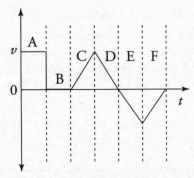

The area under an acceleration-versus-time graph gives the average velocity.

Example 17 The velocity of an object as a function of time is given by the following graph:

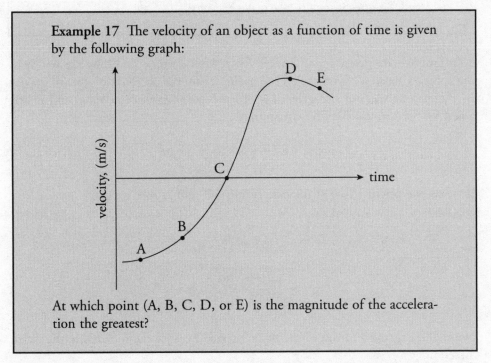

At which point (A, B, C, D, or E) is the magnitude of the acceleration the greatest?

One Slope to Rule Them All
The greater the slope at a point of a velocity-versus-time graph, the greater the acceleration is.

The greater the slope at a point of a position-versus-time graph, the greater the velocity is.

Solution. The acceleration is the slope of the velocity-versus-time graph. Although this graph is not composed of straight lines, the concept of slope still applies; at each point, the slope of the curve is the slope of the tangent line to the curve. The slope is essentially zero at Points A and D (where the curve is flat), small and positive at B, and small and negative at E. The slope at Point C is large and positive, so this is where the object's acceleration is the greatest.

FREE FALL

The simplest real-life example of motion under pretty constant acceleration is the motion of objects in Earth's gravitational field, near the surface of the Earth and ignoring any effects due to the air (mainly air resistance). With these effects ignored, an object can fall *freely*. That is, it can fall experiencing only acceleration due to gravity. Near the surface of the Earth, the gravitational acceleration has a constant magnitude of about 9.8 m/s²; this quantity is denoted g (for *gravitational* acceleration). On the AP Physics 1 Exam, you may use $g = 10$ m/s^2 as a simple approximation to $g = 9.8$ m/s^2. Even though you can use the exact value on your calculator, this approximation helps to save time, so in this book, we will always use $g = 10$ m/s^2. And, of course, the gravitational acceleration vector, **g**, points *downward*.

Since gravity points downward, for the sake of keeping things consistent throughout the rest of this chapter, we will set gravity as $g = -10$ m/s^2.

> **Example 18** A rock is dropped from a cliff 80 m above the ground. How long does it take to reach the ground?

Solution. We are given v_o, and asked for t. Unless you specifically see words to the contrary (such as, "you are on the moon where the acceleration due to gravity is…"), assume you are also given $a = -10$ m/s^2. Because v is missing, and it isn't asked for, we can use Big Five equation #3.

$$y = y_0 + v_0 t + \frac{1}{2}at^2 \Rightarrow y = \frac{1}{2}at^2$$

A Note About Gravity
We don't always have to set gravity as negative. You are allowed to choose any coordinate system you wish as long as you match the positive or negative with gravity correctly.

If we set the origin $y_0 = 0$ at the base of the cliff and $v_0 = 0$, we get:

$$t = \sqrt{\frac{2y}{a}}$$

$$t = \sqrt{\frac{2(-80 \text{ m})}{(-10 \text{ m/s}^2)}} = 4 \text{ s}$$

Note: The negative in front of the 80 is inserted because the rock fell in the down direction.

> **Example 19** A baseball is thrown straight upward with an initial speed of 20 m/s. How high will it go?

Solution. We are given v_0, $a = -10$ m/s^2 is implied, and we are asked for y. Now, neither t nor v is expressly given; however, we know the vertical velocity at the top is 0 (otherwise the baseball would still rise). Consequently, we use Big Five equation #5.

$$v^2 = v_0^2 + 2a(y - y_0) \Rightarrow -2ay = v_0^2$$

The time it takes for an object to be thrown straight up is the same as the time it takes for an object to fall down from the peak back into your hand.

We set $y_0 = 0$ and we know that $v = 0$, so that leaves us with:

$$y = -\frac{v_0^2}{2a}$$

$$y = -\frac{(20 \text{ m/s})^2}{2(-10 \text{ m/s}^2)} = 20 \text{ m}$$

Example 20 One second after being thrown straight down, an object is falling with a speed of 20 m/s. How fast will it be falling 2 seconds later?

Solution. We're given v_0, a, and t, and asked for v. Since x is missing, we use Big Five #2:

$$v = v_0 + at = (-20 \text{ m/s}) + (-10 \text{ m/s}^2)(2 \text{ s}) = -40 \text{ m/s}$$

The negative sign in front of the 40 simply indicates that the object is traveling in the down direction.

Example 21 If an object is thrown straight upward with an initial speed of 8 m/s and takes 3 seconds to strike the ground, from what height was the object thrown?

Solution. We're given a, v_0, and t, and we need to find y_0. Because v is missing, we use Big Five #3:

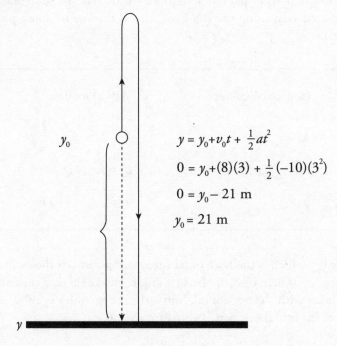

$$y = y_0 + v_0 t + \frac{1}{2}at^2$$

$$0 = y_0 + (8)(3) + \frac{1}{2}(-10)(3^2)$$

$$0 = y_0 - 21 \text{ m}$$

$$y_0 = 21 \text{ m}$$

PROJECTILE MOTION

In general, an object that moves near the surface of the Earth will not follow a straight-line path (for example, a baseball hit by a bat, a golf ball struck by a club, or a tennis ball hit from the baseline). If we launch an object at an angle other than straight upward and consider only the effect of acceleration due to gravity, then the object will travel along a parabolic trajectory.

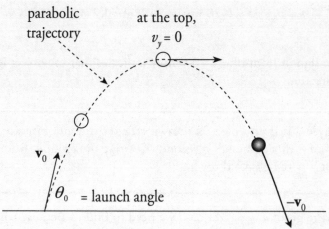

A Note About Vectors
Vectors that are perpendicular to each other do not affect each other's magnitudes, only their directions.

To simplify the analysis of parabolic motion, we analyze the horizontal and vertical motions separately, using the Big Five. This is the key to doing projectile motion problems.

Horizontal motion:	Vertical motion:
$\Delta x = v_{0x}t$	$y = y_0 + v_{0y}t - \dfrac{1}{2}gt^2$
$v_x = v_{0x}$ (constant!)	$v_y = v_{0y} - gt$
$a_x = 0$	$a_y = -g = -10 \text{ m/s}^2$

The quantity v_{0x}, which is the horizontal (or x) component of the initial velocity, is equal to $v_0 \cos \theta_0$, where θ_0 is the **launch angle**, the angle that the initial velocity vector, $\mathbf{v}_0$, makes with the horizontal. Similarly, the quantity v_{0y}, the vertical (or y) component of the initial velocity, is equal to $v_0 \sin \theta_0$.

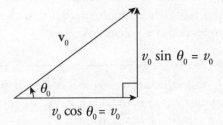

$$v_0 \sin \theta_0 = v_0$$

$$\theta_0$$

$$v_0 \cos \theta_0 = v_0$$

Example 22 An object is thrown horizontally with an initial speed of 10 m/s. It hits the ground 4 seconds later. How far did it drop in 4 seconds?

Solution. The first step is to decide whether this is a *horizontal* question or a *vertical* question, since you must consider these motions separately. The question *How far did it drop?* is a *vertical* question, so the set of equations we will consider are those listed above under *vertical motion*. Next, *How far...?* implies that we will use the first of the vertical-motion equations, the one that gives vertical displacement, Δy.

Now, since the object is thrown horizontally, there is no vertical component to its initial velocity vector $\mathbf{v}_0$; that is, $v_{0y} = 0$. Therefore,

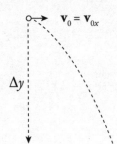

$$\Delta y = v_{0y}t - \frac{1}{2}gt^2 \rightarrow \Delta y = -\frac{1}{2}gt^2 \quad (\text{because } v_{0y} = 0)$$

$$= -\frac{1}{2}(10)(4^2)$$

$$= -80 \text{ m}$$

The fact that Δy is negative means that the displacement is *down*. Also, notice that the information given about v_{0x} is irrelevant to the question.

Example 23 From a height of 100 m, a ball is thrown horizontally with an initial speed of 15 m/s. How far does it travel horizontally in the first 2 seconds?

Solution. The question, *How far does it travel horizontally...?*, immediately tells us that we should use the first of the horizontal-motion equations:

$$\Delta x = v_{0x}t = (15 \text{ m/s})(2 \text{ s}) = 30 \text{ m}$$

The information that the initial vertical position is 100 m above the ground is irrelevant (except for the fact that it's high enough that the ball doesn't strike the ground before the two seconds have elapsed).

> **Example 24** A projectile is traveling in a parabolic path for a total of 6 seconds. How does its horizontal velocity 1 s after launch compare to its horizontal velocity 4 s after launch?

Solution. The only acceleration experienced by the projectile is due to gravity, which is purely vertical, so that there is no horizontal acceleration. If there's no horizontal acceleration, then the horizontal velocity cannot change during flight, and the projectile's horizontal velocity 1 s after it's launched is the same as its horizontal velocity 3 s later.

> **Example 25** An object is projected upward with a 30° launch angle and an initial speed of 40 m/s. How long will it take for the object to reach the top of its trajectory? How high is this?

For the vertical component of parabolic motion, you can treat it as throwing an object straight up and it falling back down. Just remember that the object traveling to its peak is only a part of its time of travel. It still needs to fall.

Solution. When the projectile reaches the top of its trajectory, its velocity vector is momentarily horizontal; that is, $v_y = 0$. Using the vertical-motion equation for v_y, we can set it equal to 0 and solve for t:

$$v_y \overset{set}{=} 0 \Rightarrow v_{0y} - gt = 0$$

$$t = \frac{v_{0y}}{g} = \frac{v_0 \sin \theta_0}{g} = \frac{\left(40 \text{ m/s}\right) \sin 30°}{10 \text{ m/s}^2} = 2 \text{ s}$$

At this time, the projectile's vertical displacement is

$$\Delta y = v_{0y}t - \frac{1}{2}\left(g\right)t^2 = \left(v_0 \sin \theta_0\right)t - \frac{1}{2}\left(g\right)t^2$$

$$= \left[\left(40 \text{ m/s}\right) \sin 30°\right]\left(2 \text{ s}\right) - \frac{1}{2}\left(10 \text{ m/s}^2\right)\left(2 \text{ s}\right)^2$$

$$= 20 \text{ m}$$

> **Example 26** An object is projected upward with a 30° launch angle from the ground and an initial speed of 60 m/s. For how many seconds will it be in the air? How far will it travel horizontally? Assume it returns to its original height.

Solution. The total time the object spends in the air is equal to twice the time required to reach the top of the trajectory (because the parabola is symmetrical). So, as we did in the previous example, we find the time required to reach the top by setting v_y equal to 0, and now double that amount of time:

$$v_y \overset{set}{=} 0 \implies v_{0y} - gt = 0$$

$$t = \frac{v_{0y}}{g} = \frac{v_0 \sin \theta_0}{g} = \frac{(60 \text{ m/s}) \sin 30°}{10 \text{ m/s}^2} = 3 \text{ s}$$

Therefore, the *total* flight time (that is, up and down) is $t_t = 2t = 2 \times (3 \text{ s}) = 6 \text{ s}$.

Now, using the first horizontal-motion equation, we can calculate the horizontal displacement after 6 seconds:

$$\Delta x = v_{0x} t_t = (v_0 \cos \theta_0) t_t = [(60 \text{ m/s}) \cos 30°](6 \text{ s}) = 312 \text{ m}$$

By the way, assuming it lands back at its original height, the full horizontal displacement of a projectile is called the projectile's **range**.

Because the motion in this example is parabolic, $v_f = -v_0$. If you use that value for v_f in the equations here, you can solve for the entire flight time (making it unnecessary to multiply by 2 at the end). Both methods give the correct solution, so use whichever you find easier.

Chapter 4 Review Questions

Answers and Explanations can be found in Chapter 13.

Section I: Multiple-Choice

[handwritten: acceleration]

1. An object that's moving with constant speed travels once around a circular path. Which of the following statements are true concerning this motion? Select two answers.

   (A) The displacement is zero.
   (B) The average speed is zero. ✗
   (C) The acceleration is zero. ✓
   (D) The velocity is changing.

2.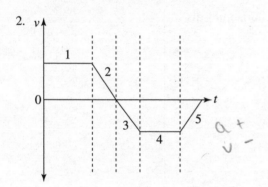

   In section 5 of the velocity-time graph, the object is

   (A) speeding up moving in the positive direction
   (B) slowing down moving in the positive direction
   (C) speeding up moving in the negative direction
   (D) slowing down moving in the negative direction

3. Which of the following statements are true about uniformly accelerated motion? Select two answers.

   (A) If an object's acceleration is constant, then it must move in a straight line. ✗
   (B) If an object's acceleration is zero, then it's speed must remain constant. ✓
   (C) If an object's speed remains constant, then its acceleration must be zero. ✓
   (D) If the object's direction of motion is changing, then its acceleration is not zero.

4. A baseball is thrown straight upward. What is the ball's acceleration at its highest point?

   (A) $\frac{1}{2} g$, downward
   (B) g, downward
   (C) $\frac{1}{2} g$, upward
   (D) g, upward

   [handwritten: $\Delta x = v_{ix}t + \frac{1}{2}at^2$]
   [handwritten: $v_f = v_0 + at$]
   [handwritten: $v^2 = v_0^2 + 2a(\Delta x)$]

5. How long would it take a car, starting from rest and accelerating uniformly in a straight line at 5 m/s², to cover a distance of 200 m ?

   (A) 9.0 s
   (B) 10.5 s
   (C) 12.0 s
   (D) 15.5 s

   [handwritten: $200 = \frac{1}{2}(5)t^2$]
   [handwritten: $200 = 2.5t^2$]

6. A rock is dropped off a cliff and strikes the ground with an impact velocity of 30 m/s. How high was the cliff?

   (A) 20 m
   (B) 30 m
   (C) 45 m
   (D) 60 m

   [handwritten: $30^2 = 0^2 + 2(-9.8)(\Delta y)$]
   [handwritten: $900 = -19.6(\Delta y)$]

7. A stone is thrown horizontally with an initial speed of 10 m/s from a bridge. Assuming that air resistance is negligible, how long would it take the stone to strike the water 80 m below the bridge?

   [handwritten: 10 m/s →]

   (A) 1 s
   (B) 2 s
   (C) 4 s
   (D) 8 s

   [handwritten: $v_{0y} = 0$]
   [handwritten: $-80 = \frac{1}{2}(-9.8)t^2$]
   [handwritten: $-80 = -4.9t^2$]

8. A soccer ball, at rest on the ground, is kicked with an initial velocity of 10 m/s at a launch angle of 30°. Calculate its total flight time, assuming that air resistance is negligible.

   (A) 0.5 s
   (B) 1 s
   (C) 2 s
   (D) 4 s

   [handwritten: $10 \quad 30 \quad v_x = \cos 30(10) = 8.11$]
   [handwritten: $v_y = \sin 30(10) = 5$]
   [handwritten: $0 = 5t + \frac{1}{2}(-9.8)t^2$]
   [handwritten: $-5 = -4.9t^2 \quad 0 = 5t - 4.9t^2$]
   [handwritten: $0 = t(5 - 4.9t)$]

9. A stone is thrown horizontally with an initial speed of 30 m/s from a bridge. Find the stone's total speed when it enters the water 4 seconds later, assuming that air resistance is negligible. → 30 m/s

   (A) 30 m/s
   (B) 40 m/s
   (C) 50 m/s
   (D) 60 m/s

10. Which one of the following statements is true concerning the motion of an ideal projectile launched at an angle of 45° to the horizontal?

    (A) The acceleration vector points opposite to the velocity vector on the way up and in the same direction as the velocity vector on the way down.
    (B) The speed at the top of the trajectory is zero.
    (C) The object's total speed remains constant during the entire flight.
    (D) The vertical speed decreases on the way up and increases on the way down.

11. A stone is thrown vertically upwards with an initial speed of 5 m/s. What is the velocity of the stone 3 seconds later?

    (A) 25 m/s, upward
    (B) 25 m/s, downward
    (C) 35 m/s, upward
    (D) 35 m/s, downward

12. A car travelling at a speed of v_0 applies its brakes, skidding to a stop over a distance of x m. Assuming that the deceleration due to the brakes is constant, what would be the skidding distance of the same car if it were travelling with twice the initial speed?

    (A) $2x$ m
    (B) $3x$ m
    (C) $4x$ m
    (D) $8x$ m

Section II: Free-Response

1. This question concerns the motion of a car on a straight track; the car's velocity as a function of time is plotted below.

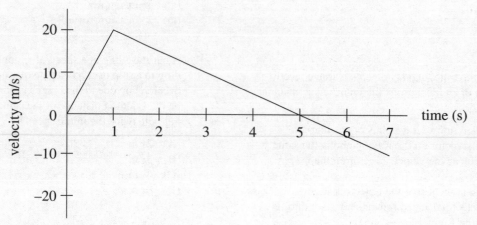

    (a)   Describe what happened to the car at time $t = 1$ s.

    (b)   How does the car's average velocity between time $t = 0$ and $t = 1$ s compare to its average velocity between times $t = 1$ s and $t = 5$ s?

    (c)   What is the displacement of the car from time $t = 0$ to time $t = 7$ s?

    (d)   Plot the car's acceleration during this interval as a function of time.

    (e)   Make a sketch of the object's position during this interval as a function of time. Assume that the car begins at $x = 0$.

2. Consider a projectile moving in a parabolic trajectory under constant gravitational acceleration. Its initial velocity has magnitude v_0, and its launch angle (with the horizontal) is θ_0.

    (a)   Calculate the maximum height, H, of the projectile.

    (b)   Calculate the (horizontal) range, R, of the projectile.

    (c)   For what value of θ_0 will the range be maximized?

    (d)   If $0 < h < H$, compute the time that elapses between passing through the horizontal line of height h in both directions (ascending and descending); that is, compute the time required for the projectile to pass through the two points shown in this figure:

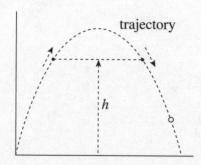

3. A cannonball is shot with an initial speed of 50 m/s at a launch angle of 40° toward a castle wall 220 m away. The height of the wall is 30 m. Assume that effects due to the air are negligible. (For this problem, use $g = 9.8$ m/s².)

    (a)   Show that the cannonball will strike the castle wall.

    (b)   How long will it take for the cannonball to strike the wall?

    (c)   At what height above the base of the wall will the cannonball strike?

4. A cannonball is fired with an initial speed of 40 m/s and a launch angle of 30° from a cliff that is 25 m tall.

    (a)   What is the flight time of the cannonball?

    (b)   What is the range of the cannonball?

    (c)   In the two graphs below, plot the horizontal speed of the projectile versus time and the vertical speed of the projectile versus time (from the initial launch of the projectile to the instant it strikes the ground).

Horizontal Speed vs. Time

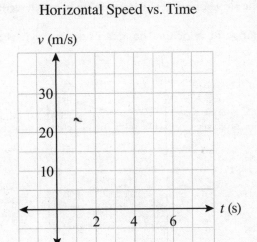

Vertical Speed vs. Time

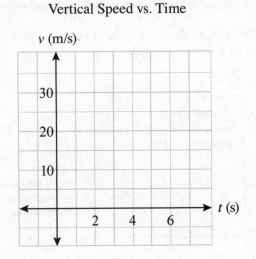

Summary

Graphs are very useful tools to help visualize the motion of an object. They can also help you solve problems once you learn how to translate from one graph to another. As you work with graphs, keep the following things in mind:

o Always look at a graph's axes first. This sounds obvious, but one of the most common mistakes students make is looking at a velocity-versus-time graph, thinking about it as if it were a position-versus-time graph.

o Don't ever assume one box is one unit. Look at the numbers on the axes.

o Lining up position-versus-time graphs *directly above* velocity-versus-time graphs and *directly above* acceleration-versus-time graphs is a must. This way you can match up key points from one graph to the next.

o The slope of an *x* versus *t* graph gives velocity. The slope of a *v* versus *t* graph gives acceleration.

o The area under an *a* versus *t* graph gives the change in velocity. The area under a *v* versus *t* graph gives the displacement.

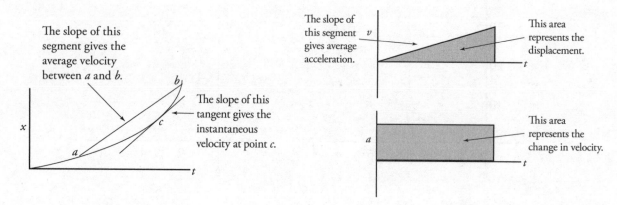

o   The motion of an object in one dimension can be described using the Big Five equations. Look for what is given, determine what you're looking for, and use the equation that has those variables in them. Remember that sometimes there is hidden (assumed) information in the problem, such as $a = -10 \text{ m/s}^2$.

Name of Equation	Equation	Missing Variable
Big Five #1:	$\Delta x = \dfrac{1}{2}(v_0 + v)t$	a
Big Five #2:	$v = v_0 + at$	x
Big Five #3:	$x = x_0 + v_0 t + \dfrac{1}{2}at^2$	v
Big Five #4:	$x = x_0 + vt - \dfrac{1}{2}at^2$	v_0
Big Five #5:	$v^2 = v_0^{\,2} + 2a(x - x_0)$	t

o   For projectiles, it is important to separate the horizontal and vertical components.

Horizontal Motion: Vertical Motion:

$x = v_x t$ $y = y_0 + v_0 t + \dfrac{1}{2}gt^2$

$v_x = v_{0x} = \text{constant}$ $v_y = v_{0y} + gt$

$a_x = 0$ $a = g = -10 \text{ m/s}^2$

o   At any given moment the relationship between v, v_x, and v_y is given by

$v_x = v\cos\theta$ $v_y = v\sin\theta$

$v^2 = v_x^{\,2} + v_y^{\,2}$ and $\theta = \tan^{-1}\left(\dfrac{v_y}{v_x}\right)$

Chapter 5
Newton's Laws

"Nature is pleased with simplicity. And nature is no dummy."

—Sir Isaac Newton

The Englishman Sir Isaac Newton published a book in 1687 called *Philosphiae Naturalis Principa Mathematica* (The Mathematical Principles of Natural Philosophy)—referred to nowadays as simply The Principia—which began the modern study of physics as a scientific discipline. Three of the laws that Newton stated in The Principia form the basis for dynamics and are known simply as Newton's Laws of Motion.

In kinematics we discovered the nature of how objects move, but there still existed the question of why objects move the way they do? For this fundamental task of understanding the cause of motion, we must turn our attention to dynamics.

INTRODUCTION TO FORCES

An interaction between two bodies—a push or a pull—is force. If an apple falls from a tree, it falls to the ground. If you pull on a door handle, it opens a door. If you push on a crate, you move it. In all these cases, some force is required for these actions to happen. In the first case of an apple falling from a tree, the Earth is exerting a downward pull called gravitational force. When you stand on the floor, the floor provides an upward force called the normal force. When you slide the crate across the floor, the floor exerts a frictional force against the crate.

NEWTON'S FIRST LAW

Newton's First Law says that *an object will continue in its state of motion unless compelled to change by a force impressed upon it.* That is, unless an unbalanced force acts on an object, the object's velocity will not change: If the object is at rest, then it will stay at rest; if it is moving, then it will continue to move at a constant speed in a straight line.

This property of objects, their natural resistance to changes in their state of motion, is called **inertia**. In fact, the First Law is often referred to as the **Law of Inertia**.

NEWTON'S SECOND LAW

Newton's Second Law predicts what will happen when an unbalanced force *does* act on an object: The object's velocity will change; the object will accelerate. More precisely, it says that its acceleration, **a**, will be directly proportional to the strength of the total—or *net*—force ($\mathbf{F}_{net}$) and inversely proportional to the object's mass, m:

$$\mathbf{a} = \mathbf{F}/\mathbf{m} \text{ or}$$

$$\mathbf{F}_{net} = ma \text{ or } \Sigma F = ma$$

$$\text{Units} = \text{Newtons} = \text{kg} \cdot \text{m/s}^2$$

This is the most important equation in mechanics!

Two identical boxes, one empty and one full, have different masses. The box that's full has the greater mass, because it contains more stuff; more stuff, more mass. Mass (m) is measured in *kilograms*, abbreviated as kg. (Note: An object whose mass is 1 kg weighs about 2.2 pounds.) If a given force produces some change in velocity for a 2 kg object, then a 1 kg object would experience twice that change. Note that mass is a proxy for the extent of inertia inherent in an object and thus inertia is a reflection of an object's mass.

Forces are represented by vectors; they have magnitude and direction. If several different forces act on an object simultaneously, then the net force, F_{net}, is the vector sum of all these forces. (The phrase *resultant force* is also used to mean *net force*.)

Since $F_{net} = ma$, and m is a *positive* scalar, the direction of **a** always matches the direction of F_{net}.

Finally, since $F = ma$, the units for F equal the units of m times the units of a:

$$[F] = [m][a]$$
$$= kg \cdot m/s^2$$

A force of 1 kg·m/s² is renamed 1 **newton** (abbreviated as N). A medium-size apple weighs about 1 N.

NEWTON'S THIRD LAW

This is the law that's commonly remembered as *to every action, there is an equal, but opposite, reaction*. More precisely, if Object 1 exerts a force on Object 2, then Object 2 exerts a force back on Object 1, equal in strength but opposite in direction. These two forces, $F_{1\text{-on-}2}$ and $F_{2\text{-on-}1}$, are called an **action/reaction pair**.

> **Example 1** What net force is required to maintain a 5,000 kg object moving at a constant velocity of magnitude 7,500 m/s?

Solution. The First Law says that any object will continue in its state of motion unless an unbalanced force acts on it. Therefore, no net force is required to maintain a 5,000 kg object moving at a constant velocity of magnitude 7,500 m/s.

You might be asking, "If no net force is needed to keep a car moving at a constant speed, why does the driver need to press down on the gas pedal in order to maintain a constant speed?" There is a big difference between *force* and *net force*. As the car moves forward, there is a frictional force (more on that shortly) opposite the direction of motion that would be slowing the car down. The gas supplies energy to the engine to spin the tires to exert a forward force on the car to counteract friction and make the net force zero, which maintains a constant speed.

Here's another way to look at it: Constant velocity means $\mathbf{a} = 0$, so the equation $F_{net} = ma$ immediately gives $F_{net} = 0$.

If an object is experiencing zero acceleration (its velocity is not increasing or decreasing), then the object is experiencing zero net force.

> **Example 2** How much force is required to cause an object of mass 2 kg to have an acceleration of 4 m/s²?

Solution. According to the Second Law, $F_{net} = ma = (2 \text{ kg})(4 \text{ m/s}^2) = 8$ N.

> **Example 3** An object feels two forces; one of strength 8 N pulling to the left and one of strength 20 N pulling to the right. If the object's mass is 4 kg, what is its acceleration?

Solution. Forces are represented by vectors and can be added and subtracted. Therefore, an 8 N force to the left added to a 20 N force to the right yields a net force of 20 – 8 = 12 N to the right. Then Newton's Second Law gives $\mathbf{a} = F_{net}/m = (12$ N to the right$)/(4$ kg$) = 3$ m/s² to the right.

NEWTON'S LAWS: A SUMMARY

These three laws will appear time and time again in later concepts. Let's sum up those three laws in easier terms.

Newton's First Law

- Moving stuff keeps moving; resting stuff keeps resting
- Law of Inertia: Objects naturally resist changes in their velocities
- Measure of inertia is mass

Newton's Second Law

$$F_{net} = ma$$

- Force (**F**) acting on an object is equal to the mass (*m*) of an object times its acceleration (*a*)
- Forces are vectors
- Units: Newton (N) = kg·m/s²

Newton's Third Law

For action/reaction pairs:

$$F_{(1 \text{ on } 2)} = -F_{(2 \text{ on } 1)}$$

the forces are equal but in opposite directions

WEIGHT

Mass and weight are not the same thing—there is a clear distinction between them in physics—but they are often used interchangeably in everyday life. The **weight** of an object is the gravitational force exerted on it by the Earth (or by whatever planet it happens to be on). Mass, by contrast, is a measure of the quantity of matter that comprises an object. An object's mass does not change with location. Weight changes depending on location. For example, you weigh less on the Moon than you do on Earth.

Since weight is a force, we can use $F = ma$ to compute it. What acceleration would the gravitational force impose on an object? The gravitational acceleration, of course! Therefore, setting $a = g$, the equation $F = ma$ becomes

$$F_w = mg \text{ or } F_g = mg$$

This is the equation for the weight of an object of mass m. (Weight is often symbolized as F_g, rather than F_w.) Notice that mass and weight are proportional but not identical. Furthermore, mass is measured in kilograms, while weight is measured in newtons.

Example 4 What is the mass of an object that weighs 500 N?

Solution. Since weight is m multiplied by g, mass is F_w (weight) divided by g. Therefore,

$$m = F_w/g = (500 \text{ N})/(10 \text{ m/s}^2) = 50 \text{ kg}$$

Amass Your Mass Facts (Groan)
Also, remember that mass is a proxy for inertia.

Big Idea #1
Per the College Board's AP Physics 1 Course and Exam information. Check out all of the Big Ideas at apstudent. collegeboard.org/ apcourse/ap-physics-1

> **Example 5** A person weighs 150 pounds. Given that a pound is a unit of weight equal to 4.45 N, what is this person's mass?

Solution. This person's weight in newtons is (150 lb)(4.45 N/lb) = 667.5 N, so his mass is

$$m = F_w/g = (667.5 \text{ N})/(10 \text{ m/s}^2) = 66.75 \text{ kg}$$

> **Example 6** A book whose mass is 2 kg rests on a table. Find the magnitude of the force exerted by the table on the book.

Solution. The book experiences two forces: the downward pull of the Earth's gravity and the upward, supporting force exerted by the table. Since the book is at rest on the table, its acceleration is zero, so the net force on the book must be zero. Therefore, the magnitude of the support force must equal the magnitude of the book's weight, which is $F_w = mg = (2 \text{ kg})(10 \text{ m/s}^2) = 20$ N.

NORMAL FORCE

When an object is in contact with a surface, the surface exerts a contact force on the object. The component of the contact force that's *perpendicular* to the surface is called the **normal force** on the object. (In physics, the word *normal* means *perpendicular*.) The normal force is what prevents objects from falling through table-tops or you from falling through the floor. The normal force is denoted by $\mathbf{F_N}$, or simply by **N**. (If you use the latter notation, be careful not to confuse it with N, the abbreviation for the newton.)

> **Example 7** A book whose mass is 2 kg rests on a table. Find the magnitude of the normal force exerted by the table on the book.

Solution. The book experiences two forces: the downward pull of Earth's gravity and the upward, supporting force exerted by the table. Since the book is at rest on the table, its acceleration is zero, so the net force on the book must be zero. Therefore, the magnitude of the support force must equal the magnitude of the book's weight, which is $F_w = mg = (2)(10) = 20$ N. This means the normal force must be 20 N as well: $\mathbf{F_N} = 20$ N. (Note that this is a repeat of Example 6, except now we have a name for the "upward, supporting force exerted by the table"; it's called the normal force.)

AN OVERALL STRATEGY

The previous examples are the lowest level of understanding Newton's laws. They are pretty straightforward thinking sometimes referred to as "plug and chug." Most of physics is not that simple. Frequently there is more than one force acting on an object, and many times angles are involved. Following the below strategy can greatly increase your chance of success for all but the most trivial of Newton's Second Law problems.

I. You must be able to visualize what's going on. Make a sketch if it helps, but definitely make a free-body diagram by doing the following:

    A. Draw a dot to represent the object. Draw arrows going away from the dot to represent any (all) forces acting on the object.

        i. Anything touching the object exerts a force.

            a. If the thing touching the object is a rope, ropes can only pull. Draw the force accordingly.

            b. If the thing touching the object is a table, ramp, floor, or some other flat surface, a surface can exert two forces.

                1. The surface exerts a force perpendicular to itself toward the object. This force is always present if two things are in contact, and it is called the normal force.

                2. If there is kinetic friction present, then the surface exerts a force on the object that is parallel to the surface and opposite to the direction of motion.

        ii. Some things can exert a force without touching an object. For example, the Earth pulls down on everything via the mystery of gravity. Electricity and magnetism also exert their influences without actually touching. Unless you hear otherwise, gravity points down!

        iii. If you know one force is bigger than another, you should draw that arrow longer than the smaller force's arrow.

        iv. Don't draw a velocity and mistake it for a force. No self-respecting velocity vector hangs out in a free-body diagram! Oh, and there is no such thing as the force of inertia.

II. Clearly define an appropriate coordinate system. Be sure to break up each force that does not lie on an axis into its x- and y-components.

III. Write out Newton's Second Law in the form of $\sum F_x = ma_x$ and/or $\sum F_y = ma_y$, using the forces identified in the free-body diagram to fill in the appropriate forces.

IV. Do the math.

As you go though the following examples, notice how this strategy is used.

Example 8 Draw a free-body diagram for each of the following situations:

a) A ball sits at rest.	b) Stickman's foot kicks the ball.	c) The ball rolls at a constant velocity across level ground.
d) The ball rolls through a rough patch and slows down due to friction.	e) The ball rolls up a ramp with friction.	f) The ball rolls back down the ramp with friction.

Solution. Notice that gravity always points down (even on ramps). Also the normal force is perpendicular to the surface (even on ramps). Do not always put the normal opposite the direction of gravity, because the normal is relative to the surface, which may be tilted. Finally, friction is always parallel to the surface (or perpendicular to the normal) and tends to point in the opposite direction from motion.

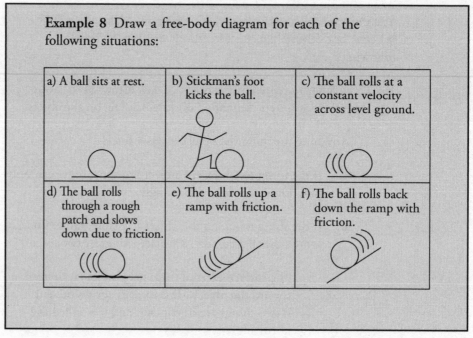

a) A ball sits at rest.	b) Stickman's foot kicks the ball.	c) The ball rolls at a constant velocity across level, frictionless ground.
d) The ball rolls through a rough patch and slows down due to friction.	e) The ball rolls up a ramp with friction.	f) The ball rolls back down the ramp with friction.

> **Example 9** A can of paint with a mass of 6 kg hangs from a rope. If the can is to be pulled up to a rooftop with an acceleration of 1 m/s^2, what must the tension in the rope be?

Solution. First draw a picture. Represent the object of interest (the can of paint) as a heavy dot, and draw the forces that act on the object as arrows connected to the dot. This is called a **free-body** (or force) **diagram**.

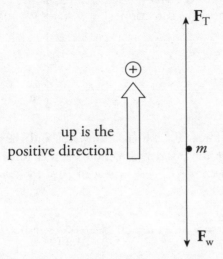

We have the tension force in the rope, F_T (also symbolized merely by **T**), which is upward, and the weight, F_w, which is downward. Calling *up* the positive direction, the net force is $F_T - F_w$. The Second Law, $F_{net} = ma$, becomes $F_T - F_w = ma$, so

$$F_T = F_w + ma = mg + ma = m(g + a) = 6(10 + 1) = 66 \text{ N}$$

> **Example 10** A can of paint with a mass of 6 kg hangs from a rope. If the can is pulled up to a rooftop with a constant velocity of 1 m/s, what must the tension in the rope be?

Solution. The phrase "constant velocity" automatically means $a = 0$ and, therefore, $F_{net} = 0$. In the diagram above, F_T would need to have the same magnitude as F_w in order to keep the can moving at a constant velocity. Thus, in this case, $F_T = F_w = mg = (6)(10) = 60 \text{ N}$.

> **Example 11** How much tension must a rope have to lift a 50 N object with an acceleration of 10 m/s^2?

Solution. First draw a free-body diagram:

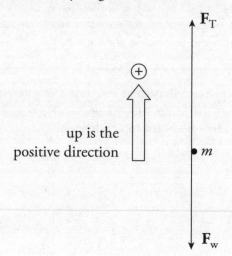

We have the tension force, $\mathbf{F}_T$, which is upward, and the weight, $\mathbf{F}_w$, which is downward. Calling *up* the positive direction, the net force is $F_T - F_w$. The Second Law, $F_{net} = ma$, becomes $F_T - F_w = ma$, so $F_T = F_w + ma$. Remembering that $m = F_w/g$, we find that

$$F_T = F_w + ma = F_w + \frac{F_w}{g}a = 50 \text{ N} + \frac{50 \text{ N}}{10 \text{ m/s}^2}\left(10 \text{ m/s}^2\right) = 100 \text{ N}$$

FRICTION

When an object is in contact with a surface, the surface exerts a contact force on the object. The component of the contact force that's *parallel* to the surface is called the **friction force** on the object. Friction, like the normal force, arises from electrical interactions between atoms of which the object is composed and those of which the surface is composed.

We'll look at two main categories of friction: (1) **static friction** and (2) **kinetic (sliding) friction**. Static friction results from the weak electrostatic bonds formed between the surfaces when the object is at rest. These bonds need to be broken before the object will slide. As a result, the static friction force is higher than the kinetic friction force. When the object begins to slide, the weak electronegative bonds cannot form fast enough, so the kinetic friction is lower. It is easier to keep an object moving than it is to start an object moving because the Kinetic μ is lower than the static μ. Static friction occurs when there is no relative motion between the object and the surface (no sliding); kinetic friction occurs when there *is* relative motion (when there's sliding).

The strength of the friction force depends, in general, on two things: the nature of the surfaces and the strength of the normal force. The nature of the surfaces is represented by the **coefficient of friction**, which is denoted by μ (*mu*) and has no units. The greater this number is, the stronger the friction force will be. For example, the coefficient of friction between rubber-soled shoes and a wooden floor is 0.7, but between rubber-soled shoes and ice, it's only 0.1. Also, since kinetic friction is generally weaker than static friction (it's easier to keep an object sliding once it's sliding than it is to start the object sliding in the first place), there are two coefficients of friction: one for static friction (μ_s) and one for kinetic friction (μ_k). For a given pair of surfaces, it's virtually always true that $\mu_k < \mu_s$. The strengths of these two types of friction forces are given by the following equations:

$$F_{\text{static friction, max}} = \mu_s F_N$$

$$F_{\text{kinetic friction}} = \mu_k F_N$$

Note that the equation for the strength of the static friction force is for the *maximum* value only. This is because static friction can vary, precisely counteracting weaker forces that attempt to move an object. For example, suppose an object experiences a normal force of $F_N = 100$ N and the coefficient of static friction between it and the surface it's on is 0.5. Then, the *maximum* force that static friction can exert is (0.5)(100 N) = 50 N. However, if you push on the object with a force of, say, 20 N, then the static friction force will be 20 N (in the opposite direction), *not* 50 N; the object won't move. The net force on a stationary object must be zero. Static friction can take on all values, up to a certain maximum, and you must overcome the maximum static friction force to get the object to slide. The direction of $F_{\text{kinetic friction}} = F_{f\,(\text{kinetic})}$ is opposite to that of motion (sliding), and the direction of $F_{\text{static friction}} = F_{f\,(\text{static})}$ is usually, but not always, opposite to that of the intended motion.

Example 12 A crate of mass 20 kg is sliding across a wooden floor. The coefficient of kinetic friction between the crate and the floor is 0.3.

    (a) Determine the strength of the friction force acting on the crate.

    (b) If the crate is being pulled by a force of 90 N (parallel to the floor), find the acceleration of the crate.

Solution. First draw a free-body diagram:

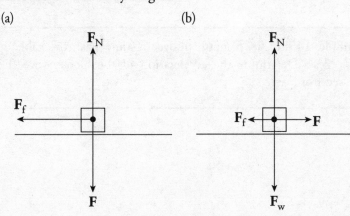

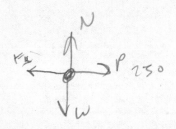

$F_s = .4 (100 \cdot 20)$

$F_s = 400$ = max force
before it moves

Push = 250N

$250 < 400$, won't
slide

(a) The normal force on the object balances the object's weight, so $F_N = mg =$ (20 kg)(10 m/s²) = 200 N. Therefore, $F_{(kinetic)} = \mu_k F_N = (0.3)(200 \text{ N}) = 60 \text{ N}$.

(b) The net horizontal force that acts on the crate is $F - F_f = 90 \text{ N} - 60 \text{ N} = 30 \text{ N}$, so the acceleration of the crate is $a = F_{net}/m = (30 \text{ N})/(20 \text{ kg}) = 1.5 \text{ m/s}^2$.

Example 13 A crate of mass 100 kg rests on the floor. The coefficient of static friction is 0.4. If a force of 250 N (parallel to the floor) is applied to the crate, what's the magnitude of the force of static friction on the crate?

Good to Know
This example illustrates how static friction can vary.

Solution. The normal force on the object balances its weight, so $F_N = mg = $ (100 kg) (10 m/s²) = 1,000 N. Therefore, $F_{\text{static friction, max}} = F_{\text{f (static), max}} = \mu_s F_N = (0.4)(1,000 \text{ N})$ = 400 N. This is the *maximum* force that static friction can exert, but in this case it's not the actual value of the static friction force. Since the applied force on the crate is only 250 N, which is less than the $F_{\text{f (static), max}}$, the force of static friction will be less also: $F_{\text{f (static)}} = 250 \text{ N}$, and the crate will not slide.

PULLEYS

Pulleys are devices that change the direction of the tension force in the cords that slide over them. Pulley systems multiply the force by however many strings are pulling on the object

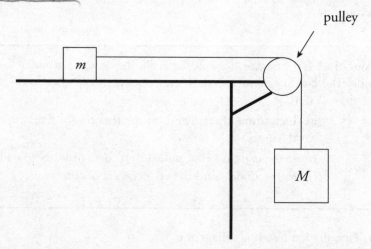

pulley

Example 14 In the diagram above, assume that the tabletop is frictionless. Determine the acceleration of the blocks once they're released from rest.

Solution. There are two blocks, so we draw two free-body diagrams:

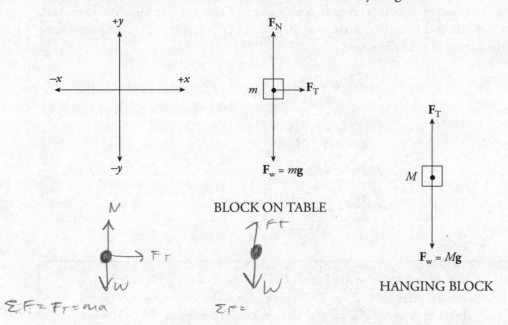

BLOCK ON TABLE

HANGING BLOCK

$$\Sigma F = F_T = ma \qquad \Sigma F =$$

To get the acceleration of each one, we use Newton's Second Law, $\mathbf{F}_{net} = ma$.

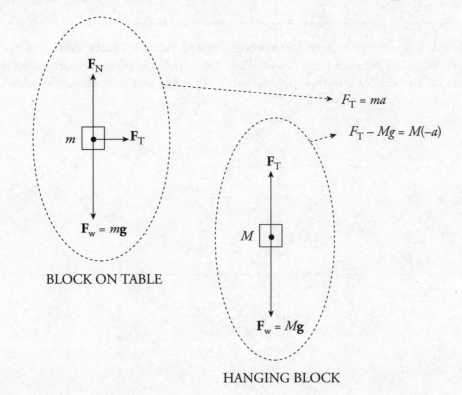

BLOCK ON TABLE

HANGING BLOCK

$$F_T = ma$$

$$F_T - Mg = M(-a)$$

Note that there are two unknowns, F_T and a, but we can eliminate F_T by adding the two equations, and then we can solve for a.

Newton's Second Law is defined as $\mathbf{F} = ma$, whereas the Force required to accelerate an object is equal to the mass of the object multiplied by the magnitude of the desired acceleration. If the net force in a system is not zero, then the object in the system is accelerating. Newtons second law can be used to calculate the acceleration of objects with unbalanced forces. Once the net force is determined, substitute this force and mass into the $\mathbf{F} = ma$ ($\mathbf{F}_{net} = ma$) equation to determine the acceleration of the object.

$$F_T = ma$$

$$Mg - F_T = Ma$$

Add the equations to eliminate F_T.

$$Mg = ma + Ma$$

$$= a(m + M)$$

$$\frac{Mg}{m + M} = a$$

Example 15 Using the same diagram as in the previous example, assume that $m = 2$ kg, $M = 10$ kg, and the coefficient of kinetic friction between the small block and the tabletop is 0.5. Compute the acceleration of the blocks.

Solution. Once again, draw a free-body diagram for each object. Note that the only difference between these diagrams and the ones in the previous example is the inclusion of the force of (kinetic) friction, $\mathbf{F}_f$, that acts on the block on the table.

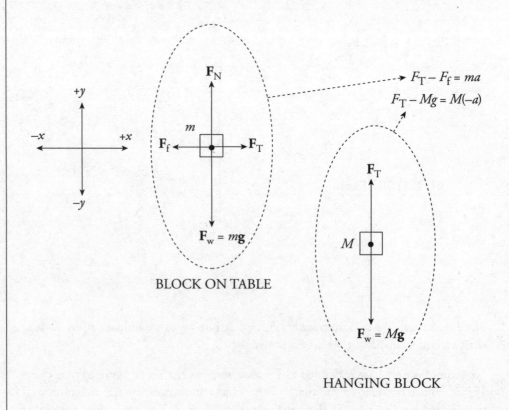

BLOCK ON TABLE

HANGING BLOCK

As before, we have two equations that contain two unknowns (a and F_T):

$$F_T - F_f = ma \quad (1)$$
$$F_T - Mg = M(-a) \quad (2)$$

Subtract the equations (thereby eliminating F_T) and solve for a. Note that, by definition, $F_f = \mu F_N$, and from the free-body diagram for m, we see that $F_N = mg$, so $F_f = \mu mg$:

$$Mg - F_f = ma + Ma$$
$$Mg - \mu mg = a(m + M)$$
$$\frac{M - \mu m}{m + M} g = a$$

Substituting in the numerical values given for m, M, and μ, we find that $a = \dfrac{3}{4} g$ (or 7.5 m/s²).

Example 16 In the previous example, calculate the tension in the cord.

Solution. Since the value of a has been determined, we can use either of the two original equations to calculate F_T. Using Equation (2), $F_T - Mg = M(-a)$ (because it's simpler), we find

$$F_T = Mg - Ma = Mg - M \cdot \frac{3}{4} g = \frac{1}{4} Mg = \frac{1}{4}(10)(10) = 25 \text{ N}$$

As you can see, we would have found the same answer if Equation (1) had been used:

$$F_T - F_f = ma \implies F_T = F_f + ma = \mu mg + ma = \mu mg + m \cdot \frac{3}{4} g = mg \left(\mu + \frac{3}{4} \right)$$
$$= (2)(10)(0.5 + 0.75)$$
$$= 25 \text{ N}$$

INCLINED PLANES

An inclined plane is basically a ramp. If you look at the forces acting on a block that sits on a ramp using a standard coordinate system, it initially looks straightforward. However, part of the normal force acts in the x direction, part acts in the y direction, and the block has acceleration in both the x and y directions. If friction is present, it also has components in both the x and y directions. The math has the potential to be quite cumbersome.

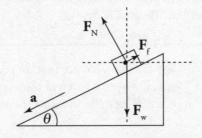

This is a non-rotated coordinate system—notice $\mathbf{F}_f$, $\mathbf{F}_N$, and $\mathbf{a}$ will each have to be broken into x- and y-components.

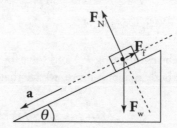

This is a rotated coordinate system—$\mathbf{F}_N$ acts only in the y direction, $\mathbf{F}_f$ and the acceleration act only in the x direction, and only $\mathbf{F}_w$ must be broken into components. If an object of mass m is on the ramp, then the force of gravity on the object, $\mathbf{F}_w = m\mathbf{g}$, has two components: One that's parallel to the ramp ($mg \sin \theta$) and one that's normal to the ramp ($mg \cos \theta$), where θ is the incline angle. The force driving the block down the inclined plane is the component of the block's weight that's parallel to the ramp: $mg \sin \theta$.

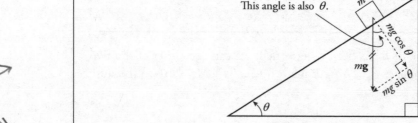

> **Example 17** A block slides down a frictionless, inclined plane that makes a 30° angle with the horizontal. Find the acceleration of this block.

Solution. Let m denote the mass of the block, so the force that pulls the block down the incline is $mg \sin\theta$, and the block's acceleration down the plane is

$$a = \frac{F}{m} = \frac{mg \sin \theta}{m} = g \sin \theta = g \sin 30° = \frac{1}{2}g = 5 \text{ m/s}^2$$

> **Example 18** A block slides down an inclined plane that makes a 30° angle with the horizontal. If the coefficient of kinetic friction is 0.3, find the acceleration of the block.

Solution. First draw a free-body diagram. Notice that, in the diagram shown below, the weight of the block, $F_w = mg$, has been written in terms of its scalar components: $F_w \sin \theta$ parallel to the ramp and $F_w \cos \theta$ normal to the ramp:

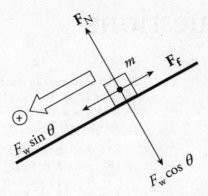

The force of friction, F_f, that acts up the ramp (opposite to the direction in which the block slides) has magnitude $F_f = \mu F_N$. But the diagram shows that $F_N = F_w \cos\theta$, so $F_f = \mu(mg\cos\theta)$. Therefore, the net force down the ramp is

$$F_w \sin\theta - F_f = mg\sin\theta - \mu mg\cos\theta = mg(\sin\theta - \mu\cos\theta)$$

Then, setting F_{net} equal to ma, we solve for a:

$$
\begin{aligned}
a = \frac{F_{net}}{m} &= \frac{mg(\sin\theta - \mu\cos\theta)}{m} \\
&= g(\sin\theta - \mu\cos\theta) \\
&= (10 \text{ m/s}^2)(\sin 30° - 0.3\cos 30°) \\
&= 2.4 \text{ m/s}^2
\end{aligned}
$$

Chapter 5 Review Questions

Answers and Explanations can be found in Chapter 13.

Section I: Multiple-Choice

1. A person standing on a horizontal floor feels two forces: the downward pull of gravity and the upward supporting force from the floor. These two forces

    (A) have equal magnitudes and form an action/reaction pair
    (B) have equal magnitudes but do not form an action/reaction pair
    (C) have unequal magnitudes and form an action/reaction pair
    (D) have unequal magnitudes and do not form an action/reaction pair

2. A person who weighs 800 N steps onto a scale that is on the floor of an elevator car. If the elevator accelerates upward at a rate of 5 m/s², what will the scale read?

    (A) 400 N
    (B) 800 N
    (C) 1,000 N
    (D) 1,200 N

3. A frictionless inclined plane of length 20 m has a maximum vertical height of 5 m. If an object of mass 2 kg is placed on the plane, which of the following best approximates the net force it feels?

    (A) 5 N
    (B) 10 N
    (C) 15 N
    (D) 20 N

4. A 20 N block is being pushed across a horizontal table by an 18 N force. If the coefficient of kinetic friction between the block and the table is 0.4, find the acceleration of the block.

    (A) 0.5 m/s²
    (B) 1 m/s²
    (C) 5 m/s²
    (D) 7.5 m/s²

5. The coefficient of static friction between a box and a ramp is 0.5. The ramp's incline angle is 30°. If the box is placed at rest on the ramp, the box will do which of the following?

    (A) Accelerate down the ramp
    (B) Accelerate briefly down the ramp but then slow down and stop
    (C) Move with constant velocity down the ramp
    (D) Not move

6.

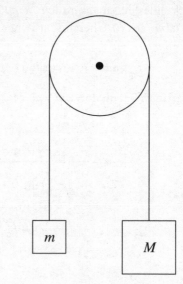

Assuming a frictionless, massless pulley, determine the acceleration of the blocks once they are released from rest.

(A) $\dfrac{m}{M+m}g$

(B) $\dfrac{M}{m}g$

(C) $\dfrac{M+m}{M-m}g$

(D) $\dfrac{M-m}{M+m}g$

7. If all of the forces acting on an object balance so that the net force is zero, then

   (A) the object must be at rest
   (B) the object's speed will decrease
   (C) the object's direction of motion can change, but not its speed
   (D) None of the above will occur.

8. A block of mass m is at rest on a frictionless, horizontal table placed in a laboratory on the surface of the Earth. An identical block is at rest on a frictionless, horizontal table placed on the surface of the Moon. Let $\mathbf{F}$ be the net force necessary to give the Earth-bound block an acceleration of $\mathbf{a}$ across the table. Given that g_{Moon} is one-sixth of g_{Earth}, the force necessary to give the Moon-bound block the same acceleration $\mathbf{a}$ across the table is

   (A) $\mathbf{F}/6$
   (B) $\mathbf{F}/3$
   (C) $\mathbf{F}$
   (D) $6\mathbf{F}$

9. A crate of mass 100 kg is at rest on a horizontal floor. The coefficient of static friction between the crate and the floor is 0.4, and the coefficient of kinetic friction is 0.3. A force $\mathbf{F}$ of magnitude 344 N is then applied to the crate, parallel to the floor. Which of the following is true?

   (A) The crate will accelerate across the floor at 0.5 m/s².
   (B) The static friction force, which is the reaction force to $\mathbf{F}$ as guaranteed by Newton's Third Law, will also have a magnitude of 344 N.
   (C) The crate will slide across the floor at a constant speed of 0.5 m/s.
   (D) The crate will not move.

10. Two crates are stacked on top of each other on a horizontal floor; Crate #1 is on the bottom, and Crate #2 is on the top. Both crates have the same mass. Compared to the strength of the force $\mathbf{F}_1$ necessary to push only Crate #1 at a constant speed across the floor, the strength of the force $\mathbf{F}_2$ necessary to push the stack at the same constant speed across the floor is greater than F_1 because

   (A) the normal force on Crate #1 is greater
   (B) the coefficient of kinetic friction between Crate #1 and the floor is greater
   (C) the coefficient of static friction between Crate #1 and the floor is greater
   (D) the weight of Crate #1 is greater

Section II: Free-Response

1. This question concerns the motion of a crate being pulled across a horizontal floor by a rope. In the diagram below, the mass of the crate is m, the coefficient of kinetic friction between the crate and the floor is μ, and the tension in the rope is $\mathbf{F}_T$.

    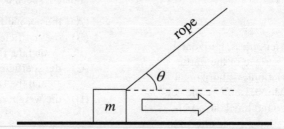

    (a)   Draw and label all of the forces acting on the crate.

    (b)   Compute the normal force acting on the crate in terms of m, F_T, θ, and g.

    (c)   Compute the acceleration of the crate in terms of m, F_T, θ, μ, and g.

2. In the diagram below, a massless string connects two blocks—of masses m_1 and m_2, respectively—on a flat, frictionless tabletop. A force $\mathbf{F}$ pulls on Block #2, as shown:

    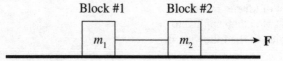

    (a)   Draw and label all of the forces acting on Block #1.

    (b)   Draw and label all of the forces acting on Block #2.

    (c)   What is the acceleration of Block #1? Please state your answer in terms of F, m_1, and m_2.

    (d)   What is the tension in the string connecting the two blocks? Please state your answer in terms of F, m_1, and m_2.

    (e)   If the string connecting the blocks were not massless, but instead had a mass of m, determine

    (i)   the acceleration of Block #1, in terms of F, m_1, and m_2.

    (ii) the difference between the strength of the force that the connecting string exerts on Block #2 and the strength of the force that the connecting string exerts on Block #1. Please state your answer in terms of F, m_1, and m_2.

3. In the figure shown, assume that the pulley is frictionless and massless.

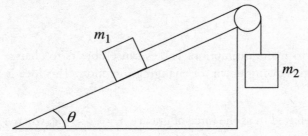

(a) If the surface of the inclined plane is frictionless, determine what value(s) of θ will cause the box of mass m_1 to

    (i) accelerate up the ramp
    (ii) slide up the ramp at constant speed

(b) If the coefficient of kinetic friction between the surface of the inclined plane and the box of mass m_1 is μ_k, derive (but do not solve) an equation satisfied by the value of θ, which will cause the box of mass m_1 to slide up the ramp at constant speed.

4. A sky diver is falling with speed v_0 through the air. At that moment (time $t = 0$), she opens her parachute and experiences the force of air resistance whose strength is given by the equation $F = kv$, in which k is a proportionality constant and v is her descent speed. The total mass of the sky diver and equipment is m. Assume that g is constant throughout her descent.

(a) Draw and label all the forces acting on the sky diver after her parachute opens.

(b) Determine the sky diver's acceleration in terms of m, v, k, and g.

(c) Determine the sky diver's terminal speed (that is, the eventual constant speed of descent).

(d) Sketch a graph of v as a function of time, being sure to label important values on the vertical axis.

Summary

- Forces are not needed to maintain motion. Forces cause objects to change their motion, whether that means speeding up, slowing down, or changing direction. This idea is expressed by Newton's Second Law: $F_{net} = ma$.

- Weight is commonly referred to as the force of gravity F_w or $F_g = mg$, where g is -10 m/s^2.

- Friction is defined by $F_f = \mu N$ and comes in two types—static and kinetic where $\mu_{kinetic} < \mu_{static}$.

- The normal force (N or sometimes F_N) is frequently given by N or $F_N = mg \cos \theta$, where θ is the angle between the horizontal axis and the surface on which the object rests.

- To solve almost any Newton's Second Law problem, use the following strategy:

    I.   You must be able to visualize what's going on. Make a sketch if it helps. Make sure to make a free-body diagram (or FBD).

    II. Clearly define an appropriate coordinate system.

    III. Write out Newton's Second Law in the form of $\sum F_x = ma_x$ and/or $\sum F_y = ma_y$, using the forces identified in the FBD to fill in the appropriate forces.

    IV. Do the math.

Chapter 6
Work, Energy, and Power

"Energy cannot be created or destroyed, it can only be changed from one form to another."

—Albert Einstein

Kinematics and dynamics are about change. Simple observations of our environment show us that change is occurring all around us. But what is needed to make an object change, and where did that change go to? It wasn't until more than one hundred years after Newton that the idea of energy became incorporated into physics, but today it permeates every branch of the subject.

ENERGY: AN OVERVIEW

So what is energy? How do we determine the energy of a system? These are not easy questions. It is difficult to give a precise definition of energy; there are different forms of energy as a result of different kinds of forces. Energy can come as a result of gravitational force, the speed of an object, stored energy in springs, heat loss to nuclear energy. But one truth remains the same for all of them: the Law of Conservation of Energy. Energy can not just appear out of nowhere nor can it dissapear in a closed system; it must always take on another form. Force is the agent for this change, energy is the measure of that change, and work is the method of transferring energy from one system to another.

WORK

Work Units
The unit for work, the newton-meter (N·m), is renamed a joule and abbreviated as J.

Named after English physicist James Prescott Joule, one joule is the work required to produce one watt of power for one second.

When you lift a book from the floor, you exert a force on it over a distance, and when you push a crate across a floor, you also exert a force on it over a distance. The application of force over a distance and the resulting change in energy of the system that the force acted on, give rise to the concept of **work**.

Definition. If a constant force **F** acts over a distance **d** and **F** is parallel to **d**, then the work done by **F** is the product of force and distance:

$$W = \mathbf{Fd}$$

Notice that, although work depends on two vectors (**F** and **d**), work itself is *not* a vector. *Work is a scalar quantity.* However, even being a scalar quantity there exists positive, negative, and zero work.

> **Example 1** You slowly lift a book of mass 2 kg at constant velocity a distance of 3 m. How much work did you do on the book?

When the Formula for Work Works
$W = Fd\cos\theta$ only works when the forces do not change as the object moves. This also means a constant acceleration is delivered on the mass.

Solution. In this case, the force you exert must balance the weight of the book (otherwise the velocity of the book wouldn't be constant), so $F = mg = (2 \text{ kg})(10 \text{ m/s}^2) = 20$ N. Since this force is straight upward and the displacement of the book is also straight upward, **F** and **d** are parallel, so the work done by your lifting force is $W = Fd = (20 \text{ N})(3 \text{ m}) = 60$ N·m, or 60 J.

WORK AT AN ANGLE

The previous formula works only when work is done completely parallel to the intended distance of travel. What happens when the force is done at an angle? The formula becomes:

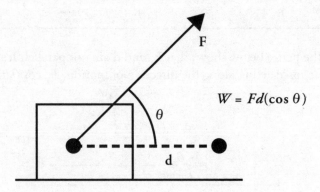

$$W = Fd(\cos \theta)$$

$$W = Fd (\cos \theta)$$

Let's compare the work done in a few instances with a force being applied between certain angles:

Angle (θ)	$0 \leq \theta < 90$	$\theta = 90$	$90 < \theta \leq 180$
Cos (θ)	positive	ZERO	negative
Work	positive	ZERO	negative
Speed of Object	increases	constant speed	decreases

Note that when we do positive work we increase the speed of an object, whereas when we do negative work we slow an object down. This will be important to note when we relate work with kinetic energy and potential energy.

Don't Be Tricked!
A force applied perpendicular to the intended direction of motion always does ZERO work!

Example 2 A 15 kg crate is moved along a horizontal floor by a warehouse worker who's pulling on it with a rope that makes a 30° angle with the horizontal. The tension in the rope is 69 N, and the crate slides a distance of 10 m. How much work is done on the crate by the worker?

Solution. The figure below shows that $\mathbf{F_T}$ and $\mathbf{d}$ are not parallel. It's only the component of the force acting along the direction of motion, $F_T \cos \theta$, that does work.

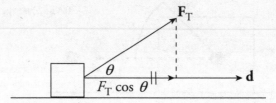

Therefore,

$$W = (F_T \cos \theta)d = (69 \text{ N} \cdot \cos 30°)(10 \text{ m}) = 600 \text{ J}$$

Example 3 In the previous example, assume that the coefficient of kinetic friction between the crate and the floor is 0.4.

  (a) How much work is done by the normal force?
  (b) How much work is done by the friction force?

Solution.

  (a) Clearly, the normal force is not parallel to the motion, so we use the general definition of work. Since the angle between $\mathbf{F_N}$ and $\mathbf{d}$ is 90° (by definition of *normal*) and cos 90° = 0, the normal force does zero work.

  (b) The friction force, $\mathbf{F_f}$ is also not parallel to the motion; it's *antiparallel*. That is, the angle between $\mathbf{F_f}$ and $\mathbf{d}$ is 180°. Since cos 180° = −1, and since the strength of the normal force is $F_n = F_w - F_{T,y} = mg - F_T \cdot \sin(\theta) = (15 \text{ kg})(10 \text{ m/s}^2) - (69 \text{ N})(1/2) = 115.5 \text{ N}$, the work done by the friction force is:

$$W = -F_f d = -\mu_k F_N \cdot d = -(0.4)(115.5 \text{ N})(10 \text{ m}) = -462 \text{ J}$$

Antiparallel
When vectors that are parallel but pointing in opposite directions, if these vectors are joined at the tail, they form an angle of 180 degrees.

Example 4 A box slides down an inclined plane (incline angle = 37°). The mass of the block, m, is 35 kg, the coefficient of kinetic friction between the box and the ramp, μ_k, is 0.3, and the length of the ramp, d, is 8 m.

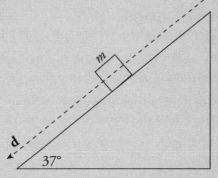

    (a) How much work is done by gravity?

    (b) How much work is done by the normal force?

    (c) How much work is done by friction?

    (d) What is the total work done?

Solution.

(a) Recall that the force that's directly responsible for pulling the box down the plane is the component of the gravitational force that's parallel to the ramp: $F_w \sin \theta = mg \sin \theta$ (where θ is the incline angle). This component is parallel to the motion, so the work done by gravity is

$$W_{\text{by gravity}} = (mg \sin \theta)d = (35 \text{ kg})(10 \text{ N/kg})(\sin 37°)(8 \text{ m}) = 1690 \text{ J}$$

Note that the work done by gravity is positive, as we would expect it to be, since gravity is helping the motion. Also, be careful with the angle θ. The general definition of work reads $W = (F \cos \theta)d$, where θ is the angle between **F** and **d**. However, the angle between **F**$_w$ and **d** is *not* 37° here, so the work done by gravity is not $(mg \cos 37°)d$. The angle θ used in the calculation above is the incline angle.

(b) Since the normal force is perpendicular to the motion, the work done by this force is zero.

Zero Work
Part (b) of this question is a typical trick question. Just remember, a force applied perpendicular to direction of travel does zero work.

(c) The strength of the normal force is $F_w \cos \theta$ (where θ is the incline angle), so the strength of the friction force is $F_f = \mu_k F_N = \mu_k F_w \cos \theta = \mu_k mg \cos \theta$. Since $\mathbf{F}_f$ is antiparallel to $\mathbf{d}$, the cosine of the angle between these vectors (180°) is –1, so the work done by friction is

$$W_{\text{by friction}} = -F_f d = -(\mu_k mg \cos \theta)(d) =$$
$$-(0.3)(35 \text{ kg})(10 \text{ N/kg})(\cos 37°)(8 \text{ m}) = -671 \text{ J}$$

Note that the work done by friction is negative, as we expect it to be, since friction is opposing the motion.

(d) The total work done is found simply by adding the values of the work done by each of the forces acting on the box:

$$W_{\text{total}} = \Sigma W = W_{\text{by gravity}} + W_{\text{by normal force}} + W_{\text{by friction}} = 1{,}690 + 0 + (-671) = 1{,}019 \text{ J}$$

WORK DONE BY A VARIABLE FORCE

If a force remains constant over the distance through which it acts, then the work done by the force is simply the product of force and distance. However, if the force does not remain constant, then the work done by the force is given by the area under the curve of a force versus displacement graph. In physics language, the term "under the curve" really means between the line itself and zero.

Big Idea #3

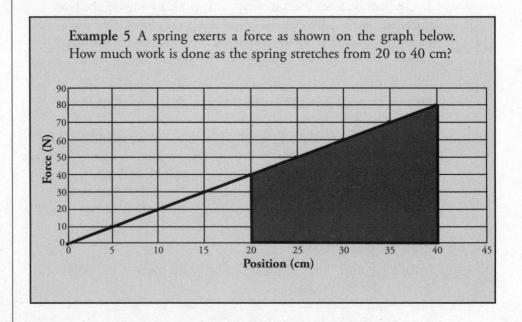

Example 5 A spring exerts a force as shown on the graph below. How much work is done as the spring stretches from 20 to 40 cm?

Solution. The area under the curve will be equal to the work done. In this case, we have some choices. You may recognize this shape as a trapezoid (it might help to momentarily rotate your head, or this book, 90 degrees) to see this.

$$A = \frac{1}{2}(b_1 + b_2)h$$

$$A = \frac{1}{2}(40\,\text{N} + 80\,\text{N})(0.20\,\text{m})$$

$$= 12\,\text{N} \cdot \text{m or } 12\,\text{J}$$

Similar to the previous chapter, units for work and energy should be confined to kg, m, and s, which is why we converted here.

An alternative choice is to recognize this shape as a triangle sitting on top of a rectangle. The total area is simply the area of the rectangle plus the area of the triangle.

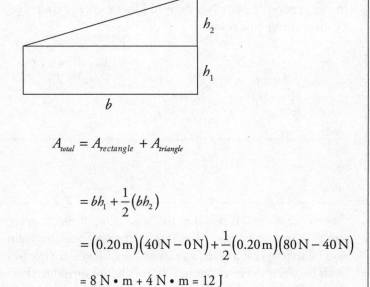

$$A_{total} = A_{rectangle} + A_{triangle}$$

$$= bh_1 + \frac{1}{2}(bh_2)$$

$$= (0.20\,\text{m})(40\,\text{N} - 0\,\text{N}) + \frac{1}{2}(0.20\,\text{m})(80\,\text{N} - 40\,\text{N})$$

$$= 8\,\text{N} \cdot \text{m} + 4\,\text{N} \cdot \text{m} = 12\,\text{J}$$

KINETIC ENERGY

Consider an object at rest ($v_0 = 0$), and imagine that a steady force is exerted on it, causing it to accelerate. Let's be more specific; let the object's mass be m, and let $\mathbf{F}$ be the force acting on the object, pushing it in a straight line. The object's acceleration is $a = F/m$, so after the object has traveled a distance Δx under the action of this force, its final speed, v, is given by Big Five #5:

$$v^2 = v_0^2 + 2a(x - x_0) = 2a(x - x_0) = 2\frac{F}{m}(x - x_0) \quad \Rightarrow \quad F(x - x_0) = \frac{1}{2}mv^2$$

But the quantity $F(x-x_0)$ is the work done by the force, so $W = \frac{1}{2}mv^2$. The work done on the object has transferred energy to it, in the amount $\frac{1}{2}mv^2$. The energy an object possesses by virtue of its motion is therefore defined as $\frac{1}{2}mv^2$ and is called **kinetic energy**:

$$K = \frac{1}{2}mv^2$$

THE WORK–ENERGY THEOREM

Need for Speed
As we mentioned before, when we do positive work we increase the speed of an object. As a result, we also increase its kinetic energy. Doing positive work means making a positive change in kinetic energy.

In the previous section, we derived kinetic energy from Big Five #5. Let's solve it a different way this time:

$v^2 = v_0^2 + 2ad$ Recall: $a = F/m$
$2(F/m)d = v^2 - v_0^2$ Recall: $W = Fd$
$W = (1/2)mv^2 - (1/2)mv_0^2$
$W = K_{final} - K_{initial}$

$$W_{total} = \Delta K$$

Work is Less Work
In some situations using the Big Five Equations is easier. However, in several instances skipping the Big Five and using work and energy can make problems much simpler. This is because we do not have to worry about the directional part of vectors and only deal with scalar quantities.

One of the questions posed at the beginning of this chapter was "How does a system gain or lose energy?" The Work–Energy Theorem begins to answer that question by stating that a system gains or loses kinetic energy by transferring it through work between the environment (forces being introduced into the system) and the system. In basic terms, doing positive work means increasing kinetic energy.

Example 6 What is the kinetic energy of a ball (mass = 0.10 kg) moving with a speed of 30 m/s?

Solution. From the definition,

$$K = \frac{1}{2}mv^2 = \frac{1}{2}(0.10 \text{ kg})(30 \text{ m/s})^2 = 45 \text{ J}$$

Example 7 A tennis ball (mass = 0.06 kg) is hit straight upward with an initial speed of 50 m/s. How high would it go if air resistance were negligible?

Solution. This could be done using the Big Five, but let's try to solve it using the concepts of work and energy. As the ball travels upward, gravity acts on it by doing negative work. [The work is negative because gravity is opposing the upward motion. F_w and d are in opposite directions, so $\theta = 180°$, which tells us that $W = (F_w \cos \theta)d = -F_w d$.] At the moment the ball reaches its highest point, its speed is 0, so its kinetic energy is also 0. The Work–Energy Theorem says

$$W = \Delta K \quad \Rightarrow \quad -F_w d = 0 - \frac{1}{2}mv_0^2 \quad \Rightarrow \quad d = \frac{\frac{1}{2}mv_0^2}{F_w} = \frac{\frac{1}{2}mv_0^2}{mg} = \frac{\frac{1}{2}v_0^2}{g} = \frac{\frac{1}{2}(50 \text{ m/s})^2}{10 \text{ m/s}^2} = 125 \text{ m}$$

Example 8 Consider the box sliding down the inclined plane in Example 4. If it starts from rest at the top of the ramp, with what speed does it reach the bottom?

Solution. It was calculated in Example 4 that W_{total} = 1,019 J. According to the Work–Energy Theorem,

$$W_{total} = \Delta K \quad \Rightarrow \quad W_{total} = K_f - K_i = K_f = \frac{1}{2}mv^2 \quad \Rightarrow \quad v = \sqrt{\frac{2W_{total}}{m}} = \sqrt{\frac{2(1{,}019 \text{ J})}{35 \text{ kg}}} = 7.6 \text{ m/s}$$

Example 9 A pool cue striking a stationary billiard ball (mass = 0.25 kg) gives the ball a speed of 2 m/s. If the average force of the cue on the ball was 200 N, over what distance did this force act?

Solution. The kinetic energy of the ball as it leaves the cue is

$$K = \frac{1}{2}mv^2 = \frac{1}{2}(0.25 \text{ kg})(2 \text{ m/s})^2 = 0.50 \text{ J}$$

The work (W) done by the cue gave the ball this kinetic energy, so

$$W = \Delta K \quad \Rightarrow \quad W = K_f \quad \Rightarrow \quad Fd = K \quad \Rightarrow \quad d = \frac{K}{F} = \frac{0.50 \text{ J}}{200 \text{ N}} = 0.0025 \text{ m} = 0.25 \text{ cm}$$

POTENTIAL ENERGY

A Simple Analogy
Energy is the ability to do work, NOT doing work.

Kinetic energy is the energy an object has by virtue of its motion. Potential energy is independent of motion; it arises from the object's position (or the system's configuration). For example, a ball at the edge of a tabletop has energy that could be transformed into kinetic energy if it falls off. An arrow in an archer's pulled-back bow has energy that could be transformed into kinetic energy if the archer releases the arrow. Both of these examples illustrate the concept of **potential energy**, the energy an object or system has by virtue of its position or configuration. In each case, work was done on the object to put it in the given configuration (the ball was lifted to the tabletop, the bowstring was pulled back), and since work is the means of transferring energy, these things have stored energy that can be retrieved as kinetic energy. Also remember that potential energy can be found in multiple sources such as chemical sources, mechanical sources, and objects in gravitational fields. This is **potential energy**, denoted by U.

Because there are different types of forces, there are different types of potential energy. The ball at the edge of the tabletop provides an example of **gravitational potential energy**, U_g, which is the energy stored by virtue of an object's position in a gravitational field. This energy would be converted to kinetic energy as gravity pulled the ball down to the floor. For now, let's concentrate on gravitational potential energy.

Assume the ball has a mass m of 2 kg and that the tabletop is h = 1.5 m above the floor. How much work did gravity do as the ball was lifted from the floor to the table? The strength of the gravitational force on the ball is F_w = mg = (2 kg)(10 N/kg) = 20 N. The force $\mathbf{F}_w$ points downward, and the ball's motion was upward, so the work done by gravity during the ball's ascent was:

$$W_{\text{by gravity}} = -F_w h = -mgh = -(20 \text{ N})(1.5 \text{ m}) = -30 \text{ J}$$

Who Cares About Potential Energy?
What if an object were to never fall? Does it even matter if we calculate its potential energy? Calculating potential energy at a point does not tell us much. We only care about potential energy when we make it change from one point to another point. This in turn translates into doing work (increasing kinetic energy).

So someone performed +30 J of work to raise the ball from the floor to the tabletop. That energy is now stored and, if the ball were given a push to send it over the edge, by the time the ball reached the floor it would acquire a kinetic energy of 30 J. We therefore say that the change in the ball's gravitational potential energy in moving from the floor to the table was +30 J. That is,

$$\Delta U_g = -W_{\text{by gravity}}$$

Note that potential energy, like work (and kinetic energy), is expressed in joules.

In general, if an object of mass m is raised a height h (which is small enough that g stays essentially constant over this altitude change), then the increase in the object's gravitational potential energy is

$$\Delta U_g = mgh$$

An important fact that makes the above equation possible is that the work done by gravity as the object is raised does not depend on the path taken by the object. The ball could be lifted straight upward or in some curvy path; it would make no difference. Gravity is said to be a **conservative** force because of this property. Conversely, any work done by a nonconservative force is path-dependent. Such is the case for friction and air resistance, as different paths taken may require more or less work to get from the initial to final positions.

If we decide on a reference level to call $h = 0$, then we can say that the gravitational potential energy of an object of mass m at a height h is $U_g = mgh$. In order to use this last equation, it's essential that we choose a reference level for height. For example, consider a passenger in an airplane reading a book. If the book is 1 m above the floor of the plane, then, to the passenger, the gravitational potential energy of the book is mgh, where $h = 1$ m. However, to someone on the ground looking up, the floor of the plane may be, say, 9,000 m above the ground. So, to this person, the gravitational potential energy of the book is mgH, where $H = 9,001$ m. What both would agree on, though, is that the difference in potential energy between the floor of the plane and the position of the book is $mg \times (1 \text{ m})$, since the airplane passenger would calculate the difference as $mg \times (1 \text{ m} - 0 \text{ m})$, while the person on the ground would calculate it as $mg \times (9{,}001 \text{ m} - 9{,}000 \text{ m})$. Differences, or changes, in potential energy are unambiguous, but values of potential energy are relative.

Example 10 A stuntwoman (mass = 60 kg) scales a 40-meter-tall rock face. What is her gravitational potential energy (relative to the ground)?

Solution. Calling the ground $h = 0$, we find

$$U_g = mgh = (60 \text{ kg})(10 \text{ m/s}^2)(40 \text{ m}) = 24{,}000 \text{ J}$$

> **Example 11** If the stuntwoman in the previous example were to jump off the cliff, what would be her final speed as she landed on a large, air-filled cushion lying on the ground?

Solution. The gravitational potential energy would be transformed into kinetic energy. So

$$U \to K \quad \Rightarrow \quad U \to \frac{1}{2}mv^2 \quad \Rightarrow \quad v = \sqrt{\frac{2 \cdot U}{m}} = \sqrt{\frac{2(24{,}000 \text{ J})}{60 \text{ kg}}} = 28 \text{ m/s}$$

CONSERVATION OF MECHANICAL ENERGY

We have seen energy in its two basic forms: Kinetic energy (K) and potential energy (U). The sum of an object's kinetic and potential energies is called its **total mechanical energy**, E.

$$E = K + U$$

(Note that because U is relative, so is E.) Assuming that no nonconservative forces (friction, for example) act on an object or system while it undergoes some change, then mechanical energy is conserved. Mechanical energy, that is the sum of potential and kinetic energies, is dissipated or converted into other energy forms such as heat, by nonconservative forces. So, if there are no nonconservative forces acting on the system, the initial mechanical energy, E_i, is equal to the final mechanical energy, E_f, or

$$K_i + U_i = K_f + U_f$$

This is the simplest form of the **Law of Conservation of Total Energy**.

Let's evaluate a situation in which an object, initially at rest, is falling from a building and calculate the total mechanical energy in each situation:

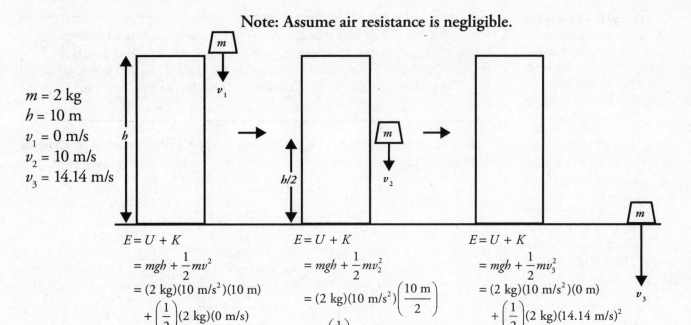

Note: Assume air resistance is negligible.

$m = 2$ kg
$h = 10$ m
$v_1 = 0$ m/s
$v_2 = 10$ m/s
$v_3 = 14.14$ m/s

$E = U + K$

$= mgh + \dfrac{1}{2}mv^2$

$= (2 \text{ kg})(10 \text{ m/s}^2)(10 \text{ m})$

$+ \left(\dfrac{1}{2}\right)(2 \text{ kg})(0 \text{ m/s})$

$= 200 \text{ J} + 0 \text{ J}$

$= 200 \text{ J}$

$E = U + K$

$= mgh + \dfrac{1}{2}mv_2^2$

$= (2 \text{ kg})(10 \text{ m/s}^2)\left(\dfrac{10 \text{ m}}{2}\right)$

$+ \left(\dfrac{1}{2}\right)(2 \text{ kg})(10 \text{ m/s})^2$

$= 100 \text{ J} + 100 \text{ J}$

$= 200 \text{ J}$

$E = U + K$

$= mgh + \dfrac{1}{2}mv_3^2$

$= (2 \text{ kg})(10 \text{ m/s}^2)(0 \text{ m})$

$+ \left(\dfrac{1}{2}\right)(2 \text{ kg})(14.14 \text{ m/s})^2$

$= 0 \text{ J} + 200 \text{ J}$

$= 200 \text{ J}$

What can you say about the calculated total mechanical energy for each situation? They all are equal. An independent system obeys the Law of Conservation of Energy. Since the total mechanical energy remains the same throughout its travel, sometimes using energy to solve problems is a lot simpler than using kinematics.

Example 12 A ball of mass 2 kg is gently pushed off the edge of a tabletop that is 5.0 m above the floor. Find the speed of the ball as it strikes the floor.

Solution. Ignoring the friction due to the air, we can apply Conservation of Mechanical Energy. Calling the floor our $h = 0$ reference level, we write

$$K_i + U_i = K_f + U_f$$

$$0 + mgh = \frac{1}{2}mv^2 + 0$$

$$v = \sqrt{2gh}$$

$$= \sqrt{2(10 \text{ m/s}^2)(5.0 \text{ m})}$$

$$= 10 \text{ m/s}$$

Note that the ball's potential energy decreased, while its kinetic energy increased. This is the basic idea behind Conservation of Mechanical Energy: One form of energy decreases while the other increases.

Example 13 A box is projected up a long ramp (incline angle with the horizontal = 37°) with an initial speed of 10 m/s. If the surface of the ramp is very smooth (essentially frictionless), how high up the ramp will the box go? What distance along the ramp will it slide?

Solution. Because friction is negligible, we can apply Conservation of Mechanical Energy. Calling the bottom of the ramp our $h = 0$ reference level, we write

$$K_i + U_i = K_f + U_f$$

$$\frac{1}{2}mv_0^2 + 0 = 0 + mgh$$

$$h = \frac{\frac{1}{2}v_0^2}{g}$$

$$= \frac{\frac{1}{2}(10 \text{ m/s})^2}{10 \text{ m/s}^2}$$

$$= 5 \text{ m}$$

Since the incline angle is $\theta = 37°$, the distance d it slides up the ramp is found in this way:

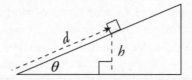

$$h = d \sin\theta$$

$$d = \frac{h}{\sin\theta} = \frac{5 \text{ m}}{\sin 37°} = \frac{25}{3} \text{ m} = 8.3 \text{ m}$$

Example 14 A skydiver jumps from a hovering helicopter that's 3,000 meters above the ground. If air resistance can be ignored, how fast will he be falling when his altitude is 2,000 m?

Solution. Ignoring air resistance, we can apply Conservation of Mechanical Energy. Calling the ground our $h = 0$ reference level, we write

$$K_i + U_i = K_f + U_f$$

$$0 + mgH = \frac{1}{2}mv^2 + mgh$$

$$v = \sqrt{2g(H - h)}$$

$$= \sqrt{2(10 \text{ m/s}^2)(3,000 \text{ m} - 2,000 \text{ m})}$$

$$= 140 \text{ m/s}$$

CONSERVATION OF ENERGY WITH NONCONSERVATIVE FORCES

The equation $K_i + U_i = K_f + U_f$ holds if no nonconservative forces are doing work. However, if work is done by such forces during the process under investigation, then the equation needs to be modified to account for this work as follows:

$$K_i + U_i + W_{other} = K_f + U_f$$

Why on the Left and not the Right?
Work done by nonconservative forces is placed on the initial energy side because the final energy accounts for both the initial energy plus the energy that is dissipated by the object as it overcomes non-conservative forces.

Example 15 Wile E. Coyote (mass = 40 kg) falls off a 50-meter-high cliff. On the way down, the force of air resistance has an average strength of 100 N. Find the speed with which he crashes into the ground.

Solution. The force of air resistance opposes the downward motion, so it does negative work on the coyote as he falls: $W_r = -F_r h$. Calling the ground $h = 0$, we find that

$$K_i + U_i + W_r = K_f + U_f$$
$$0 + mgh + (-F_r h) = \frac{1}{2} mv^2 + 0$$
$$v = \sqrt{2h(g - F_r/m)} = \sqrt{2(50)(10 - 100/40)} = 27 \text{ m/s}$$

Example 16 A skier starts from rest at the top of a 20° incline and skis in a straight line to the bottom of the slope, a distance d (measured along the slope) of 400 m. If the coefficient of kinetic friction between the skis and the snow is 0.2, calculate the skier's speed at the bottom of the run.

Solution. The strength of the friction force on the skier is $F_f = \mu_k F_N = \mu_k (mg \cos \theta)$, so the work done by friction is $-F_f d = \mu_k (mg \cos \theta) \cdot d$. The vertical height of the slope above the bottom of the run (which we designate the $h = 0$ level) is $h = d \sin \theta$. Therefore, Conservation of Mechanical Energy (including the negative work done by friction) gives

$$K_i + U_i + W_{friction} = K_{ff} + U$$
$$0 + mgh + (-\mu_k mg \cos \theta \cdot d) = \frac{1}{2} mv^2 + 0$$
$$mg(d \sin \theta) + (-\mu_k mg \cos \theta \cdot d) = \frac{1}{2} mv^2$$
$$gd(\sin \theta - \mu_k \cos \theta) = \frac{1}{2} v^2$$
$$v = \sqrt{2gd(\sin \theta - \mu_k \cos \theta)}$$
$$= \sqrt{2(10)(400)[\sin 20° - (0.2) \cos 20°]}$$
$$= 35 \text{ m/s}$$

Working Backward to Go Forward
Working with energy does not take vectors into account. With many problems, we can solve them backwards with Conservation of Energy in order to find out the initial energy, work, or final energy; and in turn, we can solve for other useful information such as height or speed.

So far, any of the problems we have solved in this chapter could have been solved using the kinematics equations and Newton's laws. The truly powerful thing about energy is that in a closed system, changes in energy are independent of the path you take. This allows you to solve many problems you would not otherwise be able to solve. With many energy problems you do not need to measure time with a stopwatch, you do not need to know the mass of the object, you do not need a constant acceleration (remember that is required for our Big Five equations from kinematics), and you do not need to know the path the object takes.

Example 17 A roller coaster at an amusement park is at rest on top of a 30 m hill (point A). The car starts to roll down the hill and reaches point B, which is 10 m above the ground, and then rolls up the track to point C, which is 20 m above the ground.

(a) A student assumes no energy is lost, and solves for how fast is the car moving at point C using energy arguments. What answer does he get?

(b) If the final speed at C is actually measured to be 2 m/s, where did the lost energy go?

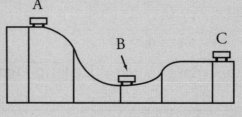

Solution.

(a) Our standard energy equation states

$$K_i + U_i = K_f + U_f$$

or

$$\frac{1}{2}mv_i^2 + mgh_i = \frac{1}{2}mv_f^2 + mgh_f$$

Canceling the mass, setting $v_i = 0$ m/s, and rearranging terms, we get

$$v_f = \sqrt{2g\Delta h}$$

$$v_f = \sqrt{21(10 \text{ m/s}^2)(30 \text{ m} - 20 \text{ m})}$$

$$v_f = \sqrt{200 \text{ m}^2/\text{s}^2} = 10\sqrt{2} \text{ m/s}$$

(b) The lost energy was likely lost as heat.

POWER

Simply put, **power** is the rate at which work gets done (or energy gets transferred, which is the same thing). Suppose Scott and Jean each do 1,000 J of work, but Scott does the work in 2 minutes, while Jean does it in 1 minute. They both did the same amount of work, but Jean did it more quickly; thus Jean was more powerful. Here's the definition of power:

$$\text{Power} = \frac{\text{Work}}{\text{time}} \qquad \text{—in symbols} \rightarrow \qquad P = \frac{W}{t}$$

The unit of power is the joule per second (J/s), which is renamed the **watt**, and symbolized W (not to be confused with the symbol for work, W). One watt is 1 joule per second: 1 W = 1 J/s. Here in the United States, which still uses older units like inches, feet, yards, miles, ounces, pounds, and so forth, you still hear of power ratings (particularly of engines) expressed in horsepower. One horsepower is defined as 1 hp = 746 W.

Note that this conversion will be provided on the test.

$$P = W/t \qquad\qquad \text{Recall: } W = Fd$$
$$P = Fd/t \qquad\qquad \text{Recall: } v = d/t$$

$$P = W/t = Fd/t = Fv$$

This equation only applies for a constant force parallel to a constant velocity. Remember to check that your equation fits the given circumstances!

> **Example 18** A mover pushes a large crate (mass m = 75 kg) from the inside of the truck to the back end (a distance of 6 m), exerting a steady push of 300 N. If he moves the crate this distance in 20 s, what is his power output during this time?

Solution. The work done on the crate by the mover is $W = Fd$ = (300 N)(6 m) = 1,800 J. If this much work is done in 20 s, then the power delivered is $P = W/t$ = (1,800 J)/(20 s) = 90 W.

Example 19 What must be the power output of an elevator motor that can lift a total mass of 1,000 kg and give the elevator a constant speed of 8.0 m/s?

Solution. The equation $P = Fv$, with $F = mg$, yields

$$P = mgv = (1{,}000 \text{ kg})(10 \text{ N/kg})(8.0 \text{ m/s}) = 80{,}000 \text{ W} = 80 \text{ kW}$$

Chapter 6 Review Questions

Answers and Explanations can be found in Chapter 13.

Section I: Multiple-Choice

1. A force **F** of strength 20 N acts on an object of mass 3 kg as it moves a distance of 4 m. If **F** is perpendicular to the 4 m displacement, the work it does is equal to

    (A) 0 J
    (B) 60 J
    (C) 80 J
    (D) 600 J

2. Under the influence of a force, an object of mass 4 kg accelerates from 3 m/s to 6 m/s in 8 s. How much work was done on the object during this time?

    (A) 27 J
    (B) 54 J
    (C) 72 J
    (D) 96 J

3. A box of mass m slides down a frictionless inclined plane of length L and vertical height h. What is the change in its gravitational potential energy?

    (A) $-mgL$
    (B) $-mgh$
    (C) $-mgL/h$
    (D) $-mgh/L$

4. While a person lifts a book of mass 2 kg from the floor to a tabletop, 1.5 m above the floor, how much work does the gravitational force do on the book?

    (A) −30 J
    (B) −15 J
    (C)   0 J
    (D)   15 J

5. A block of mass 3.5 kg slides down a frictionless inclined plane of length 6.4 m that makes an angle of 30° with the horizontal. If the block is released from rest at the top of the incline, what is its speed at the bottom?

    (A) 5.0 m/s
    (B) 6.4 m/s
    (C) 8.0 m/s
    (D) 10 m/s

6. A block of mass m slides from rest down an inclined plane of length s and height h. If F is the magnitude of the force of kinetic friction acting on the block as it slides, then the kinetic energy of the block when it reaches the bottom of the incline will be equal to

    (A) mgh
    (B) $mgs - Fh$
    (C) $mgh - Fs$
    (D) $mgs - Fs$

7. As a rock of mass 4 kg drops from the edge of a 40-meter-high cliff, it experiences air resistance, whose average strength during the descent is 20 N. At what speed will the rock hit the ground?

    (A) 10 m/s
    (B) 12 m/s
    (C) 16 m/s
    (D) 20 m/s

8. An astronaut drops a rock from the top of a crater on the Moon. When the rock is halfway down to the bottom of the crater, its speed is what fraction of its final impact speed?

    (A) $\dfrac{1}{4}$

    (B) $\dfrac{1}{2\sqrt{2}}$

    (C) $\dfrac{1}{2}$

    (D) $\dfrac{1}{\sqrt{2}}$

9. A force of 200 N is required to keep an object sliding at a constant speed of 2 m/s across a rough floor. How much power is being expended to maintain this motion?

    (A) 50 W
    (B) 100 W
    (C) 200 W
    (D) 400 W

Section II: Free-Response

1. A box of mass m is released from rest at Point A, the top of a long, frictionless slide. Point A is at height H above the level of Points B and C. Although the slide is frictionless, the horizontal surface from Point B to C is not. The coefficient of kinetic friction between the box and this surface is μ_k, and the horizontal distance between Point B and C is x.

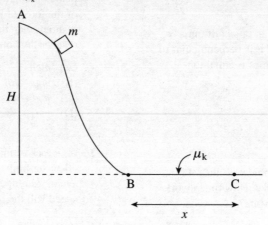

(a) Find the speed of the box when its height above Point B is $\frac{1}{2}H$.

(b) Find the speed of the box when it reaches Point B.

(c) Determine the value of μ_k so that the box comes to rest at Point C.

(d) Now assume that Points B and C were not on the same horizontal level. In particular, assume that the surface from B to C had a uniform upward slope so that Point C were still at a horizontal distance of x from B but now at a vertical height of y above B. Answer the question posed in part (c).

(e) If the slide were not frictionless, determine the work done by friction as the box moved from Point A to Point B if the speed of the box as it reached Point B were half the speed calculated in part (b).

2. A student uses a digital camera and computer to collect the following data about a ball as it slides down a curved frictionless track. The initial release point is 1.5 meters above the ground and the ball is released from rest. He prints up the following data and then tries to analyze it.

time (s)	velocity (m/s)
0.00	0.00
0.05	1.41
0.10	2.45
0.15	3.74
0.20	3.74
0.25	3.46
0.30	3.16
0.35	2.83
0.40	3.46
0.45	4.24
0.50	4.47

(a) Based on the data, what are the corresponding heights for each data point?

(b) What time segment experiences the greatest acceleration and what is the value of this acceleration?

(c) How would the values change if the ball were replaced by an identical ball with double the mass?

3. A car with a mass of 800 kg is traveling with an initial speed of 10 m/s. When the brakes of the car are applied, the car starts to skid, and it experiences a frictional force with $\mu_k = 0.2$.

(a) What is the skidding distance of the car?

(b) How would the skidding distance change if the initial speed of the car were doubled?

(c) How would the skidding distance change if the initial mass of the car were doubled?

Summary

- Work is force applied across a displacement. Work can cause a change in energy. Positive work puts energy into a system, while negative work takes energy out of a system. Basic equations for work include:

  $W = Fd \cos\theta$

  $\text{Work} = \Delta KE$

  $W = $ area under an F versus d graph

- Energy is a conserved quantity. By that we mean the total initial energy is equal to the total final energy. Basic equations with energy include:

  $K = \dfrac{1}{2}mv^2$

  $U_g = mgh$

- Often, we limit ourselves to mechanical energy with no heat lost or gained. In this case:

  $K_i + U_i \pm W = K_f + U_f$

- Power is the rate at which one does work and is given by:

  $P = \dfrac{W}{t}$ or $P = Fv$

Chapter 7
Linear Momentum

"Nothing will happen until something moves."

—Albert Einstein

Previously we discovered the nature of objects, the reason why objects move the way they do, and the energy required to make objects move. Now we will predict the nature of objects when they interact with other objects and the resulting outcome.

INTRODUCTION TO MOMENTUM

A pool stick hitting the cue ball, a car collision, the Death Star exploding—physics is about the interaction of objects. A collision is a complex interaction between two objects. But what happens after two objects interact with each other? If we use Newton's Second Law, it will prove to be quite a tedious challenge. The forces occurring during a car collision are unimaginably complex. And not all collisions end with the same results: sometimes they move away from each other, sometimes they move in the same direction, and sometimes the objects stick together. If we were to measure their speeds before the collision and their speeds after the collision, we would also get different results. Using our observations and a general understanding of Newton's Second Law, we could predict that the smaller car in a collision (head on and both cars traveling with the same speed) would end up having the greater speed in the end. How can such a complex interaction give rise to a simple outcome?

WHAT IS MOMENTUM?

Momentum Units
$p = mv$ so the units for momentum is a kg · m/s. There is no special term for this unit.

When Newton first expressed his Second Law, he didn't write $\mathbf{F}_{net} = m\mathbf{a}$. Instead, he expressed the law in the words, *The alteration of motion is...proportional to the... force impressed....* By "motion," he meant the product of mass and velocity, a vector quantity known as **linear momentum** and denoted by **p**:

$$p = mv$$

Momentum is a Vector
Remember that a vector has magnitude and direction. In collision problems, be aware of orientation and assign negative values to negative velocities and positive values for positive velocities.

So Newton's original formulation of the Second Law read $\Delta\mathbf{p} \propto \mathbf{F}$, or, equivalently, $\mathbf{F} \propto \Delta\mathbf{p}$. But a large force that acts for a short period of time can produce the same change in linear momentum as a small force acting for a greater period of time. Knowing this, we can turn the proportion above into an equation, if we take the average force that acts over the time interval Δt:

$$\mathbf{F} = \frac{\Delta\mathbf{p}}{\Delta t}$$

This equation becomes $\mathbf{F} = m\mathbf{a}$, since $\Delta\mathbf{p}/\Delta t = \Delta(m\mathbf{v})/\Delta t = m(\Delta\mathbf{v}/\Delta t) = m\mathbf{a}$.

Example 1 A golfer strikes a golf ball of mass 0.05 kg, and the time of impact between the golf club and the ball is 1 ms. If the ball acquires a velocity of magnitude 70 m/s, calculate the average force exerted on the ball.

Solution. Using Newton's Second Law, we find

$$\overline{F} = \frac{\Delta p}{\Delta t} = \frac{\Delta (mv)}{\Delta t} = m \frac{v - 0}{\Delta t} = (0.05 \text{ kg}) \frac{70 \text{ m/s}}{10^{-3} \text{ s}} = 3{,}500 \text{ N}$$

IMPULSE

The product of force and the time during which it acts is known as **impulse**; it's a vector quantity that's denoted by **J**:

$$\mathbf{J} = \mathbf{F}\Delta t$$

Impulse is equal to change in linear momentum. In terms of impulse, Newton's Second Law can be written in yet another form:

$$\mathbf{J} = \Delta \mathbf{p}$$

Sometimes this is referred to as the **Impulse-Momentum Theorem**, but it's just another way of writing Newton's Second Law. The impulse delivered to an object may be found by taking the area under a force-versus-time graph.

The Impulse-Momentum Theorem basically states that an impulse that is delivered on an object changes its momentum. The momentum "after" the collision is equal to the momentum "before" the collision added in with the impulse required to get your final outcome. In equation form, this states:

$$p_{\text{final}} = p_{\text{initial}} + J$$

Impulse eliminates the need to use $F = ma$ and simplifies the intricate forces delivered into mass, initial, and final velocities.

Reminder!
Remember that momentum and Kinetic Energy are not the same thing.

Newton's First Law
Objects naturally resist changes in their motion. In order for us to change an object's speed or direction, we must induce some kind of force over a period of time.

Get Used to It

The question you see in Example 2 here is a common type of problem that you will see on the AP Physics 1 Exam.

Example 2 A football team's kicker punts the ball (mass = 0.4 kg) and gives it a launch speed of 30 m/s. Find the impulse delivered to the football by the kicker's foot and the average force exerted by the kicker on the ball, given that the impact time is 8 ms.

Solution. As we know, impulse is equal to change in linear momentum, so

$$J = \Delta p = p_f - p_i = p_f = mv = (0.4 \text{ kg})(30 \text{ m/s}) = 12 \text{ kg·m/s}$$

Using the equation $\overline{F} = J/\Delta t$, we find that the average force exerted by the kicker is

$$\overline{F} = J/\Delta t = (12 \text{ kg} \cdot \text{m/s})/(8 \times 10^{-3} \text{ s}) = 1,500 \text{ N}$$

Example 3 An 80 kg stuntman jumps out of a window that's 45 m above the ground.

    (a) How fast is he falling when he reaches ground level?

    (b) He lands on a large, air-filled target, coming to rest in 1.5 s. What average force does he feel while coming to rest?

    (c) What if he had instead landed on the ground (impact time = 10 ms)?

Solution.

    (a) His gravitational potential energy turns into kinetic energy:

$$mgh = \frac{1}{2}mv^2, \text{ so}$$
$$v = \sqrt{2gh} = \sqrt{2(10)(45)} = 30 \text{ m/s}$$

    (You could also have answered this question using Big Five #5.)

    (b) Using $\mathbf{F} = \Delta \mathbf{p}/\Delta t$, we find that

$$\mathbf{F} = \frac{\Delta \mathbf{p}}{\Delta t} = \frac{\mathbf{p}_f - \mathbf{p}_i}{\Delta t} = \frac{0 - m\mathbf{v}_i}{\Delta t} = \frac{-(80 \text{ kg})(30 \text{ m/s})}{1.5 \text{ s}} = -1,600 \text{ N} \Rightarrow F = 1,600 \text{ N}$$

    (c) In this case,

$$\mathbf{F} = \frac{\Delta \mathbf{p}}{\Delta t} = \frac{\mathbf{p}_f - \mathbf{p}_i}{\Delta t} = \frac{0 - m\mathbf{v}_i}{\Delta t} = \frac{-(80 \text{ kg})(30 \text{ m/s})}{10 \times 10^{-3} \text{ s}} = -240,000 \text{ N} \Rightarrow F = 240,000 \text{ N}$$

The negative signs in the vector answers to (b) and (c) simply tell you that the forces are acting in the opposite direction of motion and will cause the object to slow down. This force is equivalent to about 27 tons (!), more than enough to break bones and cause fatal brain damage. Notice how crucial impact time is: Increasing the slowing-down time reduces the acceleration and the force, ideally enough to prevent injury. This is the purpose of air bags in cars, for instance.

Example 4 A small block of mass $m = 0.07$ kg, initially at rest, is struck by an impulsive force F of duration 10 ms whose strength varies with time according to the following graph:

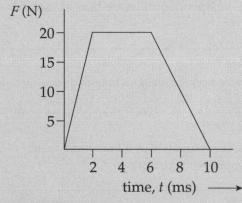

What is the resulting speed of the block?

Curved Graphs?
Calculus will be needed to find the true area under curved graphs. But, the AP Physics 1 Exam does not require you to know calculus. You can approximate areas under curved graphs by using basic shapes you have studied before in geometry.

Solution. The impulse delivered to the block is equal to the area under the F-versus-t graph. The region is a trapezoid, so its area, $\frac{1}{2}(\text{base}_1 + \text{base}_2) \times \text{height}$, can be calculated as follows:

$$J = \frac{1}{2}[(10 \text{ ms} - 0) + (6 \text{ ms} - 2 \text{ ms})] \times (20 \text{ N}) = 0.14 \text{ N} \cdot \text{s}$$

Now, by the Impulse-Momentum Theorem,

$$J = \Delta p = p_f - p_i = mv \quad \Rightarrow \quad v_f = \frac{J}{m} = \frac{0.14 \text{ N} \cdot \text{s}}{0.07 \text{ kg}} = 2 \text{ m/s}$$

CONSERVATION OF LINEAR MOMENTUM

Newton's Third Law states that when one object exerts a force on a second object, the second object exerts an equal but opposite force on the first. Newton's Third Law combines with Newton's Second Law when two objects interact with each other.

In the previous section, we redefined Newton's Second Law as Impulse-Momentum Theorem, $J = \Delta p$. If we combine the laws and interpret them in terms of momentum, it states that two interacting objects experience equal but opposite momentum changes (assuming we have an isolated system, meaning no external forces).

> *The total linear momentum of an isolated system remains constant.*

The momentum "before" equals the momentum "after." This is the Law of Conservation of Momentum, which states:

> total $\mathbf{p}_{initial}$ = total $\mathbf{p}_{final}$

Example 5 An astronaut is floating in space near her shuttle when she realizes that the cord that's supposed to attach her to the ship has become disconnected. Her total mass (body + suit + equipment) is 91 kg. She reaches into her pocket, finds a 1 kg metal tool, and throws it out into space with a velocity of 9 m/s directly away from the ship. If the ship is 10 m away, how long will it take her to reach it?

Solution. Here, the astronaut + tool are the system. Because of Conservation of Linear Momentum,

$$m_{astronaut}\,\mathbf{v}_{astronaut} + m_{tool}\,\mathbf{v}_{tool} = 0$$

$$m_{astronaut}\,\mathbf{v}_{astronaut} = -m_{tool}\mathbf{v}_{tool}$$

$$\mathbf{v}_{astronaut} = -\frac{m_{tool}}{m_{astronaut}}\mathbf{v}_{tool}$$

$$= -\frac{1\text{ kg}}{90\text{ kg}}(-9\text{ m/s}) = +0.1\,\text{m/s}$$

Using *distance = rate × time*, we find

$$t = \frac{d}{v} = \frac{10 \text{ m}}{0.1 \text{ m/s}} = 100 \text{ s}$$

COLLISIONS

Conservation of Linear Momentum is routinely used to analyze **collisions**. The objects whose collision we will analyze form the *system*, and although the objects exert forces on each other during the impact, these forces are only *internal* (they occur within the system), and the system's total linear momentum is conserved.

Let's break down the collision types:

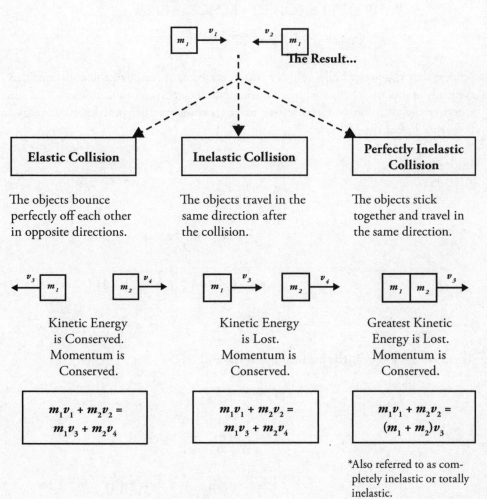

Two objects collide with one another.

The Result...

Elastic Collision	Inelastic Collision	Perfectly Inelastic Collision
The objects bounce perfectly off each other in opposite directions.	The objects travel in the same direction after the collision.	The objects stick together and travel in the same direction.
Kinetic Energy is Conserved. Momentum is Conserved.	Kinetic Energy is Lost. Momentum is Conserved.	Greatest Kinetic Energy is Lost. Momentum is Conserved.
$m_1 v_1 + m_2 v_2 = m_1 v_3 + m_2 v_4$	$m_1 v_1 + m_2 v_2 = m_1 v_3 + m_2 v_4$	$m_1 v_1 + m_2 v_2 = (m_1 + m_2) v_3$

*Also referred to as completely inelastic or totally inelastic.

> **Example 6** Two balls roll toward each other. The red ball has a mass of 0.5 kg and a speed of 4 m/s just before impact. The green ball has a mass of 0.2 kg and a speed of 2 m/s. After the head-on collision, the red ball continues forward with a speed of 2 m/s. Find the speed of the green ball after the collision. Was the collision elastic?

Remember Vectors!
Momentum is a vector. If you set an object traveling to the right as the positive momentum, you must give an object going to the left a negative momentum.

Solution. First, remember that momentum is a vector quantity, so the direction of the velocity is crucial. Since the balls roll toward each other, one ball has a positive velocity while the other has a negative velocity. Let's call the red ball's velocity before the collision positive; then $v_{red} = +4$ m/s, and $v_{green} = -2$ m/s. Using a prime to denote *after the collision*, Conservation of Linear Momentum gives us the following:

$$\text{total } \mathbf{p}_{before} = \text{total } \mathbf{p}_{after}$$

$$m_{red}\mathbf{v}_{red} + m_{green}\mathbf{v}_{green} = m_{red}\mathbf{v}'_{red} + m_{green}\mathbf{v}'_{green}$$

$$(0.5)(+4) + (0.2)(-2) = (0.5)(+2) + (0.2)\mathbf{v}'_{green}$$

$$\mathbf{v}'_{green} = +3.0 \text{ m/s}$$

Notice that the green ball's velocity was reversed as a result of the collision; this typically happens when a lighter object collides with a heavier object. To see whether the collision was elastic, we need to compare the total kinetic energies before and after the collision.

Initially

Elastic versus Inelastic
All collisions conserve momentum. In order to determine whether an unknown collision was elastic or inelastic, we must check if kinetic energy was lost. If kinetic energy was lost, then the collision was inelastic.

$$K_t = K_1 + K_2$$

$$= \frac{1}{2}m_1 v_{i1}^2 + \frac{1}{2}m_2 v_{i2}^2$$

$$= \frac{1}{2}(0.5)(+4)^2 + \frac{1}{2}(0.2)(-2)^2$$

$$= 4.4 \text{ J}$$

There are 4.4 joules at the beginning. At the end,

$$K_t = K_1 + K_2$$

$$= \frac{1}{2}m_1 v_{f1}^2 + \frac{1}{2}m_2 v_{f2}^2$$

$$= \frac{1}{2}(0.5)(+2)^2 + \frac{1}{2}(0.2)(3)^2$$

$$= 1.9 \text{ J}$$

So, there is less kinetic energy at the end compared to the beginning. Kinetic energy was lost (so the collision was inelastic). Most of the lost energy was transferred as heat; the two objects are both slightly warmer as a result of the collision.

Example 7 Two balls roll toward each other. The red ball has a mass of 0.5 kg and a speed of 4 m/s just before impact. The green ball has a mass of 0.3 kg and a speed of 2 m/s. If the collision is completely inelastic, determine the velocity of the composite object after the collision.

Solution. If the collision is completely inelastic, then, by definition, the masses stick together after impact, moving with a velocity, v'. Applying Conservation of Linear Momentum, we find

$$\text{total } \mathbf{p}_{before} = \text{total } \mathbf{p}_{after}$$
$$m_{red}\,\mathbf{v}_{red} + m_{green}\,\mathbf{v}_{green} = (m_{red} + m_{green})\mathbf{v}'$$
$$(0.5)(4) + (0.3)(-2) = (0.5 + 0.3)\mathbf{v}'$$
$$\mathbf{v}' = +1.8 \text{ m/s}$$

Example 8 A 500 kg car travels 20 m/s due north. It hits a 500 kg car traveling due west at 30 m/s. The cars lock bumpers and stick together. What is the velocity the instant after impact?

Solution. This problem illustrates the vector nature of numbers. First, look only at the x (east-west) direction. There is only one car moving west and its momentum is given by:

$$\mathbf{p} = m\mathbf{v} \Rightarrow (500 \text{ kg})(30 \text{ m/s}) = 15,000 \text{ kg} \cdot \text{m/s west}$$

Next, look only at the y (north-south) direction. There is only one car moving north and its momentum is given by:

$$\mathbf{p} = m\mathbf{v} \Rightarrow (500 \text{ kg})(20 \text{ m/s}) = 10,000 \text{ kg} \cdot \text{m/s north}$$

The total final momentum is the resultant of these two vectors. Use the Pythagorean Theorem:

$$(10,000)^2 + (15,000)^2 = \mathbf{p}_f^2$$

$$\mathbf{p} = 18,027 \text{ kg} \cdot \text{m/s}$$

To find the total velocity you need to solve for v using

$$\mathbf{p_f} = mv \Rightarrow 18{,}027 \text{ kg} \cdot \text{m/s} = (500 \text{ kg} + 500 \text{ kg})(v)$$

$$v = 18 \text{ m/s}$$

However, we are not done yet because velocity has both a magnitude (which we now know is 18 m/s) and a direction. The direction can be expressed using $\tan \theta = 10{,}000/15{,}000$ so $\theta = \tan^{-1}(10{,}000/15{,}000)$, or 33.7 degrees north of west. Again, most of the AP Physics 1 Exam is in degrees, not radians.

Example 9 An object of mass m moves with velocity $\mathbf{v}$ toward a stationary object of mass $2m$. After impact, the objects move off in the directions shown in the following diagram:

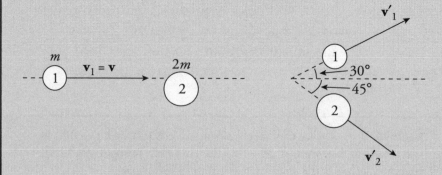

Before the collision   After the collision

(a) Determine the magnitudes of the velocities after the collision (in terms of v).
(b) Is the collision elastic? Explain your answer.

Solution.

(a) Conservation of Linear Momentum is a principle that establishes the equality of two vectors: $\mathbf{p}_{total}$ before the collision and $\mathbf{p}_{total}$ after the collision. Writing this single vector equation as two equations, one for the x-component and one for the y, we have

x component: $mv = mv_1' \cos 30° + 2mv_2' \cos 45°$ (1)

y component: $0 = mv_1' \sin 30° - 2mv_2' \sin 45°$ (2)

Adding these equations eliminates v_2', because $\cos 45° = \sin 45°$.

$$mv = mv_1'(\cos 30° + \cos 30°)$$

and lets us determine v_1':

$$v_1' = \frac{v}{\cos 30° + \sin 30°} = \frac{2v}{1 + \sqrt{3}}$$

Substituting this result into Equation (2) gives us

$$0 = m\frac{2v}{1 + \sqrt{3}}\sin 30° - 2mv_2'\sin 45°$$

$$2mv_2'\sin 45° = m\frac{2v}{1 + \sqrt{3}}\sin 30°$$

$$v_2' = \frac{\dfrac{2v}{1+\sqrt{3}}\sin 30°}{2\sin 45°} = \frac{v}{\sqrt{2}(1 + \sqrt{3})}$$

(b) The collision is elastic only if kinetic energy is conserved. The total kinetic energy after the collision, K', is calculated as follows:

$$K' = \frac{1}{2} \cdot mv_1'^2 + \frac{1}{2} \cdot 2mv_2'^2$$

$$= \frac{1}{2}m\left(\frac{2v}{1+\sqrt{3}}\right)^2 + m\left(\frac{v}{\sqrt{2}(1+\sqrt{3})}\right)^2$$

$$= mv^2\left[\frac{2}{(1+\sqrt{3})^2} + \frac{1}{2(1+\sqrt{3})^2}\right]$$

$$= \frac{5}{2(1+\sqrt{3})^2}mv^2$$

However, the kinetic energy before the collision is just $K = \frac{1}{2}mv^2$, so the fact that

$$\frac{5}{2(1+\sqrt{3})^2} < \frac{1}{2}$$

tells us that K' is less than K, so some kinetic energy is lost; the collision is inelastic.

Chapter 7 Review Questions

Answers and Explanations can be found in Chapter 13.

Section I: Multiple-Choice

1. An object of mass 2 kg has a linear momentum of magnitude 6 kg · m/s. What is this object's kinetic energy?

   (A) 3 J
   (B) 6 J
   (C) 9 J
   (D) 12 J

2. A ball of mass 0.5 kg, initially at rest, acquires a speed of 4 m/s immediately after being kicked by a force of strength 20 N. For how long did this force act on the ball?

   (A) 0.01 s
   (B) 0.1 s
   (C) 0.2 s
   (D) 1 s

3. A box with a mass of 2 kg accelerates in a straight line from 4 m/s to 8 m/s due to the application of a force whose duration is 0.5 s. Find the average strength of this force.

   (A) 4 N
   (B) 8 N
   (C) 12 N
   (D) 16 N

4. A ball of mass m traveling horizontally with velocity **v** strikes a massive vertical wall and rebounds back along its original direction with no change in speed. What is the magnitude of the impulse delivered by the wall to the ball?

   (A) $\frac{1}{2}m\mathbf{v}$

   (B) $m\mathbf{v}$

   (C) $2m\mathbf{v}$

   (D) $4m\mathbf{v}$

5. Two objects, one of mass 3 kg moving with a speed of 2 m/s and the other of mass 5 kg and speed 2 m/s, move toward each other and collide head-on. If the collision is perfectly inelastic, find the speed of the objects after the collision.

   (A) 0.25 m/s
   (B) 0.5 m/s
   (C) 0.75 m/s
   (D) 1 m/s

6. Object 1 moves toward Object 2, whose mass is twice that of Object 1 and which is initially at rest. After their impact, the objects lock together and move with what fraction of Object 1's initial kinetic energy?

   (A) 1/18
   (B) 1/9
   (C) 1/6
   (D) 1/3

7. Two objects move toward each other, collide, and separate. If there was no net external force acting on the objects, but some kinetic energy was lost, then

   (A) the collision was elastic and total linear momentum was conserved
   (B) the collision was elastic and total linear momentum was not conserved
   (C) the collision was not elastic and total linear momentum was conserved
   (D) the collision was not elastic and total linear momentum was not conserved

8. Two frictionless carts (mass = 500 g each) are sitting at rest on a perfectly level table. The teacher taps the release so that one cart pushes off the other. If one of the carts has a speed of 2 m/s, then what is the final momentum of the system (in kg · m/s)?

   (A) 2,000
   (B) 1,000
   (C) 2
   (D) 0

9. A wooden block of mass M is moving at speed V in a straight line.

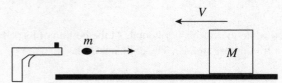

How fast would the bullet of mass m need to travel to stop the block (assuming that the bullet became embedded inside)?

(A) $mV/(m + M)$
(B) mV/M
(C) MV/m
(D) $(m + M)V/m$

10. Which of the following best describes a perfectly inelastic collision free of external forces?

(A) Total linear momentum is never conserved.
(B) Total linear momentum is sometimes conserved.
(C) Kinetic energy is never conserved.
(D) Kinetic energy is always conserved.

11. Object 1 moves with an initial speed of v_0 toward Object 2, which has a mass half that of Object 1. If the final speed of both objects after colliding is 0, what must have been the initial speed of Object 2 (assuming no external forces)?

(A) v_0
(B) $2v_0$
(C) $\frac{1}{2}v_0$
(D) $4v_0$

Section II: Free-Response

1. A steel ball of mass m is fastened to a light cord of length L and released when the cord is horizontal. At the bottom of its path, the ball strikes a hard plastic block of mass $M = 4m$, initially at rest on a frictionless surface. The collision is elastic.

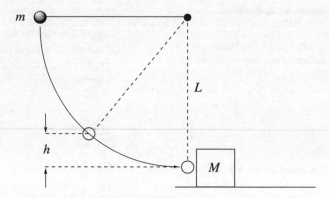

(a) Find the tension in the cord when the ball's height above its lowest position is $\frac{1}{2}L$. Write your answer in terms of m and g.

(b) Find the speed of the block immediately after the collision.

(c) To what height h will the ball rebound after the collision?

2. A *ballistic pendulum* is a device that may be used to measure the muzzle speed of a bullet. It is composed of a wooden block suspended from a horizontal support by cords attached at each end. A bullet is shot into the block, and as a result of the perfectly inelastic impact, the block swings upward. Consider a bullet (mass m) with velocity v as it enters the block (mass M). The length of the cords supporting the block each have length L. The maximum height to which the block swings upward after impact is denoted by y, and the maximum horizontal displacement is denoted by x.

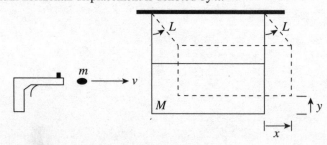

(a) In terms of m, M, g, and y, determine the speed v of the bullet.

(b) What fraction of the bullet's original kinetic energy is lost as a result of the collision? What happens to the lost kinetic energy?

(c) If y is very small (so that y^2 can be neglected), determine the speed of the bullet in terms of m, M, g, x, and L.

(d) Once the block begins to swing, does the momentum of the block remain constant? Why or why not?

Summary

- Momentum is a vector quantity given by $p = mv$. If you push on an object for some amount of time we call that an impulse (J). Impulses cause a change in momentum. Impulse is also a vector quantity and these ideas are summed up in the equations $J = F\Delta t$ or $J = \Delta p$.

- Momentum is a conserved quantity in a closed system (that is, a system with no external forces). That means:

  total p_i = total p_f.

- Overall strategy for conservation of momentum problems:
  I. Create a coordinate system.
  II. Break down each object's momentum into x- and y-components. That is $p_x = p \cos\theta$ and $p_y = p \sin\theta$ for any given object.
  III. $\sum p_{xi} = \sum p_{xf}$ and $\sum p_{yi} = \sum p_{yf}$
  IV. Sometimes you end up rebuild-ing vectors in the end. Remember total $p = \sqrt{p_x^2 + p_y^2}$ and the angle is given by $\theta = \tan^{-1}\left(\dfrac{p_y}{p_x}\right)$.

Chapter 8
Uniform Circular Motion, Newton's Law of Gravitation, and Rotational Motion

"I can calculate the motion of heavenly bodies but not the madness of people."

—Sir Isaac Newton

Since man has looked up to the stars, we have always tried to find reason as to why they move the way they do. The stars and planets make elliptical orbits, but this violates Newton's First Law in which objects will continue to move straight. What was making these orbits elliptical? Our first answers explained motion in a linear fashion. Then we added a second dimension and our motion became parabolic. Now we will explore when objects begin to undergo circular motion. Then we will have a better understanding of the orbit of our Moon around the Earth and the Earth around the Sun.

UNIFORM CIRCULAR MOTION

Let's simplify matters and consider the object's speed around its path to be constant. This is called **uniform circular motion**. You should remember that although the speed may be constant, the velocity is not, because the direction of the velocity is always changing. Since the velocity is changing, there must be acceleration. This acceleration does not change the speed of the object; it only changes the direction of the velocity to keep the object on its circular path. Also, in order to produce an acceleration, there must be a force; otherwise, the object would move off in a straight line (Newton's First Law).

Take a look at the figures below. The figure on the left shows an object moving along a circular trajectory, along with its velocity vectors at two nearby points. The vector v_1 is the object's velocity at time $t = t_1$, and v_2 is the object's velocity vector a short time later (at time $t = t_2$). The velocity vector is always tangential to the object's path (whatever the shape of the trajectory). Notice that since we are assuming constant speed, the lengths of v_1 and v_2 (their magnitudes) are the same.

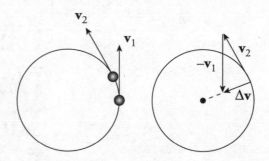

Non-Uniform Circular Objects
Planets and most objects do not undergo uniform circular motion. They usually follow elliptical orbits with varying speeds. For the purposes of the AP Physics 1 Exam, this will not be tested.

Since $\Delta v = v_2 - v_1$ points toward the center of the circle (see the figure on the right), so does the acceleration, since $a = \Delta v/\Delta t$. Because the acceleration vector points toward the center of the circle, it's called **centripetal acceleration**, or a_c. The centripetal acceleration is what turns the velocity vector to keep the object traveling in a circle. The magnitude of the centripetal acceleration depends on the object's speed, v, and the radius, r, of the circular path according to the equation

$$a_c = \frac{v^2}{r}$$

Example 1 An object of mass 5 kg moves at a constant speed of 6 m/s in a circular path of radius 2 m. Find the object's acceleration and the net force responsible for its motion.

Solution. By definition, an object moving at constant speed in a circular path is undergoing uniform circular motion. Therefore, it experiences a centripetal acceleration of magnitude v^2/r, always directed toward the center of the circle:

$$a_c = \frac{v^2}{r} = \frac{(6 \text{ m/s})^2}{2 \text{ m}} = 18 \text{ m/s}^2$$

The force that produces the centripetal acceleration is given by Newton's Second Law, coupled with the equation for centripetal acceleration:

$$F_c = ma_c = m\frac{v^2}{r}$$

This equation gives the magnitude of the force. As for the direction, recall that because $\mathbf{F} = m\mathbf{a}$, the directions of $\mathbf{F}$ and $\mathbf{a}$ are always the same. Since centripetal acceleration points toward the center of the circular path, so does the force that produces it. Therefore, it's called **centripetal force**. The centripetal force acting on this object has a magnitude of $F_c = ma_c = (5 \text{ kg})(18 \text{ m/s}^2) = 90 \text{ N}$.

> **Example 2** A 10.0 kg mass is attached to a string that has a breaking strength of 200 N. If the mass is whirled in a horizontal circle of radius 80 cm, what maximum speed can it have?

Solution. The first thing to do in problems like this is to identify what forces produce the centripetal acceleration. Notice that this is a horizontal circle. We can limit our examination to the horizontal (x) direction. Because gravity exerts a force in the y direction, it can be ignored. If we were given a problem with a vertical circle, we would have to include the effects of gravity, which will be demonstrated in Example 4. In this example, the tension in the string produces the centripetal force:

$$\mathbf{F}_T \text{ provides } \mathbf{F}_c \Rightarrow F_T = \frac{mv^2}{r} \Rightarrow v = \sqrt{\frac{rF_T}{m}} \Rightarrow v_{max} = \sqrt{\frac{rF_{T,\,max}}{m}}$$

$$= \sqrt{\frac{(0.80 \text{ m})(200 \text{ N})}{10 \text{ kg}}}$$

$$= 4 \text{ m/s}$$

Notice the unit change from 80 cm to 0.80 m. As a general rule, stick to kg, m, and s because the newton is composed of these units.

Centripetal Force and Centrifugal Force
Centripetal force points into the center of the circle and centrifugal points away from the circle. Centrifugal force is referred to as a fictitious force since it is not a real force. Centripetal force is the net force from the physical forces acting on the object.

Example 3 An athlete who weighs 800 N is running around a curve at a speed of 5.0 m/s in an arc whose radius of curvature, r, is 5.0 m. Find the centripetal force acting on him. What provides the centripetal force? What could happen to him if r were smaller?

Solution. Using the equation for the strength of the centripetal force, we find that

$$F_c = m\frac{v^2}{r} = \frac{F_w}{g} \cdot \frac{v^2}{r} = \frac{800 \text{ N}}{10 \text{ N/kg}} \cdot \frac{\left(5.0 \text{ m/s}\right)^2}{5.0 \text{ m}} = 400 \text{ N}$$

In this case, static friction provides the centripetal force. If the radius of curvature of the arc were smaller, then the centripetal force required to keep him running in a circle would increase. If the centripetal force increased enough, it might exceed what the force of static friction could provide, at which point he would slip.

Example 4 A roller-coaster car enters the circular-loop portion of the ride. At the very top of the circle (where the people in the car are upside down), the speed of the car is 15 m/s, and the acceleration points straight down. If the diameter of the loop is 40 m and the total mass of the car (plus passengers) is 1,200 kg, find the magnitude of the normal force exerted by the track on the car at this point.

Solution. There are two forces acting on the car at its topmost point: the normal force exerted by the track and the gravitational force, both of which point downward. At the top of the loop, the gravitational force as well as the normal force point downwards. This is because the normal force acts perpendicular to the surface of the track, and the gravitational force is always directed downward.

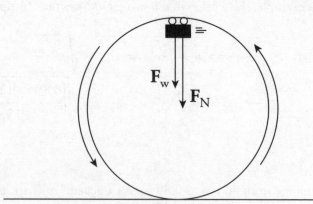

The combination of these two forces, $\mathbf{F}_N + \mathbf{F}_w$, provides the centripetal force:

$$F_N + F_w = \frac{mv^2}{r} \Rightarrow F_N = \frac{mv^2}{r} - F_w$$

$$= \frac{mv^2}{r} - mg$$

$$= m\left(\frac{v^2}{r} - g\right)$$

$$= (1,200 \text{ kg})\left[\frac{(15 \text{ m/s})^2}{\frac{1}{2}(40 \text{ m})} - 10 \text{ m/s}^2\right]$$

$$= 1,500 \text{ N}$$

Example 5 In the previous example, if the net force on the car at its topmost point is straight down, why doesn't the car fall straight down?

Solution. Remember that force tells an object how to accelerate. If the car had zero velocity at this point, then it would certainly fall straight down, but the car has a nonzero velocity (to the left) at this point. The fact that the acceleration is downward means that, at the next moment v will point down to the left at a slight angle, ensuring that the car remains on a circular path, in contact with the track.

Example 6 How would the normal force change in Example 4 if the car was at the bottom of the circle?

Solution. There are still two forces acting on the cart: The gravitational force still points downward, but the normal force pushes ninety degrees to the surface (upward). These forces now oppose one another. The combination of these two forces still provides the centripetal force. Because the centripetal acceleration points inward, we will make anything that points toward the center of the circle positive and anything that points away from the circle negative. Therefore, our equation becomes:

$$F_N - F_w = \frac{mv^2}{r} \Rightarrow F_N = \frac{mv^2}{r} + F_w$$

$$= \frac{mv^2}{r} + mg$$

$$= m\left(\frac{v^2}{r} + g\right)$$

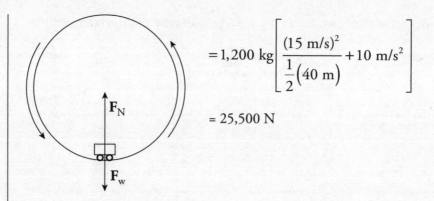

$$= 1{,}200 \text{ kg} \left[\frac{(15 \text{ m/s})^2}{\frac{1}{2}(40 \text{ m})} + 10 \text{ m/s}^2 \right]$$

$$= 25{,}500 \text{ N}$$

Notice the big difference between this answer and the answer from Example 4. This is why you would feel very little force between you and the seat at the top of the loop, but you would feel a big slam at the bottom of the loop.

NEWTON'S LAW OF GRAVITATION

Newton eventually formulated a law of gravitation: Any two objects in the universe exert an attractive force on each other—called the **gravitational force**—whose strength is proportional to the product of the objects' masses and inversely proportional to the square of the distance between them as measured from center to center. If we let G be the **universal gravitational constant**, then the strength of the gravitational force is given by the equation:

$$F_G = \frac{G m_1 m_2}{r^2}$$

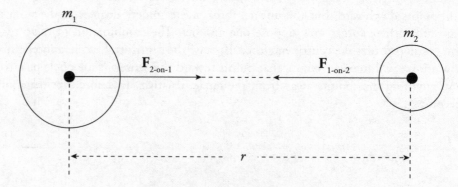

Gravity is always a pulling force.

The forces $\mathbf{F}_{\text{1-on-2}}$ and $\mathbf{F}_{\text{2-on-1}}$ act along the line that joins the bodies and form an action/reaction pair.

The first reasonably accurate numerical value for G was determined by Cavendish more than one hundred years after Newton's Law was published. To two decimal places, the currently accepted value of G is

$$G = 6.67 \times 10^{-11} \text{ N} \cdot \text{m}^2/\text{kg}^2$$

Why is gravity 9.8 m/s²? This constant is derived from this equation

$$g = Gm/r^2$$

Using the mass of the Earth and the radius of the Earth, gravity comes out to be 9.8 m/s².

Example 7 Given that the radius of the Earth is 6.37×10^6 m, determine the mass of the Earth.

Solution. Consider a small object of mass m near the surface of the Earth (mass M). Its weight is mg, but its weight is just the gravitational force it feels due to the Earth, which is GMm/R^2. Therefore,

$$mg = G\frac{Mm}{R^2} \quad \Rightarrow \quad M = \frac{gR^2}{G}$$

Since we know that $g = 10$ m/s² and $G = 6.67 \times 10^{-11}$ N·m²/kg², we can substitute to find

$$M = \frac{gR^2}{G} = \frac{(10 \text{ m/s}^2)(6.37 \times 10^6 \text{ m})^2}{6.67 \times 10^{-11} \text{ N} \cdot \text{m}^2/\text{kg}^2} = 6.1 \times 10^{24} \text{ kg}$$

Example 8 We can derive the expression $g = GM/R^2$ by equating mg and GMm/R^2 (as we did in the previous example), and this gives the magnitude of the *absolute gravitational acceleration*, a quantity that's sometimes denoted g_0. The notation g is acceleration, but with the spinning of the Earth taken into account. Show that if an object is at the equator, its *measured weight* (that is, the weight that a scale would measure), mg, is less than its *true weight*, mg_0, and compute the weight difference for a person of mass $m = 60$ kg.

Solution. Imagine looking down at the Earth from above the North Pole.

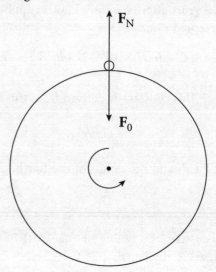

Relationships are Important
Not just in daily life but especially in physics. Recognizing the relationship between variables in formulas means at times you will not need to do much calculation.

The net force toward the center of the Earth is $\mathbf{F}_0 - \mathbf{F}_N$, which provides the centripetal force on the object. Therefore,

$$F_0 - F_N = \frac{mv^2}{R}$$

Since $v = 2\pi R/T$, where T is the Earth's rotation period, we have

$$F_0 - F_N = \frac{m}{R}\left(\frac{2\pi R}{T}\right)^2 = \frac{4\pi^2 mR}{T^2}$$

or, since $F_0 = mg_0$ and $F_N = mg$,

$$mg_0 - mg = \frac{4\pi^2 mR}{T^2}$$

Since the quantity $\dfrac{4\pi^2 mR}{T^2}$ is positive, mg must be less than mg_0. The difference between mg_0 and mg, for a person of mass $m = 60$ kg, is only:

$$\frac{4\pi^2 mR}{T^2} = \frac{4\pi^2 (60 \text{ kg})(6.37 \times 10^6 \text{ m})}{\left(24 \text{ hr} \times \frac{60 \text{ min}}{\text{hr}} \times \frac{60 \text{ s}}{\text{min}}\right)^2} = 2.0 \text{ N}$$

and the difference between g_0 and g is

$$g_0 - g = \frac{mg_0 - mg}{m} = \frac{4\pi^2 R}{T^2} = \frac{4\pi^2 (6.37 \times 10^6 \text{ m})}{\left(24 \text{ hr} \times \frac{60 \text{ min}}{\text{hr}} \times \frac{60 \text{ s}}{\text{min}}\right)^2} = 0.034 \text{ m/s}^2$$

Note that this difference is so small (< 0.3%) that it can usually be ignored.

Example 9 Communications satellites are often parked in geosynchronous orbits above Earth's surface. These satellites have orbit periods that are equal to Earth's rotation period, so they remain above the same position on Earth's surface. Determine the altitude that a satellite must have to be in a geosynchronous orbit above a fixed point on Earth's equator. (The mass of the Earth is 5.98×10^{24} kg.)

Solution. Let m be the mass of the satellite, M the mass of Earth, and R the distance from the center of Earth to the position of the satellite. The gravitational pull of Earth provides the centripetal force on the satellite, so

$$G \frac{Mm}{R^2} = \frac{mv^2}{R} \quad \Rightarrow \quad G \frac{M}{R} = v^2$$

The orbit speed of the satellite is $2\pi R/T$, so

$$G \frac{M}{R} = \left(\frac{2\pi R}{T} \right)^2$$

which implies that

$$G \frac{M}{R} = \frac{4\pi^2 R^2}{T^2} \quad \Rightarrow \quad 4\pi^2 R^3 = GMT^2 \quad \Rightarrow \quad R = \sqrt[3]{\frac{GMT^2}{4\pi^2}}$$

Now the key feature of a geosynchronous orbit is that its period matches Earth's rotation period, $T = 24$ hr. Substituting the numerical values of G, M, and T into this expression, we find that

$$R = \sqrt[3]{\frac{GMT^2}{4\pi^2}} = \sqrt[3]{\frac{(6.67 \times 10^{-11})(5.98 \times 10^{24})(24 \cdot 60 \cdot 60)^2}{4\pi^2}}$$
$$= 4.23 \times 10^7 \text{ m}$$

Therefore, if r_{E} is the radius of Earth, then the satellite's altitude above Earth's surface must be

$$h = R - r_{\text{E}} = (4.23 \times 10^7 \text{ m}) - (6.37 \times 10^6 \text{ m}) = 3.59 \times 10^7 \text{ m}$$

Example 10 The Moon orbits Earth in a (nearly) circular path at (nearly) constant speed. If M is the mass of Earth, m is the mass of the Moon, and r is the Moon's orbit radius, find an expression for the Moon's orbit speed.

Solution: We begin by answering the question, What produces the centripetal force? The answer is the gravitational pull by the Earth. We now simply translate our answer into an equation, like this:

gravitational pull	produces	the centripetal force
F_{grav}	=	$m(v^2/r)$

Since we know $F_{grav} = GMm/r^2$, we get

$$F_{grav} = F_c$$

$$G(Mm/r^2) = m(v^2/r)$$

$$G(M/r) = v^2$$

$$v = (GM/r)^{(1/2)}$$

Notice that the mass of the Moon, m, cancels out. So, any object orbiting at the same distance from the Earth as the moon must move at the same speed as the Moon.

BANKING

You may see a question or two about banking on your AP Physics 1 Exam. Banked curves are often employed by engineers in designing and constructing curved roads.

Banking allows for cars to travel around a curve at or below the posted speed limit, without relying on friction between the tires and road.

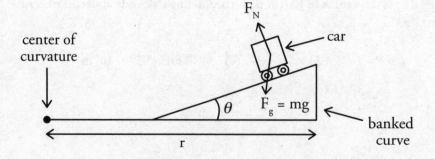

Example 11 Suppose that the radius of curvature, shown above is 60 m and the curve is banked at 11.8 degrees. What should the recommended speed be posted as?

Solution. The vertical component of the normal force must equal the downward force of gravity, *mg*. The horizontal component of the normal force is directed toward the center of curvature, producing the centripetal force that the car experiences as it rounds the curve. Thus,

$$F_g = F_N \cos\theta$$
$$mg = F_N \cos\theta$$
$$\cos\theta = \frac{mg}{F_N}$$

$$F_C = F_N \sin\theta$$
$$\frac{mv^2}{r} = F_N \sin\theta$$
$$\sin\theta = \frac{mv^2}{rF_N}$$

$$\frac{\sin\theta}{\cos\theta} = \frac{mv^2}{rF_N} \frac{F_N}{mg}$$

$$\tan\theta = \frac{v^2}{rg}$$

$$v = \sqrt{\tan\theta \cdot rg}$$

$$v = \sqrt{\tan 11.8° \cdot 60 \text{ m} \cdot 9.81 \text{ m/s}^2}$$

$$v = \sqrt{122.96 \text{ m}^2/\text{s}^2}$$

$$v \approx 11 \text{ m/s} \times \frac{1 \text{ km}}{1000 \text{ m}} \times \frac{3600 \text{ s}}{\text{h}}$$

$$v \approx 11 \frac{\text{km}}{\text{h}} \times 3.6$$

$$v \approx 39.6 \frac{\text{km}}{\text{h}}$$

Therefore, the recommended speed should be posted as 40 km/hour.

ROTATIONAL MOTION

Previously, we covered objects that undergo circular motion. The next part of this chapter focuses on taking those objects and spinning them. Previous equations involved objects moving in a linear orientation or being manipulated into circular orbit. With rotational motion, we will need to take on a new set of equations that are analogous to the physics of linear motion.

If we recall from before, an object's mass measures its inertia—its resistance to acceleration. The greater the inertia on an object, the harder it is to change its velocity. Harder to change its velocity means the object is harder to deliver an acceleration on—which in turn means the greater the inertia, the greater the force that is required in order for an object to be moved. Comparing two objects, if Object 1 has greater inertia than Object 2 and the same force is applied on both objects, Object 1 will undergo a smaller acceleration.

In the linear model, we put these in terms of force, mass, acceleration, and velocity. When it comes to rotational kinematics, we need to change up a few of these terms:

Linear Kinematics	Rotational Kinematics
Force	Torque (τ)
Mass	Moment of Inertia (I)
Acceleration	Angular Acceleration (α)
$F_{net} = ma$	$\tau_{net} = I\alpha$
Velocity	Angular Velocity (ω)

Rotational Kinematics

Like our linear equations, which are used to determine the distance (x), velocity (v), and linear acceleration, we use rotational equations to determine the same factors. In this section, we will go over angular distance (θ), angular velocity (ω), and angular acceleration (α). Finally, we will explore the relationship between these three rotational parameters and the linear parameters.

Let's start with some basic definitions.

What is angular displacement?

What is translational displacement?

If you look at a circle, you can see that 1x around the circle (1 revolution) equals 2π radians, or 360°.

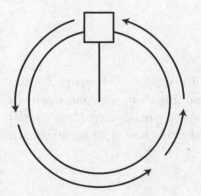 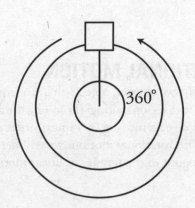

The linear position or physical distance traveled around the circle (Δs) can be related to angular position via this equation:

$$\theta = \frac{s}{r}$$

What is angular velocity ($\vec{\omega}$)?

Angular Velocity	Linear Velocity
$\vec{\omega} = \dfrac{\Delta \theta}{\Delta t}$	$\vec{V} = \dfrac{\Delta s}{\Delta t}$
Units $= \dfrac{\text{Rad}}{\text{s}}$ or $\dfrac{\text{Rev}}{\text{min}}$	Units = m/s
The angular velocity equals change in angular displacement divided by change in time.	The linear velocity equals change in distance (Δx) divided by change in time.

We can relate angular velocity to linear velocity via the following:

$$\vec{V} = r\vec{\omega}$$

Note, just like $\vec{V}$, $\vec{\omega}$ has direction!

The **Right Hand Rule** states that you must wrap your fingers around the object's path. Let's take the example of a toy car going around a circle.

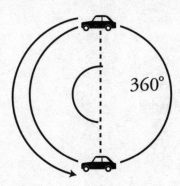

If you follow the car's path, you will find your fingers are wrapping counterclockwise. Your thumb is the direction for angular velocity. It this case it points out of the page. In physics we write the direction like this $\odot$.

What is angular acceleration (α)?

Angular acceleration, $\alpha = \dfrac{\Delta\omega}{\Delta t}$

Units: radians/s^2

Many of the rotational kinematics equations reflect linear kinematics equations.

	Rotational Motion	Linear Motion
Big Five #1:	$\Delta\theta = \dfrac{1}{2}(\omega_0 + \omega)t$	$\Delta x = \dfrac{1}{2}(v_0 + v)t$
Big Five #2:	$\omega = \omega_0 + \alpha t$	$v = v_0 + at$
Big Five #3:	$\theta = \theta_0 + \omega_0 t + \dfrac{1}{2}at^2$	$x = x_0 + v_0 t + \dfrac{1}{2}at^2$
Big Five #4:	$\theta = \theta_0 + \omega t - \dfrac{1}{2}at^2$	$x = x_0 + vt - \dfrac{1}{2}at^2$
Big Five #5:	$\omega^2 = \omega_0^2 + 2\alpha(\theta - \theta_0)$	$v^2 = v_0^2 + 2a(x - x_0)$

Example 12 Four children climb on a carousel that is initially at rest. If the carousel accelerates to 0.4 radians per second within 10 seconds, what is the angular acceleration? What is its linear rotation 3 m from axis of rotation?

Solution.

$\omega_i = 0$
$\omega_0 = 0.4$ rad/s

$\alpha = \dfrac{\omega_0 - \omega_i}{t} = \dfrac{0.4 - 0}{10} = 0.04$ rad/s^2

$\Delta\theta$
$\alpha = ?$
$T = 10$ sec

To find linear acceleration: $a = r\alpha = (3\text{ m})(0.04\text{ rad/s}^2) = .12$ m/s^2

CENTER OF MASS

In the preceding chapters, objects were treated as though they were each a single particle. In many force-diagrams we have said that all the force is being delivered at a single point on the object. What makes this point the center of mass? And why do we associate all the force being delivered at this single point on the object?

Imagine a series of experiments. We walk into a large room with a hammer and a small light that we can attach to the hammer. In the first experiment, we will attach the light to the very end of the hammer in which we hold the hammer. Then we turn off the light and throw the hammer across the room. If we trace the path of the hammer, we notice that it gives a weird spiral-shaped path as follows:

Then we repeat the experiment. This time let's attach the small light to the head of the hammer. Once again we turn off the light, throw the hammer across the room, and trace the path of the hammer. This time we notice that it follows another spiral shaped path:

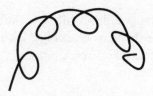

If we keep doing the experiment, after countless trials, you will notice that at a specific point, the hammer makes a parabolic path seen in the following drawing:

Apparently there was something important about that specific point. All the other points gave spiraled trajectories, but this one gave a smooth parabolic path. Upon further investigation, if we place that point on our fingers, we notice that the hammer balances nicely and is perfectly horizontal with the floor.

This certain point is the center of mass. Another way of looking at it is to say that the center of mass is the point at which we could consider all the mass of the object to be concentrated.

For a simple object such a sphere, block, or a cylinder, whose density is constant (a term in physics we call homogeneous), the center of mass is at its geometric center.

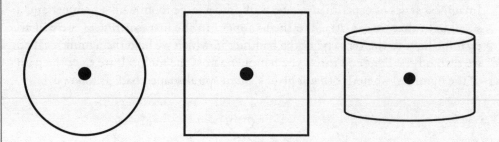

In some cases the center of mass is not located on the body of the object:

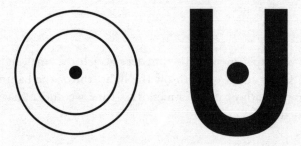

The center of mass remains motionless while every other point undergoes circular motion around it. To locate the mass of a collection of objects, we must consider each indivdual mass and a central reference point:

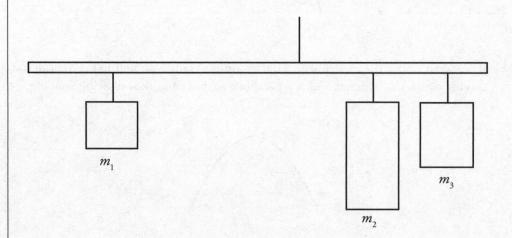

$$x(c.m.) = (x_1 m_1 + x_2 m_2 + x_3 m_3 + \dots + x_n m_n) / (m_1 + m_2 + m_3 + \dots + m_n)$$

The location of the center of mass is denoted as $x(c.m.)$. In order to use this equation we must follow a few steps:

1. Pick a convenient origin. We will call this our reference point ($x = 0$).
2. Determine the locations (x_1, x_2, x_3, ... , x_n) of the objects.
3. Calculate the center of mass by using the formula above.

Keep in mind that the stick itself matters. If the stick itself has mass, we must take that mass into account and its center position will be where its center of mass is.

TORQUE

We can tie a ball to a string and make it undergo circular motion, but how would we make that ball itself spin? We could simply palm the ball and rotate our hand or we could put our hands on opposite sides of the ball and push one hand forward and the other backward. In both cases, in order to make an object's center of mass accelerate, we need to exert a force. In order to make an object spin, we need to exert a torque.

Torque is the measure of a force's effectiveness at making an object spin or rotate. (More precisely, it's the measure of a force's effectiveness at making an object accelerate rotationally.) If an object is initially at rest, and then it starts to spin, something must have exerted a torque. And if an object is already spinning, something would have to exert a torque to get it to stop spinning.

All systems that can spin or rotate have a "center" of turning. This is the point that does not move while the remainder of the object is rotating, effectively becoming the center of the circle. There are many terms used to describe this point, including pivot point and fulcrum.

Torque has always been a topic that students have difficulty understanding. So let's go through a few examples and ask a few questions before we delve into torque:

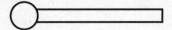

The drawing above shows a door with its hinge (pivot point) located on the left side of the door. You can try some of theses examples at home on a door in your house to get a better understanding. Let's pose two different situations for trying to close the door:

SCENARIO 1

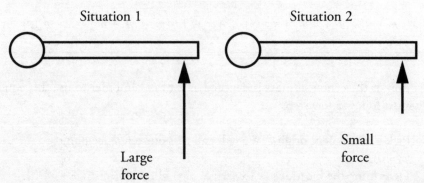

Situation 1   Situation 2

Large
force

Small
force

In Situation 1, the door will close the fastest because of the greater force used. Now in the same example, let's say that a 100 N force is going to be applied to the door but at different angles:

SCENARIO 2

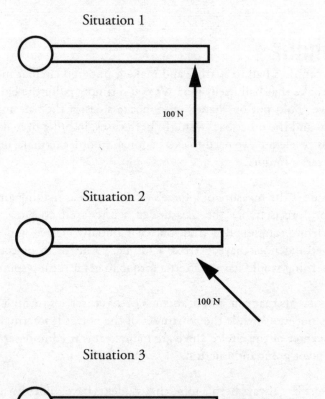

Situation 1

100 N

Situation 2

100 N

Situation 3

100 N

The door will close the fastest in Situation 1 and the second fastest in Situation 2. Situation 3 involves merely pushing on the door to no avail; the door will not close. Now let's use the same door example and apply the force to different parts of the door:

SCENARIO 3

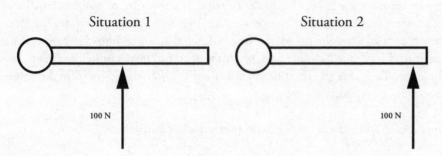

Situation 1   Situation 2

100 N   100 N

If you tried this at home, you will notice that if you push the door like Situation 2, it will be easier to close the door than if you tried to push the door like Situation 1.

If you noticed in the three scenarios there were a few points that mattered when trying to close the door. In scenario 1, the amount of force used to close the door mattered (magnitude of force). In scenario 2, the angle in which we pushed the door mattered (angle). And in scenario 3, the place in which we pushed mattered (radius).

Our force's effectiveness at making something spin or rotate was determined by three factors:

1   the magnitude of force (F)
2.   the angle (θ)
3.   the radius (r)

These three factors give us our torque equation:

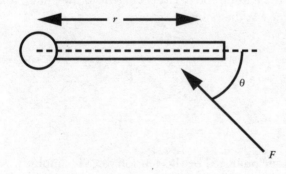

$$\tau = rF\sin\theta$$

Why Sine?
In math, this is known as a cross product between your force with your radius. It is written as $T = (r \times F)$.

There's no special name for this unit: it's just called a newton-meter. Because it is not in newtons, torque is NOT a force. (Torque is a measure of how much a force acting on an object causes that object to rotate.) In Scenario 2, Situation 3, a force was being applied straight on to the door directed straight into the pivot point. The magnitude of force did not suddenly disappear, but it was not effective at closing the door. Torque is the rotational equivalent of force in trying to make something accelerate rotationally.

Torque problems usually involve putting systems in equilibrium.

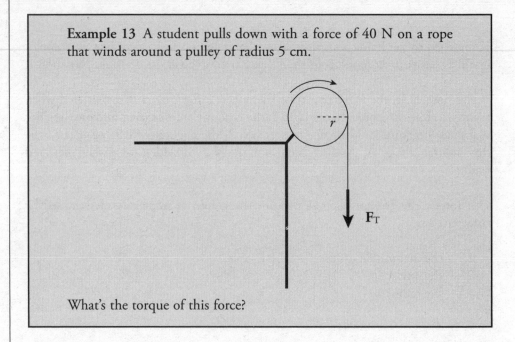

Example 13 A student pulls down with a force of 40 N on a rope that winds around a pulley of radius 5 cm.

What's the torque of this force?

Solution. Since the tension force, F_T, is tangent to the pulley, it is perpendicular to the radius vector **r** at the point of contact:

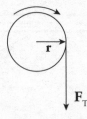

Therefore, the torque produced by this tension force is simply

$$\tau = rF_T = (0.05 \text{ m})(40 \text{ N}) = 2 \text{ N·m}$$

Example 14 What is the net torque on the cylinder shown below, which is pinned at the center?

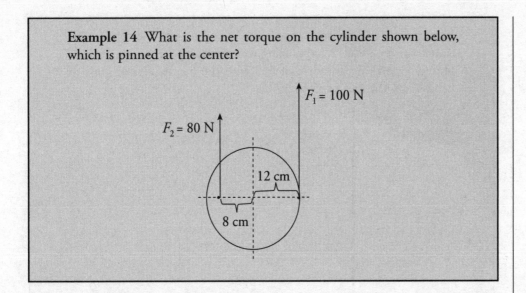

Solution. Each of the two forces produces a torque, but these torques oppose each other. The torque of F_1 is counterclockwise, and the torque of F_2 is clockwise. This can be visualized by imagining the effect of each force, assuming that the other was absent.

The **net torque** is the sum of all the torques. Counting a counterclockwise torque as positive and a clockwise torque as negative, we have

$$\tau_1 = +r_1 F_1 = +(0.12 \text{ m})(100 \text{ N}) = +12 \text{ N·m}$$

and

$$\tau_2 = -r_2 F_2 = -(0.08 \text{ m})(80 \text{ N}) = -6.4 \text{ N·m}$$

so

$$\tau_{net} = \Sigma \tau = \tau_1 + \tau_2 = (+12 \text{ N·m}) + (-6.4 \text{ N·m}) = +5.6 \text{ N·m}$$

Keep Orientation in Mind
With spinning objects, we will often be referring to directions as clockwise or counter-clockwise.

EQUILIBRIUM

An object is said to be in **translational equilibrium** if the sum of the forces acting on it is zero; that is, if $F_{net} = 0$. Similarly, an object is said to be in **rotational equilibrium** if the sum of the torques acting on it is zero; that is, if $\tau_{net} = 0$. The term *equilibrium* by itself means both translational and rotational equilibrium. A body in equilibrium may be in motion; $F_{net} = 0$ does not mean that the velocity is zero; it only means that the velocity is constant. Similarly, $\tau_{net} = 0$ does not mean that the angular velocity is zero; it only means that it's constant. If an object is at rest, then it is said to be in **static equilibrium**.

It's All About Balance
Equilibrium problems are all about balancing one force's effectiveness at turning something clockwise with another force's effectiveness at turning something counter-clockwise.

Example 15 A uniform bar of mass m and length L extends horizontally from a wall. A supporting wire connects the wall to the bar's midpoint, making an angle of 55° with the bar. A sign of mass M hangs from the end of the bar.

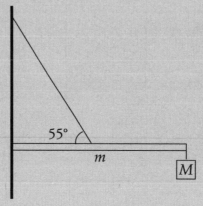

If the system is in static equilibrium and the wall has friction, determine the tension in the wire and the strength of the force exerted on the bar by the wall if m = 8 kg and M = 12 kg.

Choose Your Strategy
Some problems are easier when using center-of-mass, whereas some problems are easier using torque. Choose which one is easier for you.

Solution. Let F_C denote the (contact) force exerted by the wall on the bar. In order to simplify our work, we can write F_C in terms of its horizontal component, F_{Cx}, and its vertical component, F_{Cy}. Also, if F_T is the tension in the wire, then $F_{Tx} = F_T \cos 55°$ and $F_{Ty} = F_T \sin 55°$ are its components. This gives us the following force diagram:

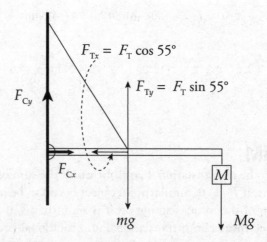

The first condition for equilibrium requires that the sum of the horizontal forces is zero and the sum of the vertical forces is zero:

$$\Sigma F_x = 0: \qquad F_{Cx} - F_T \cos 55° = 0 \qquad (1)$$

$$\Sigma F_y = 0: \qquad F_{Cy} + F_T \sin 55° - mg - Mg = 0 \qquad (2)$$

We notice immediately that we have more unknowns (F_{Cx}, F_{Cy}, F_T) than equations, so this system cannot be solved as is. The second condition for equilibrium requires that the sum of the torques about any point is equal to zero. Choosing the contact point between the bar and the wall as our pivot, only three of the forces in the diagram above produce torque: $\mathbf{F}_{Ty}$ produces a counterclockwise torque, and both $m\mathbf{g}$ and $M\mathbf{g}$ produce clockwise torques, and the sum of the three torques must equal zero. From the definition $\tau = rF \sin \theta$, and taking counterclockwise torque as positive and clockwise torque as negative, we have

$$\Sigma \tau = 0: \qquad \tfrac{1}{2}rF_{Ty} - \tfrac{1}{2}mgr - LMgr = 0 \qquad (3)$$

Note that you can divide out the r term here, so r will not factor into the final equation. The above equation contains only one unknown and can be solved immediately:

$$\tfrac{L}{2}F_{Ty} = \tfrac{L}{2}mg + LMg$$
$$F_{Ty} = mg + 2Mg = (m+2M)g$$

Since $F_{Ty} = F_T \sin 55°$, we can find that

$$F_T \sin 55° = (m+2M)g \quad \Rightarrow \quad F_T = \frac{(m+2M)g}{\sin 55°}$$
$$= \frac{(8+2\cdot 12)(10)}{\sin 55°}$$
$$= 390 \text{ N}$$

Substituting this result into Equation (1) gives us F_{Cx}:

$$F_{Cx} = F_T \cos 55° = \frac{(m+2M)g}{\sin 55°}\cos 55° = (8+2\cdot 12)(10)\cot 55° = 220 \text{ N}$$

Note: We use cot 55° here, because $\cot 55° = \dfrac{\cos 55°}{\sin 55°}$.

And finally, from Equation (2), we get

$$F_{Cy} = mg + Mg - F_T \sin 55°$$
$$= mg + Mg - \frac{(m+2M)g}{\sin 55°}\sin 55°$$
$$= -Mg$$
$$= -(12)(10)$$
$$= -120 \text{ N}$$

The fact that F_{C_y} turned out to be negative simply means that in our original force diagram, the vector $\mathbf{F}_{C_y}$ points in the direction opposite to how we drew it. That is, $\mathbf{F}_{C_y}$ points downward. Therefore, the magnitude of the total force exerted by the wall on the bar is

$$F_C = \sqrt{(F_{Cx})^2 + (F_{Cy})^2} = \sqrt{220^2 + 120^2} = 250 \text{ N}$$

ROTATIONAL INERTIA

Now that we've studied torque and rotation, we can finally put together the pieces of making an object spin. An object's rotational inertia (also known as the moment of inertia) is defined as the tendency of an object in motion to rotate until acted upon by an outside force. Think of mass as translational inertia, since it measures an object's resistance to translational acceleration, given by a in $F = ma$. Then, just as translational inertia tells us how resistant an object is to translation acceleration, an object's rotational inertia, I, tells us how resistant the object is to rotational acceleration.

Rotational acceleration or angular acceleration is the same as translational acceleration except we are taking a ball at rest and speeding up its angular velocity. In order to achieve this acceleration, a force is required. In terms of rotational inertia, torque is required to produce rotational acceleration. We need to apply a force that is effective in generating acceleration. Torque, in terms of rotational inertia, is

$$\tau = I\alpha$$

This equation makes some key relationships. The larger the rotational inertia (moment of inertia) is, the smaller the value will be for a given torque on an object. If Object 1 has a greater rotational inertia than Object 2, then it will be more difficult to rotate Object 1 than Object 2. More precisely, a greater torque will be required to give Object 1 the same rotational acceleration as Object 2, or, equivalently, if the same torque is applied to both objects, Object 1 will undergo a smaller rotational acceleration.

So, how do we find the rotational inertia of an object? It depends on the object's mass, but there is more to it than that. Two objects can have the same mass but different rotational inertias. Rotational inertia is also dependent on how the mass is distributed in an object with respect to the axis that it rotated around.

The farther away the mass is from the axis of rotation, the greater the rotational inertia will be.

Imagine a barbell with a weight near each end and an identical barbell with the weights pushed near the middle of the bar. These two barbells have the same mass, but their rotational inertias are different. If we wanted to rotate each bar around its midpoint, the first barbell has its attached masses farther away from the axis of rotation than the second one. As a result, we would find it more difficult to rotate the first barbell than the second one. The first barbell has a greater value of *I*.

Barbell 1

Barbell 2

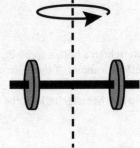

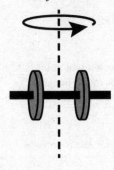

Chapter 8 Review Questions

Answers and Explanations can be found in Chapter 13.

Section I: Multiple-Choice

1. An object moves at constant speed in a circular path. Which of the following statements is true? Select two answers.

    (A) The velocity is changing.
    (B) The velocity is constant.
    (C) The magnitude of acceleration is constant.
    (D) The magnitude of acceleration is changing.

Questions 2-3:

A 60 cm rope is tied to the handle of a bucket which is then whirled in a vertical circle. The mass of the bucket is 3 kg.

2. At the lowest point in its path, the tension in the rope is 50 N. What is the speed of the bucket?

    (A) 1 m/s
    (B) 2 m/s
    (C) 3 m/s
    (D) 4 m/s

3. What is the critical speed below which the rope would become slack when the bucket reaches the highest point in the circle?

    (A) 0.6 m/s
    (B) 1.8 m/s
    (C) 2.4 m/s
    (D) 4.8 m/s

4. An object moves at a constant speed in a circular path of radius r at a rate of 1 revolution per second. What is its acceleration?

    (A) 0
    (B) $2\pi^2 r$
    (C) $2\pi^2 r^2$
    (D) $4\pi^2 r$

5.

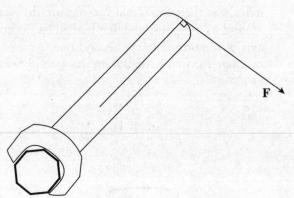

In an effort to tighten a bolt, a force **F** is applied as shown in the figure above. If the distance from the end of the wrench to the center of the bolt is 20 cm and $F = 20$ N, what is the magnitude of the torque produced by **F** ?

    (A) 1 N·m
    (B) 2 N·m
    (C) 4 N·m
    (D) 10 N·m

6.

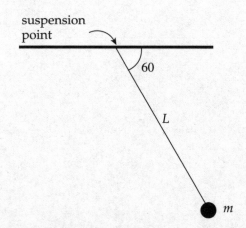

In the figure above, what is the torque about the pendulum's suspension point produced by the weight of the bob, given that the length of the pendulum, L, is 80 cm and $m = 0.50$ kg ?

    (A) 0.5 N·m
    (B) 1.0 N·m
    (C) 1.7 N·m
    (D) 2.0 N·m

7.

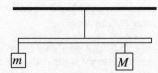

A uniform meter stick of mass 1 kg is hanging from a thread attached at the stick's midpoint. One block of mass $m = 3$ kg hangs from the left end of the stick, and another block, of unknown mass M, hangs below the 80 cm mark on the meter stick. If the stick remains at rest in the horizontal position shown above, what is M ?

(A) 4 kg
(B) 5 kg
(C) 6 kg
(D) 8 kg

8. If the distance between two point particles is doubled, then the gravitational force between them

(A) decreases by a factor of 4
(B) decreases by a factor of 2
(C) increases by a factor of 2
(D) increases by a factor of 4

9. At the surface of Earth, an object of mass m has weight w. If this object is transported to an altitude that is twice the radius of Earth, then at the new location,

(A) its mass is m and its weight is $w/2$
(B) its mass is $m/2$ and its weight is $w/4$
(C) its mass is m and its weight is $w/4$
(D) its mass is m and its weight is $w/9$

10. A moon of mass m orbits a planet of mass $100m$. Let the strength of the gravitational force exerted by the planet on the moon be denoted by F_1, and let the strength of the gravitational force exerted by the moon on the planet be F_2. Which of the following is true?

(A) $F_1 = 100F_2$
(B) $F_1 = 10F_2$
(C) $F_1 = F_2$
(D) $F_2 = 10F_1$

11. The dwarf planet Pluto has 1/500 the mass and 1/15 the radius of Earth. What is the value of g (in m/s²) on the surface of Pluto?

(A) $\dfrac{50}{225}$

(B) $\dfrac{50}{15}$

(C) $\dfrac{15}{50}$

(D) $\dfrac{225}{50}$

12. A satellite is currently orbiting Earth in a circular orbit of radius R; its kinetic energy is K_1. If the satellite is moved and enters a new circular orbit of radius $2R$, what will be its kinetic energy?

(A) $\dfrac{K_1}{4}$

(B) $\dfrac{K_1}{2}$

(C) $2K_1$

(D) $4K_1$

13. A moon of Jupiter has a nearly circular orbit of radius R and an orbit period of T. Which of the following expressions gives the mass of Jupiter?

(A) $\dfrac{4\pi^2 R}{T^2}$

(B) $\dfrac{2\pi R^3}{(GT^2)}$

(C) $\dfrac{4\pi R^2}{(GT^2)}$

(D) $\dfrac{4\pi^2 R^3}{(GT^2)}$

14. Two large bodies, Body A of mass m and Body B of mass $4m$, are separated by a distance R. At what distance from Body A, along the line joining the bodies, would the gravitational force on an object be equal to zero? (Ignore the presence of any other bodies.)

(A) $\dfrac{R}{16}$

(B) $\dfrac{R}{8}$

(C) $\dfrac{R}{4}$

(D) $\dfrac{R}{3}$

15. You are looking at a top view of a planet orbiting the Sun in a clockwise direction. Which of the following would describe the velocity, acceleration, and force acting on the planet due to the Sun's pull at point P ?

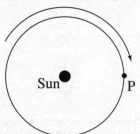

(A) $v \downarrow$ $a \uparrow$ $F \uparrow$
(B) $v \downarrow$ $a \leftarrow$ $F \leftarrow$
(C) $v \downarrow$ $a \rightarrow$ $F \rightarrow$
(D) $v \uparrow$ $a \leftarrow$ $F \leftarrow$

16. Which of the following statements are true for a satellite in outer space orbiting the Earth in uniform circular motion? Select two answers.

(A) There are no forces acting on the satellite.
(B) The force of gravity is the only force acting on the satellite.
(C) The force of gravity is balanced by outward force of the object.
(D) The mass of the satellite has no effect on the orbital speed.

17. An object of mass m is traveling at constant speed v in a circular path of radius r. How much work is done by the centripetal force during one-half of a revolution?

(A) $\pi m v^2$
(B) 0
(C) $\pi m v^2 r$
(D) $2\pi m v^2 r$

Section II: Free-Response

1. A robot probe lands on a new, uncharted planet. It has determined the diameter of the planet to be 8×10^6 m. It weighs a standard 1 kg mass and determines that 1 kg weighs only 5 newtons on this new planet.

    (a)   What must the mass of the planet be?

    (b)   What is the acceleration due to gravity on this planet? Express your answer in both m/s^2 and g's (where 1 g = 10 m/s^2).

    (c)   What is the average density of this planet?

2. The Earth has a mass of 6×10^{24} kg and orbits the Sun in 3.15×10^7 seconds at a constant radius of 1.5×10^{11} m.

    (a)   What is the Earth's centripetal acceleration around the Sun?

    (b)   What is the gravitational force acting between the Sun and Earth?

    (c)   What is the mass of the Sun?

3. An amusement park ride consists of a large cylinder that rotates around its central axis as the passengers stand against the inner wall of the cylinder. Once the passengers are moving at a certain speed, v, the floor on which they were standing is lowered. Each passenger feels pinned against the wall of the cylinder as it rotates. Let r be the inner radius of the cylinder.

    (a)   Draw and label all the forces acting on a passenger of mass m as the cylinder rotates with the floor lowered.

    (b)   Describe what conditions must hold to keep the passengers from sliding down the wall of the cylinder.

    (c)   Compare the conditions discussed in part (b) for an adult passenger of mass m and a child passenger of mass $m/2$.

4. A curved section of a highway has a radius of curvature of r. The coefficient of friction between standard automobile tires and the surface of the highway is μ_s.

    (a)   Draw and label all the forces acting on a car of mass m traveling along this curved part of the highway.

    (b)   Compute the maximum speed with which a car of mass m could make it around the turn without skidding in terms of μ_s, r, g, and m.

    City engineers are planning on banking this curved section of highway at an angle of θ to the horizontal.

    (c)   Draw and label all of the forces acting on a car of mass m traveling along this banked turn. Do not include friction.

    (d)   The engineers want to be sure that a car of mass m traveling at a constant speed v (the posted speed limit) could make it safely around the banked turn even if the road were covered with ice (that is, essentially frictionless). Compute this banking angle θ in terms of r, v, g, and m.

Summary

○ For objects undergoing uniform circular motion, the centripetal acceleration is given by $a_c = \dfrac{v^2}{r}$ and the centripetal force is given by $F_c = \dfrac{mv^2}{r}$.

○ Torque is a property of a force that makes an object rotate. The equation for torque is $\tau = rF\sin\theta$. Torques may be clockwise or counterclockwise. An object is in equilibrium if $\sum F_x = 0$, $\sum F_y = 0$, and $\sum \tau = 0$.

○ For any two masses in the universe there is a gravitational attraction given by:

$$F_g = \frac{Gm_1m_2}{r^2} \text{ where}$$

$$G = 6.67 \times 10^{-11}\ \frac{\text{Nm}^2}{\text{kg}^2}.$$

○ The acceleration due to gravity on any planet is given by:

$$g_{planet} = \frac{Gm_{planet}}{r^2}$$

○ Many times universal gravitation is linked up with circular motion (because planetary orbits are very nearly circular). Therefore, it is useful to keep the following equations for circular motion mentally linked for those questions that include orbits:

$$v = \frac{2\pi r}{T} \qquad a_c = \frac{v^2}{r} \qquad a_c = \frac{4\pi^2 r}{T^2}$$

$$F = ma_c \qquad F_c = \frac{mv^2}{r} \qquad F_c = \frac{4\pi^2 mr}{T^2}$$

Chapter 9
Oscillations

"There is nothing so small, as to escape our inquiry; hence there is a new visible World discovered to the understanding."

—Robert Hooke

Seventeenth century British physicist Robert Hooke helped pave the way for simple harmonic motion. Previously, Newton's laws of static equilibrium made it possible to show a relationship between stress and strain for complex objects. Building upon these, he developed laws on simple harmonic motion that are named after him today—Hooke's Law. Hooke is credited as the discoverer of the concise mathematical relationship of a spring.

In this section, we will focus on periodic motion that's straightforward which will help to describe oscillations and other simple harmonic motions.

SIMPLE HARMONIC MOTION

In the series of diagrams below is a fixed block on the left side of a wall. When the spring is neither stretched nor compressed (when it sits at its natural length), it is said to be in its equilibrium position. When the block is in equilibrium, the net force on the block is zero. We label this position as $x = 0$.

Let's start with a spring at rest (Diagram 1). First, we pull the block to the right, where it will experience a force pulling back toward equilibrium (Diagram 2). This force brings the block back through the equilibrium position (Diagram 3), and the block's momentum carries it past that point to location $x = -A$. At this point, the block will again be experiencing a force that pushes it toward the equilibrium position (Diagram 4). Once again, the block passes through the equilibrium position, but it's traveling to the right this time (Diagram 5). If this is taking place in ideal conditions (no friction), this back-and-forth motion will continue indefinitely and the block will oscillate from these positions in the same amount of time. The oscillations of the block at the end of this spring provide us with the physical example of simple harmonic motion (often abbreviated SHM).

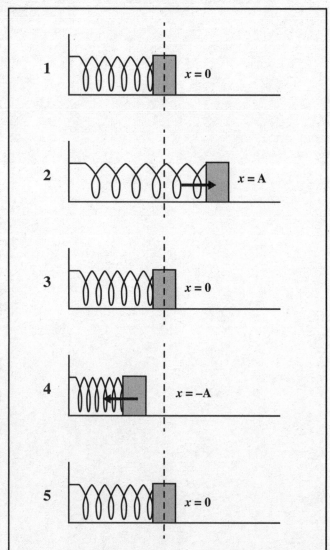

The Dynamics of SHM

Force

Since the block is accelerating and decelerating, there must be some force that is making it do so. In the case of a spring, the spring exerts a force on the block that is proportional to its displacement from its equilibrium point. Setting the equilibrium point as $x = 0$, the force exerted by the spring is

$$F = -kx$$

This is known as Hooke's Law. The proportionality constant, k, is called the spring constant and tells us how strong the spring is. The greater the value of k, the stiffer and stronger the spring is. The minus sign in Hooke's Law tells us that the force is a restoring force. A restoring force simply means that the force wants to return the object back to its equilibrium position. Hence, in the previous diagrams, when the block was on the left (when the position is negative), the force exerted was to the right; and when the block was on the right (when the position is positive), the force exerted was to the left. In all cases of the extreme left or right, the spring has a tendency to return to its original length or equilibrium position. This force helps to maintain the oscillations.

> **Example 1** A 12 cm-long spring has a force constant (k) of 400 N/m. How much force is required to stretch the spring to a length of 14 cm?

Solution. The displacement of the spring has a magnitude of $14 - 12 = 2$ cm $= 0.02$ m so, according to Hooke's Law, the spring exerts a force of magnitude $F = kx = (400 \text{ N/m})(0.02 \text{ m}) = 8$ N. Therefore, we'd have to exert this much force to keep the spring in this stretched state.

During the oscillation, the force on the block is zero when the block is at equilibrium (the point we designate as $x = 0$). This is because Hooke's Law says that the strength of the spring's restoring force is given by the equation $F = kx$, so $F = 0$ at equilibrium. The acceleration of the block is also equal to zero at $x = 0$, since $F = 0$ at $x = 0$ and $a = F/m$. At the endpoints of the oscillating region, where the block's displacement, x, has the greatest magnitude, the restoring force and the magnitude of the acceleration are both at their maximums.

Amplitude

The maximum displacement from equilibrium is called the amplitude of oscillation and is denoted by A. So instead of writing $x = x_{max}$, we write $x = A$ ($x = -x_{max}$ is $x = -A$). This number tells us how far to the left and right of equilibrium the block will travel.

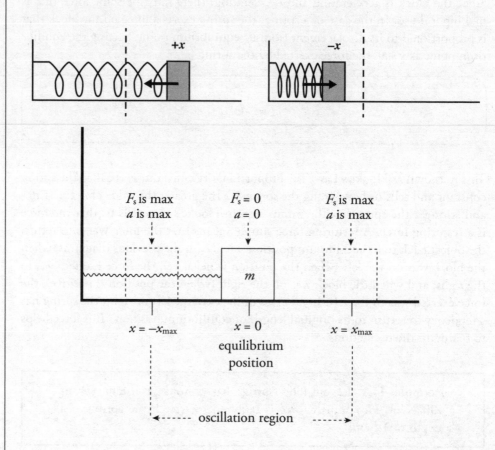

SHM in Terms of Energy

Another way to describe the block's motion is in terms of energy transfers. A stretched or compressed spring stores **elastic potential energy**, which is transformed into kinetic energy (and back again); this shuttling of energy between potential and kinetic causes the oscillations. For a spring with spring constant k, the elastic potential energy it possesses—relative to its equilibrium position—is given by the equation

$$U_s = \frac{1}{2} k x^2$$

Notice that the farther you stretch or compress a spring, the more work you have to do, and, as a result, the more potential energy that's stored.

In terms of energy transfers, we can describe the block's oscillations as follows. When you initially pull the block out, you increase the elastic potential energy of the system. Upon releasing the block, this potential energy turns into kinetic energy, and the block moves. As it passes through equilibrium, $U_S = 0$, so all the energy is kinetic. Then, as the block continues through equilibrium, it compresses the spring and the kinetic energy is transformed back into elastic potential energy.

By Conservation of Mechanical Energy, the sum $K + U_S$ is a constant. Therefore, when the block reaches the endpoints of the oscillation region (that is, when $x = \pm x_{max}$), U_S is maximized, so K must be minimized; in fact, $K = 0$ at the endpoints. As the block is passing through equilibrium, $x = 0$, so $U_S = 0$ and K is maximized.

Maximum Potential Energy
The maximum potential energy is when $x = A$ or $x = -A$.

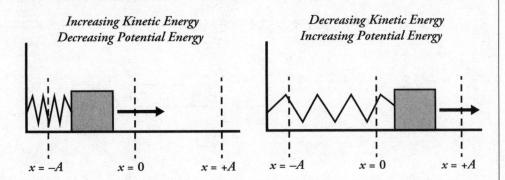

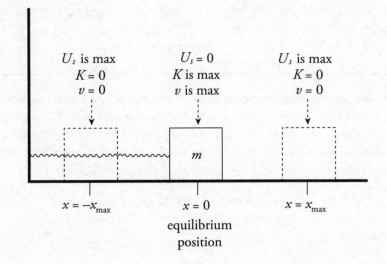

> **Example 2** A block of mass $m = 0.05$ kg oscillates on a spring whose force constant k is 500 N/m. The amplitude of the oscillations is 4.0 cm. Calculate the maximum speed of the block.

Solution. First, let's get an expression for the maximum elastic potential energy of the system:

$$U_S = \frac{1}{2}kx^2 \quad \Rightarrow \quad U_{S,\,max} = \frac{1}{2}kx_{max}^2 = \frac{1}{2}kA^2$$

Maximum Velocity
Only the maximum velocity can be calculated in a spring with this method. The AP Physics 1 Exam will not ask you to calculate other velocities at other points because this is not uniform accelerated motion.

When all this energy has been transformed into kinetic energy—which, as we discussed earlier, occurs just as the block is passing through equilibrium—the block will have a maximum kinetic energy and maximum speed of

$$U_{S,\,max} \to K_{max} \quad \Rightarrow \quad \frac{1}{2}kA^2 = \frac{1}{2}mv_{max}^2$$

$$v_{max} = \sqrt{\frac{kA^2}{m}}$$

$$= \sqrt{\frac{(500 \text{ N/m})(0.04 \text{ m})^2}{0.05 \text{ kg}}}$$

$$= 4 \text{ m/s}$$

> **Example 3** A block of mass $m = 2.0$ kg is attached to an ideal spring of force constant $k = 500$ N/m. The amplitude of the resulting oscillations is 8.0 cm. Determine the total energy of the oscillator and the speed of the block when it's 4.0 cm from equilibrium.

Solution. The total energy of the oscillator is the sum of its kinetic and potential energies. By Conservation of Mechanical Energy, the sum $K + U_S$ is a constant, so if we can determine what this sum is at some point in the oscillation region, we'll know the sum at every point. When the block is at its amplitude position, $x = 8$ cm, its speed is zero; so at this position, E is easy to figure out:

$$E = K + U_S = 0 + \frac{1}{2}kA^2 = \frac{1}{2}(500 \text{ N/m})(0.08 \text{ m})^2 = 1.6 \text{ J}$$

This gives the total energy of the oscillator at *every* position. At any position x, we have

$$\frac{1}{2}mv^2 + \frac{1}{2}kx^2 = E$$

$$v = \sqrt{\frac{E - \frac{1}{2}kx^2}{\frac{1}{2}m}}$$

so when we substitute in the numbers, we get

$$v = \sqrt{\frac{E - \frac{1}{2}kx^2}{\frac{1}{2}m}} = \sqrt{\frac{(1.6\text{ J}) - \frac{1}{2}(500\text{ N/m})(0.04\text{ m})^2}{\frac{1}{2}(2.0\text{ kg})}}$$

$$= 1.1\text{ m/s}$$

Example 4 A block of mass $m = 8.0$ kg is attached to an ideal spring of force constant $k = 500$ N/m. The block is at rest at its equilibrium position. An impulsive force acts on the block, giving it an initial speed of 2.0 m/s. Find the amplitude of the resulting oscillations.

Solution. The block will come to rest when all of its initial kinetic energy has been transformed into the spring's potential energy. At this point, the block is at its maximum displacement from equilibrium, at one of its amplitude positions, and

$$K_i + U_i = K_{ff} + U$$

$$\frac{1}{2}mv_i^2 + 0 = 0 + \frac{1}{2}kA^2$$

$$A = \sqrt{\frac{mv_i^2}{k}}$$

$$= \sqrt{\frac{(8.0\text{ kg})(2.0\text{ m/s})^2}{500\text{ N/m}}}$$

$$= 0.25\text{ m}$$

Period and Frequency

As you watch the block oscillate, you should notice that it repeats each **cycle** of oscillation in the same amount of time. A cycle is a *round-trip*: for example, from position $x = A$ over to $x = -A$ and back again to $x = A$. The amount of time it takes to complete a cycle is called the **period** of the oscillations, or T. If T is short, the block is oscillating rapidly, and if T is long, the block is oscillating slowly.

Another way of indicating the rapidity of the oscillations is to count the number of cycles that can be completed in a given time interval; the more completed cycles, the more rapid the oscillations. The number of cycles that can be completed per unit time is called the **frequency** of the oscillations, or *f*, and frequency is expressed in cycles per second. One cycle per second is one **hertz** (abbreviated **Hz**). Do not confuse lowercase *f* (frequency) with uppercase *F* (Force).

One of the most basic equations of oscillatory motion expresses the fact that the period and frequency are reciprocals of each other:

$$\text{period} = \frac{\# \text{ seconds}}{\text{cycle}} \qquad \text{while} \qquad \text{frequency} = \frac{\# \text{ cycles}}{\text{second}}$$

Two Peas in a Pod
If you have the period, you can always get the frequency and vice-versa. Note that the period and the frequency are inverses of each other.

Therefore,

$$T = \frac{1}{f} \qquad \text{and} \qquad f = \frac{1}{T}$$

Example 5 A block oscillating on the end of a spring moves from its position of maximum spring stretch to maximum spring compression in 0.25 s. Determine the period and frequency of this motion.

Solution. The period is defined as the time required for one full cycle. Moving from one end of the oscillation region to the other is only half a cycle. Therefore, if the block moves from its position of maximum spring stretch to maximum spring compression in 0.25 s, the time required for a full cycle is twice as much; $T = 0.5$ s. Because frequency is the reciprocal of period, the frequency of the oscillations is $f = 1/T = 1/(0.5 \text{ s}) = 2$ Hz.

Example 6 A student observing an oscillating block counts 45.5 cycles of oscillation in one minute. Determine its frequency (in hertz) and period (in seconds).

Solution. The frequency of the oscillations, in hertz (which is the number of cycles per second), is

$$f = \frac{45.5 \text{ cycles}}{\text{min}} \times \frac{1 \text{ min}}{60 \text{ s}} = \frac{0.758 \text{ cycles}}{\text{s}} = 0.758 \text{ Hz}$$

Therefore,

$$T = \frac{1}{f} = \frac{1}{0.758 \text{ Hz}} = 1.32 \text{ s}$$

One of the defining properties of the spring-block oscillator is that the frequency and period can be determined from the mass of the block and the force constant of the spring. The equations are as follows:

$$f = \frac{1}{2\pi}\sqrt{\frac{k}{m}} \quad \text{and} \quad T = 2\pi\sqrt{\frac{m}{k}}$$

A useful mnemonic for remembering this equation is to go in alphabetical order (clockwise) for frequency (f to k to m) and reverse alphabetical for period (T to m to k).

Let's analyze these equations. Suppose we had a small mass on a very stiff spring; then intuitively, we would expect that this strong spring would make the small mass oscillate rapidly, with high frequency and short period. Both of these predictions are substantiated by the equations above, because if m is small and k is large, then the ratio k/m is large (high frequency) and the ratio m/k is small (short period).

Example 7 A block of mass $m = 2.0$ kg is attached to a spring whose force constant, k, is 300 N/m. Calculate the frequency and period of the oscillations of this spring–block system.

Solution. According to the equations above,

$$f = \frac{1}{2\pi}\sqrt{\frac{k}{m}} = \frac{1}{2\pi}\sqrt{\frac{300 \text{ N/m}}{2.0 \text{ kg}}} = 1.9 \text{ Hz}$$

$$T = 2\pi\sqrt{\frac{m}{k}} = 2\pi\sqrt{\frac{2.0 \text{ kg}}{300 \text{ N/m}}} = 0.51 \text{ s}$$

Notice that $f \approx 2$ Hz and $T \approx 0.5$ s, and that these values satisfy the basic equation $T = 1/f$.

Example 8 A block is attached to a spring and set into oscillatory motion, and its frequency is measured. If this block were removed and replaced by a second block with 1/4 the mass of the first block, how would the frequency of the oscillations compare to that of the first block?

Solution. Since the same spring is used, k remains the same. According to the equation given on the previous page, f is inversely proportional to the square root of the mass of the block: $f \propto 1/\sqrt{m}$. Therefore, if m decreases by a factor of 4, then f increases by a factor of $\sqrt{4} = 2$.

The equations we saw on the previous page for the frequency and period of the spring-block oscillator do not contain A, the amplitude of the motion. In simple harmonic motion, *both the frequency and the period are independent of the amplitude*. The reason that the frequency and period of the spring-block oscillator are independent of amplitude is that F, the strength of the restoring force, is proportional to x, the displacement from equilibrium, as given by Hooke's Law: $F_S = -kx$.

Example 9 A student performs an experiment with a spring-block simple harmonic oscillator. In the first trial, the amplitude of the oscillations is 3.0 cm, while in the second trial (using the same spring and block), the amplitude of the oscillations is 6.0 cm. Compare the values of the period, frequency, and maximum speed of the block between these two trials.

Solution. If the system exhibits simple harmonic motion, then the period and frequency are independent of amplitude. This is because the same spring and block were used in the two trials, so the period and frequency will have the same values in the second trial as they had in the first. But the maximum speed of the block will be greater in the second trial than in the first. Since the amplitude is greater in the second trial, the system possesses more total energy ($E = \frac{1}{2} kA^2$). So when the block is passing through equilibrium (its position of greatest speed), the second system has more energy to convert to kinetic, meaning that the block will have a greater speed. In fact, from Example 2, we know that $v_{max} = A\sqrt{k/m}$ so, since A is twice as great in the second trial than in the first, v_{max} will be twice as great in the second trial than in the first.

Example 10 For each of the following arrangements of two springs, determine the **effective spring constant**, k_{eff}. This is the force constant of a single spring that would produce the same force on the block as the pair of springs shown in each case.

(a)

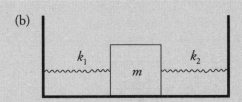

(b)

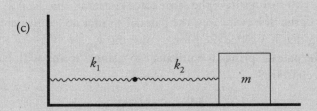

(c)

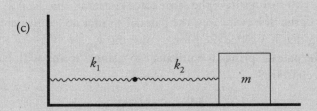

(d)   Determine k_{eff} in each of these cases if $k_1 = k_2 = k$.

Solution.

(a)   Imagine that the block was displaced a distance x to the right of its equilibrium position. Then the force exerted by the first spring would be $F_1 = -k_1 x$ and the force exerted by the second spring would be $F_2 = -k_2 x$. The net force exerted by the springs would be

$$F_1 + F_2 = -k_1 x + -k_2 x = -(k_1 + k_2)x$$

Since $F_{\text{eff}} = -(k_1 + k_2)x$, we see that $k_{\text{eff}} = k_1 + k_2$.

(b)   Imagine that the block was displaced a distance x to the right of its equilibrium position. Then the force exerted by the first spring would be $F_1 = -k_1 x$ and the force exerted by the second spring would be $F_2 = -k_2 x$. The net force exerted by the springs would be

$$F_1 + F_2 = -k_1 x + -k_2 x = -(k_1 + k_2)x$$

As in part (a), we see that, since $F_{\text{eff}} = -(k_1 + k_2)x$, we get $k_{\text{eff}} = k_1 + k_2$.

(c) Imagine that the block was displaced a distance x to the right of its equilibrium position. Let x_1 be the distance that the first spring is stretched, and let x_2 be the distance that the second spring is stretched. Then $x = x_1 + x_2$. But $x_1 = -F/k_1$ and $x_2 = -F/k_2$, so

$$\frac{-F}{k_1} + \frac{-F}{k_2} = x$$

$$-F\left(\frac{1}{k_1} + \frac{1}{k_2}\right) = x$$

$$F = -\left(\frac{1}{\frac{1}{k_1} + \frac{1}{k_2}}\right)x$$

$$F = -\frac{k_1 k_2}{k_1 + k_2}x$$

Therefore,

$$k_{\text{eff}} = \frac{k_1 k_2}{k_1 + k_2}$$

(d) If the two springs have the same force constant, that is, if $k_1 = k_2 = k$, then in the first two cases, the pairs of springs are equivalent to one spring that has twice their force constant: $k_{\text{eff}} = k_1 + k_2 = k + k = 2k$. In (c), the pair of springs is equivalent to a single spring with half their force constant:

$$k_{\text{eff}} = \frac{k_1 k_2}{k_1 + k_2} = \frac{kk}{k + k} = \frac{k^2}{2k} = \frac{k}{2}$$

Spring-Block Summary

We can summarize the dynamics of oscillations in this table:

	$x = -A$	$x = 0$	$x = +A$
Magnitude of Restoring Force	MAX	0	MAX
Magnitude of Acceleration	MAX	0	MAX
Potential Energy (U) of Spring	MAX	0	MAX
Kinetic Energy (K) of Block	0	MAX	0
Speed (v) of Block	0	MAX	0

Fortunately, this same table applies to springs, pendulums, and waves. All simple harmonic motion follows this cycle.

THE SPRING-BLOCK OSCILLATOR: VERTICAL MOTION

So far we've looked at a block sliding back and forth on a horizontal table, but the block could also oscillate vertically. The only difference would be that gravity would cause the block to move downward, to an equilibrium position at which, in contrast with the horizontal SHM we've examined, the spring would not be at its natural length. Of course, in calculating energy, the gravitational potential energy (*mgh*) must be included.

Consider a spring of negligible mass hanging from a stationary support. A block of mass m is attached to its end and allowed to come to rest, stretching the spring a distance d. At this point, the block is in equilibrium; the upward force of the spring is balanced by the downward force of gravity. Therefore,

$$kd = mg \implies d = \frac{mg}{k}$$

Another Approach
Using a force diagram let's map out the same situation:

$F_{spring} = kx$

$W = mg$

Our net force in this case is zero since it is neither moving up nor moving down. So our force, by the spring, must cancel out with our weight: $kx = mg$.

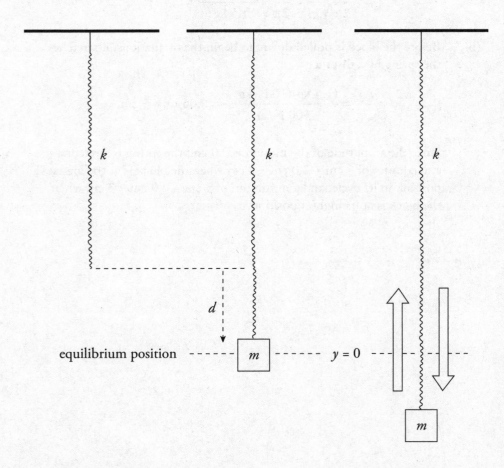

Once you have solved for the distance *d*, simply set this position as your new equilibrium position $x = 0$. At this point, you can treat the vertical spring exactly as you would a horizontal spring. Just be sure to measure distances relative to the new equilibrium when solving for any values (spring force, potential energy, and so on) that have *x* in their formulas.

Example 11 A block of mass $m = 1.5$ kg is attached to the end of a vertical spring of force constant $k = 300$ N/m. After the block comes to rest, it is pulled down a distance of 2.0 cm and released.

(a) What is the frequency of the resulting oscillations?

(b) What are the minimum and maximum amounts of stretch of the spring during the oscillations of the block?

Solution.

(a) The frequency is given by

$$f = \frac{1}{2\pi}\sqrt{\frac{k}{m}} = \frac{1}{2\pi}\sqrt{\frac{300 \text{ N/m}}{1.5 \text{ kg}}} = 2.3 \text{ Hz}$$

(b) Before the block is pulled down, to begin the oscillations, it stretches the spring by a distance

$$d = \frac{mg}{k} = \frac{(1.5 \text{ kg})(10 \text{ N/kg})}{300 \text{ N/m}} = 0.05 \text{ m} = 5 \text{ cm}$$

Since the amplitude of the motion is 2.0 cm, the spring is stretched a maximum of 5 cm + 2.0 cm = 7 cm when the block is at the lowest position in its cycle, and a minimum of 5 cm − 2.0 cm = 3 cm when the block is at its highest position.

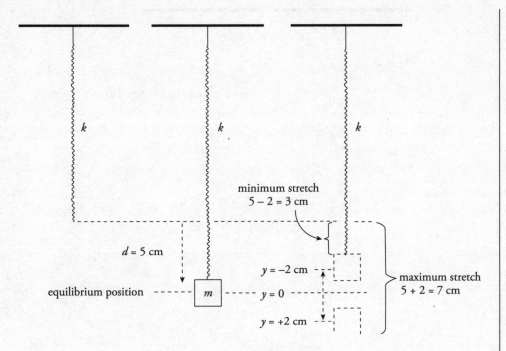

PENDULUMS

A **simple pendulum** consists of a weight of mass m attached to a string or a massless rod that swings, without friction, about the vertical equilibrium position. The restoring force is provided by gravity and, as the figure on the next page shows, the magnitude of the restoring force when the bob is θ to an angle to the vertical is given by the equation:

$$F_{restoring} = mg \sin \theta$$

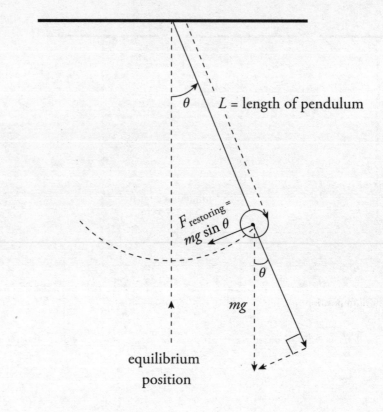

L = length of pendulum

$F_{restoring} = mg \sin \theta$

mg

equilibrium
position

Although the displacement of the pendulum is measured by the angle that it makes with the vertical, rather than by its linear distance from the equilibrium position (as was the case for the spring–block oscillator), the simple pendulum shares many of the important features of the spring–block oscillator. For example,

- Displacement is zero at the equilibrium position.
- At the endpoints of the oscillation region (where $\theta = \pm\theta_{max}$), the restoring force and the tangential acceleration (a_t) have their greatest magnitudes, the speed of the pendulum is zero, and the potential energy is maximized.
- As the pendulum passes through the equilibrium position, its kinetic energy and speed are maximized.

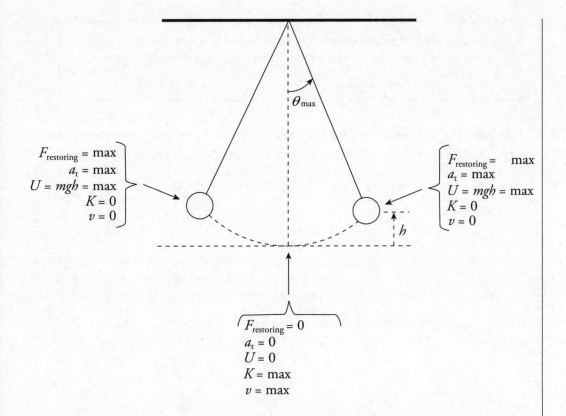

$F_{\text{restoring}} = \text{max}$
$a_t = \text{max}$
$U = mgh = \text{max}$
$K = 0$
$v = 0$

$F_{\text{restoring}} = \text{max}$
$a_t = \text{max}$
$U = mgh = \text{max}$
$K = 0$
$v = 0$

$F_{\text{restoring}} = 0$
$a_t = 0$
$U = 0$
$K = \text{max}$
$v = \text{max}$

Despite these similarities, there is one important difference. Simple harmonic motion results from a restoring force that has a strength that is proportional to the displacement. The magnitude of the restoring force on a pendulum is $mg \sin \theta$, which is *not* proportional to the displacement (θL, the arc length, with the angle measured in radians). Strictly speaking, the motion of a simple pendulum is not really simple harmonic. However, if θ is small, then $\sin \theta \approx \theta$ (measured in radians) so, in this case, the magnitude of the restoring force is approximately $mg\theta$, which *is* proportional to θ. So if θ_{max} is small, the motion can be treated as simple harmonic.

If the restoring force is given by $mg\theta$, rather than $mg \sin \theta$, then the frequency and period of the oscillations depend only on the length of the pendulum and the value of the gravitational acceleration, according to the following equations:

$$f = \frac{1}{2\pi}\sqrt{\frac{g}{L}} \quad \text{and} \quad T = 2\pi\sqrt{\frac{L}{g}}$$

A useful mnemonic for remembering these equations is to go in alphabetical order (clockwise) for frequency (f to g to L) and reverse alphabetical order (clockwise) for period (T to L to g).

Note that neither frequency nor the period depends on the amplitude (the maximum angular displacement, θ_{max}); this is a characteristic feature of simple harmonic motion. Also notice that neither the frequency nor the period depends on the mass of the weight.

Example 12 A simple pendulum has a period of 1 s on Earth. What would its period be on the Moon (where g is one-sixth of its value here)?

Solution. The equation $T = 2\pi\sqrt{L/g}$ shows that T is inversely proportional to $\sqrt{g}$, so if g decreases by a factor of 6, then T increases by factor of $\sqrt{6}$. That is,

$$T_{\text{on Moon}} = \sqrt{6} \times T_{\text{on Earth}} = (1\ \text{s})\sqrt{6} = 2.4\ \text{s}$$

Chapter 9 Review Questions

Answers and Explanations can be found in Chapter 13.

Section I: Multiple-Choice

1. Which of the following are characteristics of simple harmonic motion? Select two answers.

   (A) The acceleration is constant.
   (B) The restoring force is proportional to the displacement.
   (C) The frequency is independent of the amplitude.
   (D) The period is dependent on the amplitude.

2. A block attached to an ideal spring undergoes simple harmonic motion. The acceleration of the block has its maximum magnitude at the point where

   (A) the speed is the maximum
   (B) the speed is the minimum
   (C) the restoring force is the minimum
   (D) the kinetic energy is the maximum

3. A block attached to an ideal spring undergoes simple harmonic motion about its equilibrium position ($x = 0$) with amplitude A. What fraction of the total energy is in the form of kinetic energy when the block is at position $x = \frac{1}{2}A$?

   (A) $\frac{1}{3}$

   (B) $\frac{1}{2}$

   (C) $\frac{2}{3}$

   (D) $\frac{3}{4}$

4. A student measures the maximum speed of a block undergoing simple harmonic oscillations of amplitude A on the end of an ideal spring. If the block is replaced by one with twice its mass but the amplitude of its oscillations remains the same, then the maximum speed of the block will

   (A) decrease by a factor of 4
   (B) decrease by a factor of 2
   (C) decrease by a factor of $\sqrt{2}$
   (D) increase by a factor of 2

5. A spring–block simple harmonic oscillator is set up so that the oscillations are vertical. The period of the motion is T. If the spring and block are taken to the surface of the Moon, where the gravitational acceleration is 1/6 of its value here, then the vertical oscillations will have a period of

   (A) $\frac{T}{6}$

   (B) $\frac{T}{3}$

   (C) $\frac{T}{\sqrt{6}}$

   (D) T

6. A linear spring of force constant k is used in a physics lab experiment. A block of mass m is attached to the spring and the resulting frequency, f, of the simple harmonic oscillations is measured. Blocks of various masses are used in different trials, and in each case, the corresponding frequency is measured and recorded. If f^2 is plotted versus $1/m$, the graph will be a straight line with slope

   (A) $\frac{4\pi^2}{k^2}$

   (B) $\frac{4\pi^2}{k}$

   (C) $4\pi^2 k$

   (D) $\frac{k}{4\pi^2}$

7. A simple pendulum swings about the vertical equilibrium position with a maximum angular displacement of 5° and period T. If the same pendulum is given a maximum angular displacement of 10°, then which of the following best gives the period of the oscillations?

   (A) $\frac{T}{2}$

   (B) $\frac{T}{\sqrt{2}}$

   (C) T

   (D) $2T$

8. A block with a mass of 20 kg is attached to a spring with a force constant $k = 50$ N/m. What is the magnitude of the acceleration of the block when the spring is stretched 4 m from its equilibrium position?

   (A) 4 m/s^2
   (B) 6 m/s^2
   (C) 8 m/s^2
   (D) 10 m/s^2

9. A block with a mass of 10 kg connected to a spring oscillates back and forth with an amplitude of 2 m. What is the approximate period of the block if it has a speed of 4 m/s when it passes through its equilibrium point?

   (A) 1 s
   (B) 3 s
   (C) 6 s
   (D) 12 s

10. A block with a mass of 4 kg is attached to a spring on the wall that oscillates back and forth with a frequency of 4 Hz and an amplitude of 3 m. What would the frequency be if the block were replaced by one with one-fourth the mass and the amplitude of the block is increased to 9 m ?

    (A) 4 Hz
    (B) 8 Hz
    (C) 12 Hz
    (D) 24 Hz

Section II: Free-Response

1. The figure below shows a block of mass m (Block 1) that is attached to one end of an ideal spring of force constant k and natural length L. The block is pushed so that it compresses the spring to 3/4 of its natural length and is then released from rest. Just as the spring has extended to its natural length L, the attached block collides with another block (also of mass m) at rest on the edge of the frictionless table. When Block 1 collides with Block 2, half of its kinetic energy is lost to heat; the other half of Block 1's kinetic energy at impact is divided between Block 1 and Block 2. The collision sends Block 2 over the edge of the table, where it falls a vertical distance H, landing at a horizontal distance R from the edge.

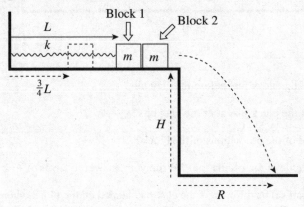

(a) What is the acceleration of Block 1 at the moment it's released from rest from its initial position? Write your answer in terms of k, L, and m.

(b) If v_1 is the velocity of Block 1 just before impact, show that the velocity of Block 1 just after impact is $\frac{1}{2}v_1$.

(c) Determine the amplitude of the oscillations of Block 1 after Block 2 has left the table. Write your answer in terms of L only.

(d) Determine the period of the oscillations of Block 1 after the collision, writing your answer in terms of T_0, the period of the oscillations that Block 1 would have had if it did not collide with Block 2.

(e) Find an expression for R in terms of H, k, L, m, and g.

2. A bullet of mass m is fired from a non-lethal pellet gun horizontally with speed v into a block of mass M initially at rest, at the end of an ideal spring on a frictionless table. At the moment the bullet hits, the spring is at its natural length, L. The bullet becomes embedded in the block, and simple harmonic oscillations result.

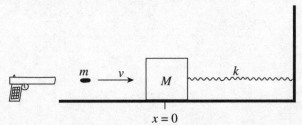

(a) Determine the speed of the block immediately after the impact by the bullet.

(b) Determine the amplitude of the resulting oscillations of the block.

(c) Compute the frequency of the resulting oscillations.

3. A block of mass M oscillating with amplitude A on a frictionless horizontal table is connected to an ideal spring of force constant k. The period of its oscillations is T. At the moment when the block is at position $x = \frac{1}{2}A$ and moving to the right, a ball of clay of mass m dropped from above lands on the block.

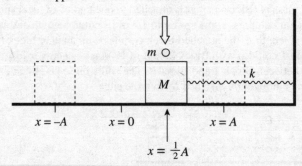

(a)   What is the velocity of the block just before the clay hits?

(b)   What is the velocity of the block just after the clay hits?

(c)   What is the new period of the oscillations of the block?

(d)   What is the new amplitude of the oscillations? Write your answer in terms of A, k, M, and m.

(e)   Would the answer to part (c) be different if the clay had landed on the block when it was at a different position? Support your answer briefly.

(f)   Would the answer to part (d) be different if the clay had landed on the block when it was at a different position? Support your answer briefly.

Summary

o $T = \dfrac{time}{\# \, cycles}$

$f = \dfrac{\# \, cycles}{time}$

$T = \dfrac{1}{f}$

o Hooke's Law holds for most springs. Formulas to keep in mind are the following:

$F_s = -kx$

$T = 2\pi \sqrt{\dfrac{m}{k}}$

$U_s = \dfrac{1}{2}kx^2$

o For small angle of a pendulum swing:

$T = 2\pi \sqrt{\dfrac{L}{g}}$

Chapter 10
Waves

"It would be possible to describe absolutely everything scientifically, but it would make no sense. It would be without meaning, as if you described a Beethoven symphony as a variation of wave pressure."

—Albert Einstein

Imagine holding the end of a long rope in your hand, with the other end attached to a wall. Move your hand up and down, and you'll create a wave that travels along the rope, from your hand to the wall. This simple example displays the basic idea of a **mechanical wave**: a disturbance transmitted by a medium from one point to another, without the medium itself being transported. In this chapter, we will discuss waves and their characteristics.

TRANSVERSE TRAVELING WAVES

Let's take a look at a long rope. Someone standing near the system would see peaks and valleys actually moving along the rope, in what's called a **traveling wave.**

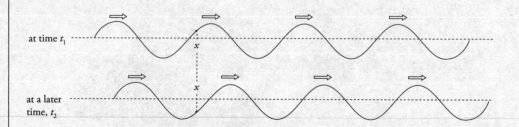

What features of the wave can we see in this point of view? Well, we can see the points at which the rope has its maximum vertical displacement above the horizontal; these points are called **crests**. The points at which the rope has its maximum vertical displacement below the horizontal are called **troughs**. These crests and troughs repeat themselves at regular intervals along the rope, and the distance between two adjacent crests (or two adjacent troughs) is the length of one wave, and is called the **wavelength** (λ, *lambda*). Also, the maximum displacement from the horizontal equilibrium position of the rope is also measurable; this is known as the **amplitude** (A) of the wave.

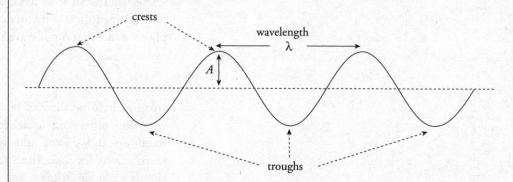

Since the direction in which the rope oscillates (vertically) is perpendicular to the direction in which the wave **propagates** (or travels, horizontally), this wave is **transverse**. The time it takes for one complete vertical oscillation of a point on the rope is called the **period**, T, of the wave, and the number of cycles it completes in one second is called its **frequency**, f. The period and frequency are established by the source of the wave and, of course, $T = 1/f$.

In addition to the amplitude, period, frequency, and wavelength, another important characteristic of a traveling wave is its speed, v. Look at the second figure on the previous page and imagine the visible point on the rope moving from its crest position, down to its trough position, and then back up to the crest position. The wave took a time period of T to move a distance of one wavelength, signified by λ. Therefore, the equation *distance = rate × time* becomes

$$\lambda = vT$$
$$\lambda \cdot \frac{1}{T} = v$$
$$\lambda f = v$$

Ride The Wave
Ocean waves are a sort of blend of transverse and compressional waves. Actually, matter moves in large circles near the surface of the ocean and progressively smaller circles deeper down as the energy in the system dissipates.

The simple equation $v = \lambda f$ shows how the wave speed, wavelength, and frequency are interconnected. It's the most basic equation in wave theory.

> **Example 1** A traveling wave on a rope has a frequency of 2.5 Hz. If the speed of the wave is 1.5 m/s, what are its period and wavelength?

Solution. The period is the reciprocal of the frequency:

$$T = \frac{1}{f} = \frac{1}{2.5 \text{ Hz}} = 0.4 \text{ s}$$

and the wavelength can be found from the equation $\lambda f = v$:

$$\lambda = \frac{v}{f} = \frac{1.5 \text{ m/s}}{2.5 \text{ Hz}} = 0.6 \text{ m}$$

> **Example 2** The period of a traveling wave is 0.5 s, its amplitude is 10 cm, and its wavelength is 0.4 m. What are its frequency and wave speed?

Solution. The frequency is the reciprocal of the period: $f = 1/T = 1/(0.5 \text{ s}) = 2$ Hz. The wave speed can be found from the equation $v = \lambda f$:

$$v = \lambda f = (0.4 \text{ m})(2 \text{ Hz}) = 0.8 \text{ m/s}$$

Note that the frequency, period, wavelength, and wave speed have nothing to do with the amplitude.

WAVE SPEED ON A STRETCHED STRING

We can also derive an equation for the speed of a transverse wave on a stretched string or rope. Let the mass of the string be m and its length be L; then its *linear mass density* (μ) is m/L. If the tension in the string is F_T, then the speed of a traveling transverse wave on this string is given by

$$v = \sqrt{\frac{F_T}{\mu}}$$

Note that v depends only on the physical characteristics of the string: its tension and linear density. So, because $v = \lambda f$ for a given stretched string, varying f will create different waves that have different wavelengths, but v will not vary.

Big Wave Rules

Notice that the equation for the wave speed on a stretched rope shows that v does not depend on f (or λ). While this may seem to contradict the first equation, $v = \lambda f$, it really doesn't. The speed of the wave depends on the characteristics of the rope; how tense it is and what it's made of. We can wiggle the end at any frequency we want, and the speed of the wave we create will be a constant. However, because $\lambda f = v$ must always be true, a higher f will mean a shorter λ and a lower f will mean a longer λ. Thus, changing f doesn't change v: it changes λ. This brings up our first big wave rule:

> **Wave Rule #1:** The speed of a wave is determined by the type of wave and the characteristics of the medium, not by the frequency.

Rule #1 deals with a wave in a single medium. What travels to you faster, a yell or a whisper? Neither does; they both travel to you at the speed of sound. In a single medium, the speed of a wave is constant and is represented by an inverse relationship between wavelength and frequency: $\lambda_1 f_1 = \lambda_2 f_2$.

Notice that two different types of waves can move with different speeds through the same medium: For example, sound and light move through air with very different speeds.

Our second wave rule addresses what happens when a wave passes from one medium into another. Because wave speed is determined by the characteristics of the medium, a change in the medium causes a change in wave speed, but the frequency won't change.

Wave Rule #2: When a wave passes into another medium, its speed changes, but its frequency does not.

Example 3 A horizontal rope with linear mass density $\mu = 0.5$ kg/m sustains a tension of 60 N. The non-attached end is oscillated vertically with a frequency of 4 Hz.

   (a) What are the speed and wavelength of the resulting wave?

   (b) How would you answer these questions if f were increased to 5 Hz?

Solution.

(a) Wave speed is established by the physical characteristics of the rope:

$$v = \sqrt{\frac{F_T}{\mu}} = \sqrt{\frac{60 \text{ N}}{0.5 \text{ kg/m}}} = 11 \text{ m/s}$$

With v, we can find the wavelength: $\lambda = v/f = (11 \text{ m/s})/(4 \text{ Hz}) = 2.8$ m.

(b) If f were increased to 5 Hz, then v would not change, but λ would; the new wavelength would be
$$\lambda' = v/f = (11 \text{ m/s})/(5 \text{ Hz}) = 2.2 \text{ m}$$

Example 4 Two ropes of unequal linear densities are connected, and a wave is created in the rope on the left, which propagates to the right, toward the interface with the heavier rope.

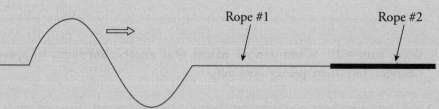

When a wave strikes the boundary to a new medium (in this case, the heavier rope), some of the wave's energy is reflected and some is transmitted. The frequency of the transmitted wave is the same, but the speed and wavelength are not. How do the speed and wavelength of the incident wave compare to the speed and wavelength of the wave as it travels through Rope #2?

Solution. Since the wave enters a new medium, it will have a new wave speed. Because Rope #2 has a greater linear mass density than Rope #1, and because v is inversely proportional to the square root of the linear mass density, the speed of the wave in Rope #2 will be less than the speed of the wave in Rope #1. Since $v = \lambda f$ must always be satisfied and f does not change, the fact that v changes means that λ must change, too. In particular, since v decreases upon entering Rope #2, so will λ.

SUPERPOSITION OF WAVES

When two or more waves meet, the displacement at any point of the medium is equal to the algebraic sum of the displacements due to the individual waves. This is **superposition**. The figure on the next page shows two wave pulses traveling toward each other along a stretched string. Note that when they meet and overlap (**interfere**), the displacement of the string is equal to the sum of the individual displacements, but after they pass, the wave pulses continue, unchanged by their meeting.

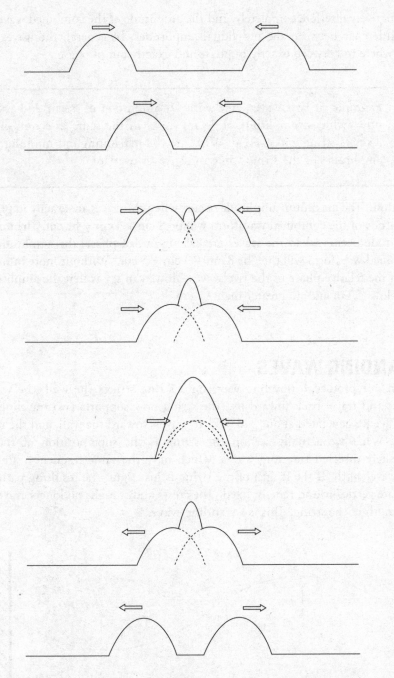

If the two waves have displacements of the same sign when they overlap, the combined wave will have a displacement of greater magnitude than either individual wave; this is called **constructive interference**. Similarly, if the waves have opposite displacements when they meet, the combined waveform will have a displacement of smaller magnitude than either individual wave; this is called **destructive interference**. If the waves travel in the same direction, the amplitude of the combined wave depends on the relative phase of the two waves. If the waves are exactly **in phase**—that is, if crest meets crest and trough meets trough—then the waves will constructively interfere completely, and the amplitude of the combined wave will be the sum of the individual amplitudes. However, if the waves are exactly **out of phase**—that is, if crest meets trough and trough meets crest—then they will

Constructive = Add

Destructive = Subtract

destructively interfere completely, and the amplitude of the combined wave will be the difference between the individual amplitudes. In general, the waves will be somewhere in between exactly in phase and exactly out of phase.

> **Example 5** Two waves, one with an amplitude of 8 cm and the other with an amplitude of 3 cm, travel in the same direction on a single string and overlap. What are the maximum and minimum amplitudes of the string while these waves overlap?

Solution. The maximum amplitude occurs when the waves are exactly in phase; the amplitude of the combined waveform will be 8 cm + 3 cm = 11 cm. The minimum amplitude occurs when the waves are exactly out of phase; the amplitude of the combined waveform will then be 8 cm − 3 cm = 5 cm. Without more information about the relative phase of the two waves, all we can say is that the amplitude will be at least 5 cm and no greater than 11 cm.

STANDING WAVES

When our prototype traveling wave on a string strikes the wall, the wave will reflect and travel back toward us. The string now supports two traveling waves; the wave we generated at our end, which travels toward the wall, and the reflected wave. What we actually see on the string is the superposition of these two oppositely directed traveling waves, which have the same frequency, amplitude, and wavelength. If the length of the string is just right, the resulting pattern will oscillate vertically and remain fixed. The crests and troughs no longer travel down the length of the string. This is a **standing wave**.

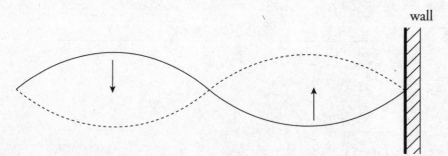

Tip
You can remember nodes as areas of "no displacement." Antinodes are the opposite of that.

The right end is fixed to the wall, and the left end is oscillated through a negligibly small amplitude so that we can consider both ends to be essentially fixed (no vertical oscillation). The interference of the two traveling waves results in complete destructive interference at some points (marked N in the figure on the next page), and complete constructive interference at other points (marked A in the figure). Other points have amplitudes between these extremes. Note another difference between a traveling wave and a standing wave: While every point on the string had the same amplitude as the traveling wave went by, each point on a string supporting a standing wave has an individual amplitude. The points marked N are called **nodes**, and those marked A are called **antinodes**.

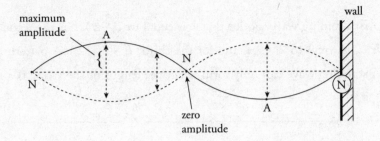

Nodes and antinodes always alternate, they're equally spaced, and the distance between two successive nodes (or antinodes) is equal to $\frac{1}{2}\lambda$. This information can be used to determine how standing waves can be generated. The following figures show the three simplest standing waves that our string can support. The first standing wave has one antinode, the second has two, and the third has three. The length of the string in all three diagrams is L.

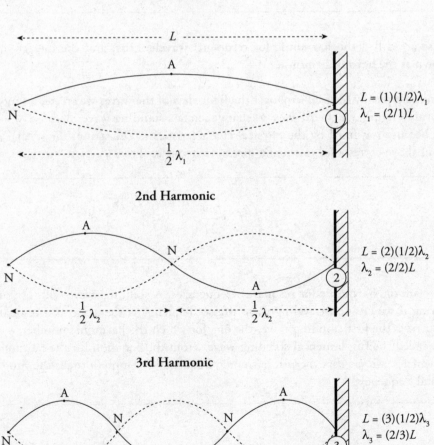

1st Harmonic

$L = (1)(1/2)\lambda_1$
$\lambda_1 = (2/1)L$

2nd Harmonic

$L = (2)(1/2)\lambda_2$
$\lambda_2 = (2/2)L$

3rd Harmonic

$L = (3)(1/2)\lambda_3$
$\lambda_3 = (2/3)L$

For the first standing wave, notice that L is equal to $1(\frac{1}{2}\lambda)$. For the second standing wave, L is equal to $2(\frac{1}{2}\lambda)$, and for the third, $L = 3(\frac{1}{2}\lambda)$. A pattern is established: A standing wave can only form when the length of the string is a multiple of $\frac{1}{2}\lambda$:

$$L = n(\frac{1}{2}\lambda)$$

Solving this for the wavelength, we get

$$\lambda_n = \frac{2L}{n}$$

These are called the **harmonic** (or **resonant**) **wavelengths**, and the integer n is known as the **harmonic number**.

Since we typically have control over the frequency of the waves we create, it's more helpful to figure out the *frequencies* that generate a standing wave. Because $\lambda f = v$, and because v is fixed by the physical characteristics of the string, the special λs found above correspond to equally special frequencies. From $f_n = v/\lambda_n$, we get

$$f_n = \frac{nv}{2L}$$

These are the **harmonic** (or **resonant**) **frequencies**. A standing wave will form on a string if we create a traveling wave whose frequency is the same as a resonant frequency. The first standing wave, the one for which the harmonic number, n, is 1, is called the **fundamental** standing wave. From the equation for the harmonic frequencies, we see that the nth harmonic frequency is simply n times the fundamental frequency:

$$f_n = nf_1$$

Similarly, the nth harmonic wavelength is equal to λ_1 divided by n. Therefore, by knowing the fundamental frequency (or wavelength), all the other resonant frequencies and wavelengths can be determined.

Example 6 A string of length 12 m that's fixed at both ends supports a standing wave with a total of 5 nodes. What are the harmonic number and wavelength of this standing wave?

Solution. First, draw a picture.

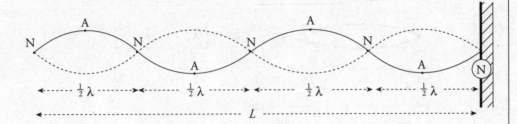

This shows that the length of the string is equal to $4(\frac{1}{2}\lambda)$, so

$$L = 4(\frac{1}{2}\lambda) \quad \Rightarrow \quad \lambda = \frac{2L}{4}$$

This is the fourth-harmonic standing wave, with wavelength λ_4 (because the expression above matches $\lambda_n = 2L/n$ for $n = 4$). Since $L = 12$ m, the wavelength is

$$\lambda_4 = \frac{2(12 \text{ m})}{4} = 6 \text{ m}$$

Example 7 A string of length 10 m and mass 300 g is fixed at both ends, and the tension in the string is 40 N. What is the frequency of the standing wave for which the distance between a node and the closest antinode is 1 m?

Solution. Because the distance between two successive nodes (or successive antinodes) is equal to $\frac{1}{2}\lambda$, the distance between a node and the closest antinode is half this, or $\frac{1}{4}\lambda$. Therefore, $\frac{1}{4}\lambda = 1$ m, so $\lambda = 4$ m. Since the harmonic wavelengths are given by the equation $\lambda_n = 2L/n$, we can find that

$$4 \text{ m} = \frac{2(10 \text{ m})}{n} \quad \Rightarrow \quad n = 5$$

The frequency of the fifth harmonic is

$$f_5 = \frac{5v}{2L} = \frac{5}{2L}\sqrt{\frac{F_T}{\mu}} = \frac{5}{2L}\sqrt{\frac{F_T}{m/L}} = \frac{5}{2(10\text{ m})}\sqrt{\frac{40\text{ N}}{(0.3\text{ kg}/10\text{ m})}} = 9.1\text{ Hz}$$

If you attached a rope or string to a ring that could slide up and down a pole without friction, you would make a rope that is fixed at one end but free at another end. This would create nodes at the closed end and antinodes at the open end. Below are some possible examples of this.

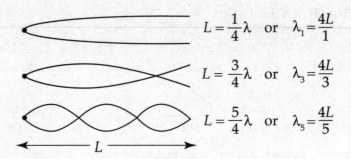

$$L = \frac{1}{4}\lambda \quad \text{or} \quad \lambda_1 = \frac{4L}{1}$$

$$L = \frac{3}{4}\lambda \quad \text{or} \quad \lambda_3 = \frac{4L}{3}$$

$$L = \frac{5}{4}\lambda \quad \text{or} \quad \lambda_5 = \frac{4L}{5}$$

If one end of the rope is free to move, standing waves can form for wavelengths of $\lambda_n = \dfrac{4L}{n}$ or by frequencies of $f_n = \dfrac{nv}{4L}$, where L is the length of the rope, v is the speed of the wave, and n must be an odd integer.

SOUND WAVES

Speed of Sound
At sea level, approximately 340 m/s.

Sound waves are produced by the vibration of an object, such as your vocal cords, a plucked string, or a jackhammer. The vibrations cause pressure variations in the conducting medium (which can be gas, liquid, or solid), and if the frequency is between 20 Hz and 20,000 Hz, the vibrations may be detected by human ears and perceived as sound. The variations in the conducting medium can be positions at which the molecules of the medium are bunched together (where the pressure is above normal), which are called **compressions**, and positions where the pressure is below normal, called **rarefactions**. In the figure on the next page, a vibrating diaphragm sets up a sound wave in an air-filled tube. Each dot represents many, many air molecules:

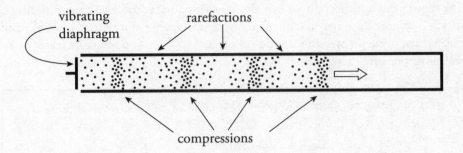

vibrating diaphragm

rarefactions

compressions

An important difference between sound waves and the waves we've been studying on stretched strings is that the molecules of the medium transmitting a sound wave move *parallel* to the direction of wave propagation, rather than perpendicular to it. For this reason, sound waves are said to be **longitudinal**. Despite this difference, all of the basic characteristics of a wave—amplitude, wavelength, period, frequency—apply to sound waves as they did for waves on a string. Furthermore, the all-important equation $\lambda f = v$ also holds true. However, because it's very difficult to draw a picture of a longitudinal wave, an alternate method is used: We can graph the pressure as a function of position:

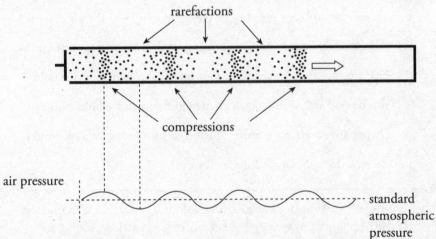

rarefactions

compressions

air pressure

standard atmospheric pressure

The speed of a sound wave depends on the medium through which it travels. In particular, it depends on the density (ρ) and on the **bulk modulus** (B), a measure of the medium's response to compression. A medium that is easily compressible, like a gas, has a low bulk modulus; liquids and solids, which are much less easily compressed, have significantly greater bulk modulus values. For this reason, sound generally travels faster through solids than through liquids and faster through liquids than through gases.

The equation that gives a sound wave's speed in terms of ρ and B is

$$v = \sqrt{\frac{B}{\rho}}$$

Logitudinal Waves
A longitudinal wave travels and oscillates in the same direction.

The speed of sound through air can also be written in terms of air's mean pressure, which is temperature dependent. At room temperature (approximately 20°C) and normal atmospheric pressure, sound travels at 343 m/s. This value increases as air warms or pressure increases.

Example 8 A sound wave with a frequency of 300 Hz travels through the air.

      (a) What is its wavelength?

      (b) If its frequency increased to 600 Hz, what are the wave's speed and wavelength?

Solution.

    (a) Using $v = 343$ m/s for the speed of sound through air, we find that

$$\lambda = \frac{v}{f} = \frac{343 \text{ m/s}}{300 \text{ Hz}} = 1.14 \text{ m}$$

    (b) Unless the ambient pressure or the temperature of the air changed, the speed of sound would not change. Wave speed depends on the characteristics of the medium, not on the frequency, so v would still be 343 m/s. However, a change in frequency would cause a change in wavelength. Since f increased by a factor of 2, λ would decrease by a factor of 2, to $\frac{1}{2}$ (1.14 m) = 0.57 m.

Example 9 A sound wave traveling through water has a frequency of 500 Hz and a wavelength of 3 m. How fast does sound travel through water?

Solution. $v = \lambda f = $ (3 m)(500 Hz) = 1,500 m/s.

Beats

If two sound waves whose frequencies are close but not identical interfere, the resulting sound modulates in amplitude, becoming loud, then soft, then loud, then soft. This is due to the fact that as the individual waves travel, they are in phase, then out of phase, then in phase again, and so on. Therefore, by superposition, the waves interfere constructively, then destructively, then constructively, and so on. When the waves interfere constructively, the amplitude increases, and the sound is loud; when the waves interfere destructively, the amplitude decreases, and the sound is soft. Each time the waves interfere constructively, producing an increase in sound level, we say that a **beat** has occurred. The number of beats per second, known as the **beat frequency**, is equal to the difference between the frequencies of the two combining sound waves:

$$f_{\text{beat}} = |f_1 - f_2|$$

If frequencies f_1 and f_2 match, then the combined waveform doesn't waver in amplitude, and no beats are heard.

Example 10 A piano tuner uses a tuning fork to adjust the key that plays the A note above middle C (whose frequency should be 440 Hz). The tuning fork emits a perfect 440 Hz tone. When the tuning fork and the piano key are struck, beats of frequency 3 Hz are heard.

    (a) What is the frequency of the piano key?

    (b) If it's known that the piano key's frequency is too high, should the piano tuner tighten or loosen the wire inside the piano in order to tune it?

Solution.

(a) Since f_{beat} = 3 Hz, the tuning fork and the piano string are off by 3 Hz. Since the fork emits a tone of 440 Hz, the piano string must emit a tone of either 437 Hz or 443 Hz. Without more information, we can't determine which.

(b) If we know that the frequency of the tone emitted by the out-of-tune string is too high (that is, it's 443 Hz), we need to find a way to lower the frequency. Remember that the resonant frequencies for a stretched string fixed at both ends are given by the equation $f = nv/2L$, and that $v = \sqrt{F_T/\mu}$. Since f is too high, v must be too high. To lower v, we must reduce F_T. The piano tuner should loosen the string and listen for beats again, adjusting the string until the beats disappear.

RESONANCE FOR SOUND WAVES

Just as standing waves can be set up on a vibrating string, standing *sound* waves can be established within an enclosure. In the figure below, a vibrating source at one end of an air-filled tube produces sound waves that travel the length of the tube.

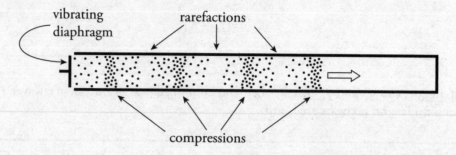

These waves reflect off the far end, and the superposition of the forward and reflected waves can produce a standing wave pattern if the length of the tube and the frequency of the waves are related in a certain way.

Notice that air molecules at the far end of the tube can't oscillate horizontally because they're up against a wall. So the far end of the tube is a displacement node. But the other end of the tube (where the vibrating source is located) is a displacement antinode. A standing wave with one antinode (A) and one node position (N) can be depicted as follows:

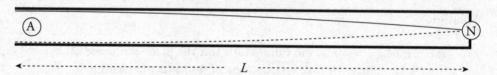

Although sound waves in air are longitudinal, here we'll show the wave as transverse so that it's easier to determine the wavelength. Since the distance between an antinode and an adjacent node is always $\frac{1}{4}$ of the wavelength, the length of the tube, L, in the figure above is $\frac{1}{4}$ the wavelength. This is the longest standing wavelength that can fit in the tube, so it corresponds to the lowest standing wave frequency, the fundamental:

$$L = \frac{\lambda_1}{4} \quad \Rightarrow \quad \lambda_1 = 4L \quad \Rightarrow \quad f_1 = \frac{v}{\lambda_1} = \frac{v}{4L}$$

Our condition for resonance was a node at the closed end and an antinode at the open end. Therefore, the next higher frequency standing wave that can be supported in this tube must have two antinodes and two nodes:

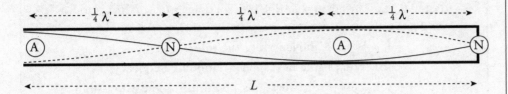

In this case, the length of the tube is equal to $3(\frac{1}{4}\lambda')$, so

$$L = \frac{3\lambda'}{4} \quad \Rightarrow \quad \lambda' = \frac{4L}{3} \quad \Rightarrow \quad f' = \frac{v}{\lambda'} = \frac{3v}{4L}$$

Here's the pattern: Standing sound waves can be established in a tube that's closed at one end if the tube's length is equal to an *odd* multiple of $\frac{1}{4}\lambda$. The resonant wavelengths and frequencies are given by the equations

For a Closed Tube, $\left.\begin{array}{l} \lambda_n = \dfrac{4L}{n} \\[2mm] f_n = n\dfrac{v}{4L} \end{array}\right\}$ for any *odd* integer n

If the far end of the tube is not sealed, standing waves can still be established in the tube, because sound waves can be reflected from the open air. A closed end is a displacement node, but an open end is a displacement antinode. In this case, then, the standing waves will have two displacement antinodes (at the ends of the tube), and the resonant wavelengths and frequencies will be given by

For an Open Tube, $\left.\begin{array}{l} \lambda_n = \dfrac{2L}{n} \\[2mm] f_n = n\dfrac{v}{2L} \end{array}\right\}$ for any integer n

Note that, while an open-ended tube can support any harmonic, a closed-end tube can only support odd harmonics.

Example 11 A closed-end tube resonates at a fundamental frequency of 440.0 Hz. The air in the tube is at a temperature of 20°C, and it conducts sound at a speed of 343 m/s.

    (a) What is the length of the tube?

    (b) What is the next higher harmonic frequency?

    (c) Answer the questions posed in (a) and (b) assuming that the tube were open at its far end.

Solution.

(a) For a closed-end tube, the harmonic frequencies obey the equation $f_n = nv/(4L)$. The fundamental corresponds to $n = 1$, so

$$f_1 = \frac{v}{4L} \quad \Rightarrow \quad L = \frac{v}{4f_1} = \frac{343 \text{ m/s}}{4(440.0 \text{ Hz})} = 0.195 \text{ m} = 19.5 \text{ cm}$$

(b) Since a closed-end tube can support only *odd* harmonics, the next higher harmonic frequency (the first **overtone**) is the *third* harmonic, f_3, which is $3f_1 = 3(440.0 \text{ Hz}) = 1{,}320 \text{ Hz}$.

(c) For an open-end tube, the harmonic frequencies obey the equation $f_n = nv/(2L)$. The fundamental corresponds to $n = 1$, so

$$f_1 = \frac{v}{2L'} \quad \Rightarrow \quad L' = \frac{v}{2f_1} = \frac{343 \text{ m/s}}{2(440.0 \text{ Hz})} = 0.390 \text{ m} = 39.0 \text{ cm}$$

And, since an open-end tube can support any harmonic, the first overtone would be the second harmonic, $f_2 = 2f_1 = 2(440.0 \text{ Hz}) = 880.0 \text{ Hz}$.

THE DOPPLER EFFECT

When a source of sound waves and a detector are not in relative motion, the frequency that the source emits matches the frequency that the detector receives.

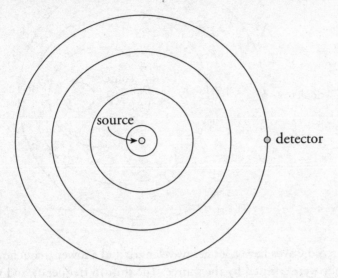

However, if there *is* relative motion between the source and the detector, then the waves that the detector receives are different in frequency (and wavelength). For example, if the detector moves toward the source, then the detector intercepts the waves at a rate higher than the one at which they were emitted; the detector hears a higher frequency than the source emitted. In the same way, if the source moves toward the detector, the wavefronts pile up, and this results in the detector receiving waves with shorter wavelengths and higher frequencies:

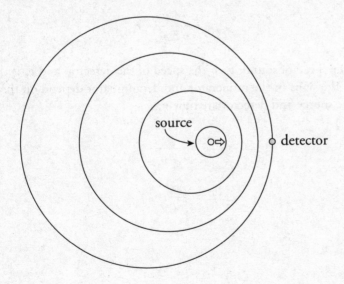

Conversely, if the detector is moving away from the source or if the source is moving away from the detector,

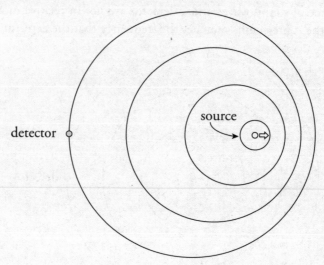

then the detected waves have a longer wavelength and a lower frequency than they had when they were emitted by the source. The shift in frequency and wavelength that occurs when the source and detector are in relative motion is known as the **Doppler effect**. In general, relative motion *toward* each other results in a frequency shift upward, and relative motion *away* from each other results in a frequency shift downward.

Let f_S be the frequency of waves that the source emits and f_D the frequency that the detector hears. To determine f_D from f_S, use the following equation

$$f_D = \frac{v \pm v_D}{v \mp v_S} \cdot f_S$$

where v is the speed of sound, v_D is the speed of the detector, and v_S is the speed of the source. The signs in the numerator and denominator depend on the directions in which the source and detector are moving.

$$f_D = \frac{v}{v - v_S} \cdot f_S$$

$$f_D = \frac{v}{v + v_S} \cdot f_S$$

The following is a chart of the four most common situations in which only one object (either the detector or source) moves.

Detector and Sound	f_D	Fraction	Velocity
D → S	↑	>1	$+v_D$
← D S	↓	<1	$-v_D$
D ← S	↑	>1	$-v_S$
D S →	↓	<1	$+v_S$

In situations where both the detector and source are moving, we can also use logic.

In a situation where both the detector and source are moving in the same direction, if the source is moving faster than the detector, we would expect the frequency heard by the detector to decrease. However, it would decrease by a factor much less than if the detector were stationary and the source was moving away from it at the same speed.

The detector is moving toward the source
(make numerator greater)

↓

$$f_D = \frac{v + v_D}{v + v_S} \cdot f_S$$

↑

The source is moving away from the detector
(make denominator greater)

We can learn something else from this. Let's use an example of a sports car and a police car. What if the sports car were moving at the same speed as the police car? Then, although both are moving relative to the road, they are not moving relative to each other. If there is no relative motion between the source and the detector, then there should be no Doppler shift. This conclusion is supported by the equation above, since $v_D = v_S$ implies that $f_D = f_S$.

Example 12 A source of 4 kHz sound waves travels at $\frac{1}{9}$ the speed of sound toward a detector that's moving at $\frac{1}{9}$ the speed of sound, toward the source.

    (a) What is the frequency of the waves as they're received by the detector?

    (b) How does the wavelength of the detected waves compare to the wavelength of the emitted waves?

Solution.

(a) Because the detector moves toward the source, we use the + sign in the numerator of the Doppler effect formula, and since the source moves toward the detector, we use the − sign in the denominator. This gives us:

$$f_D = \frac{v + v_D}{v - v_S} \cdot f_S = \frac{v + \frac{1}{9}v}{v - \frac{1}{9}v} \cdot f_S = \frac{5}{4} f_S = \frac{5}{4}(4 \text{ kHz}) = 5 \text{ kHz}$$

(b) Since the frequency shifted up by a factor of $\frac{5}{4}$, the wavelength will shift down by the same factor. That is,

$$\lambda_D = \frac{\lambda_S}{\frac{5}{4}} = \frac{4}{5}\lambda_S$$

Example 13 A person yells, emitting a constant frequency of 200 Hz, as he runs at 5 m/s toward a stationary brick wall. When the reflected waves reach the person, how many beats per second will he hear? (Use 343 m/s for the speed of sound.)

Solution. We need to determine what the frequency of the reflected waves will be when they reach the runner. First, the person is the source of the sound, and the wall is the detector. So:

$$f_{\text{wall}} = \frac{v}{v - v_{\text{runner}}} f$$

Then, as the sound waves reflect off the wall (with no change in frequency upon reflection), the wall acts as the source and the runner acts as the detector:

$$f' = \frac{v + v_{\text{runner}}}{v} f_{\text{wall}}$$

Combining these two equations gives

$$f' = \frac{v + v_{\text{runner}}}{v - v_{\text{runner}}} f$$

With $v = 343$ m/s and $v_{\text{runner}} = 5$ m/s, we get

$$f' = \frac{(343 + 5) \text{ m/s}}{(343 - 5) \text{ m/s}} (200 \text{ Hz}) = 206 \text{ Hz}$$

Therefore, the beat frequency is

$$f_{\text{beat}} = f' - f = 206 \text{ Hz} - 200 \text{ Hz} = 6 \text{ Hz}$$

Chapter 10 Review Questions

Answers and Explanations can be found in Chapter 13.

Section I: Multiple-Choice

1. What is the wavelength of a 5 Hz wave that travels with a speed of 10 m/s ?

   (A) 0.25 m
   (B) 0.5 m
   (C) 2 m
   (D) 50 m

2. A rope of length 5 m is stretched to a tension of 80 N. If its mass is 1 kg, at what speed would a 10 Hz transverse wave travel down the string?

   (A) 2 m/s
   (B) 5 m/s
   (C) 20 m/s
   (D) 200 m/s

3. A transverse wave on a long horizontal rope with a wavelength of 8 m travels at 2 m/s. At $t = 0$, a particular point on the rope has a vertical displacement of $+A$, where A is the amplitude of the wave. At what time will the vertical displacement of this same point on the rope be $-A$?

   (A) $t = \dfrac{1}{4}$ s

   (B) $t = \dfrac{1}{2}$ s

   (C) $t = 2$ s

   (D) $t = 4$ s

4. The vertical displacement, y, of a transverse traveling wave is given by the equation $y = 6 \sin (10\,t - \dfrac{1}{2}\,\pi x)$, with x and y in centimeters and t in seconds. What is the wavelength?

   (A) 0.25 cm
   (B) 0.5 cm
   (C) 2 cm
   (D) 4 cm

5. A string, fixed at both ends, supports a standing wave with a total of 4 nodes. If the length of the string is 6 m, what is the wavelength of the wave?

   (A) 0.67 m
   (B) 1.2 m
   (C) 3 m
   (D) 4 m

6. A string, fixed at both ends, has a length of 6 m and supports a standing wave with a total of 4 nodes. If a transverse wave can travel at 40 m/s down the rope, what is the frequency of this standing wave?

   (A) 6.7 Hz
   (B) 10 Hz
   (C) 20 Hz
   (D) 26.7 Hz

7. A sound wave travels through a metal rod with wavelength λ and frequency f. Which of the following best describes the wave when it passes into the surrounding air?

	Wavelength	Frequency
(A)	Less than λ	Equal to f
(B)	Less than λ	Less than f
(C)	Greater than λ	Equal to f
(D)	Greater than λ	Less than f

8. In the figure below, two speakers, S_1 and S_2, emit sound waves of wavelength 2 m, in phase with each other.

   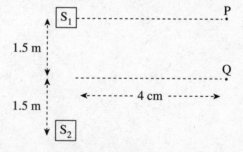

   Let A_P be the amplitude of the resulting wave at Point P, and A_Q the amplitude of the resultant wave at Point Q. How does A_P compare to A_Q ?

   (A) $A_P < A_Q$
   (B) $A_P = A_Q$
   (C) $A_P > A_Q$
   (D) $A_P < 0, A_Q > 0$

9. An organ pipe that's closed at one end has a length of 17 cm. If the speed of sound through the air inside is 340 m/s, what is the pipe's fundamental frequency?

(A) 250 Hz
(B) 500 Hz
(C) 1,000 Hz
(D) 1,500 Hz

10. A bat emits a 40 kHz "chirp" with a wavelength of 8.75 mm toward a tree and receives an echo 0.4 s later. How far is the bat from the tree?

(A) 35 m
(B) 70 m
(C) 105 m
(D) 140 m

11. A car is traveling at 20 m/s away from a stationary observer. If the car's horn emits a frequency of 600 Hz, what frequency will the observer hear? (Use $v = 340$ m/s for the speed of sound.)

(A) (34/36)(600 Hz)
(B) (34/32)(600 Hz)
(C) (36/34)(600 Hz)
(D) (32/34)(600 Hz)

Section II: Free-Response

1. A rope is stretched between two vertical supports. The points where it's attached (P and Q) are fixed. The linear density of the rope, μ, is 0.4 kg/m, and the speed of a transverse wave on the rope is 12 m/s.

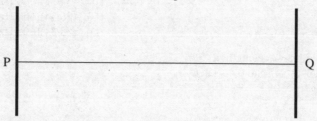

(a)   What's the tension in the rope?

(b)   With what frequency must the rope vibrate to create a traveling wave with a wavelength of 2 m?

The rope can support standing waves of lengths 4 m and 3.2 m, whose harmonic numbers are consecutive integers.

(c)   Find

    (i)   the length of the rope

    (ii) the mass of the rope

(d)   What is the harmonic number of the 4 m standing wave?

(e)   On the diagram above, draw a sketch of the 4 m standing wave, labeling the nodes and antinodes.

2. For a physics lab experiment on the Doppler effect, students are brought to a racetrack. A car is fitted with a horn that emits a tone with a constant frequency of 500 Hz, and the goal of the experiment is for the students to observe the pitch (frequency) of the car's horn.

(a) If the car drives at 50 m/s directly away from the students (who remain at rest), what frequency will they hear? (Note: speed of sound = 345 m/s)

(b) Would it make a difference if the car remained stationary and the students were driven in a separate vehicle at 50 m/s away from the car?

The following diagram shows a long straight track along which the car is driven. The students stand 40 m from the track.

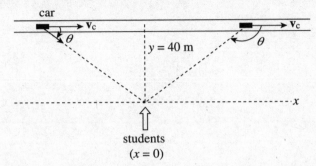

As the car drives by, the frequency which the students hear, f', is related to the car horn's frequency, f, by the equation

$$f' = \frac{v}{v - v_C \cos\theta} \cdot f$$

where v is the speed of sound, v_C is the speed of the car (50 m/s), and θ is the angle shown in the diagram.

(c) Find f' when $\theta = 60°$.

(d) Find f' when $\theta = 120°$.

(e) On the axes below, sketch the graph of f' as a function of x, the horizontal position of the car relative to the students' location.

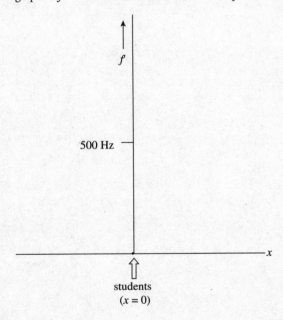

3. A group of students perform a set of physics experiments using two tuning forks, one with a frequency of 400 Hz and the other 440 Hz.

   (a)   What is the observed beat frequency when the two tuning forks are struck?

   (b)   Describe the changes to the frequency, wavelength, and speed of the sound waves from the tuning forks as they travel from air into water.

   (c)   One student strikes a tuning fork and then throws it straight up to a student on the second floor of a building. What happens to the frequency of the sound that the student on the second floor hears as the tuning fork travels upward?

Summary

○ The speed of a wave depends on the medium through which it travels. Speed can be determined by $v = f\lambda$, where f is the frequency ($f = \dfrac{\#\,cycles}{time}$) and λ is the wavelength. Because $f = \dfrac{1}{T}$, the speed can also be written $v = \dfrac{\lambda}{T}$.

○ Changing any one of the period, the frequency, or the wavelength of a wave will affect the other two quantities, but will not affect the speed as long as the medium remains the same.

○ If the wave travels through a stretched string, its speed can be determined by $v = \sqrt{\dfrac{F_T}{\mu}}$, where F_T is the tension in the spring and μ is the linear mass density (mass/length).

○ The amplitude of the waves tells how much energy is present, not how fast the wave will travel. A traveling wave may be mathematically described by

$$y = A\sin(\omega t \pm kx) \qquad \text{or} \quad y = A\sin 2\pi\left(ft \pm \frac{x}{\lambda}\right) \qquad \text{or} \quad y = A\sin\frac{2\pi}{\lambda}(vt \pm x)$$

○ Superposition is when parts of waves interact so that they constructively or destructively interfere (e.g., create larger or smaller amplitudes, respectively).

○ Standing waves on a string that is fixed at both ends can form with wavelengths of $\lambda_n = \dfrac{2L}{n}$ and frequencies of $f_n = \dfrac{nv}{2L}$, where L is the length of the string, v is the speed of the wave, and n is a whole positive number.

○ If one end of the string is free to move, standing waves can form with wavelengths of $\lambda_n = \dfrac{4L}{n}$ and frequencies of $f_n = \dfrac{nv}{4L}$, where L is the length of the string, v is the speed of the wave, and n must be a positive odd integer.

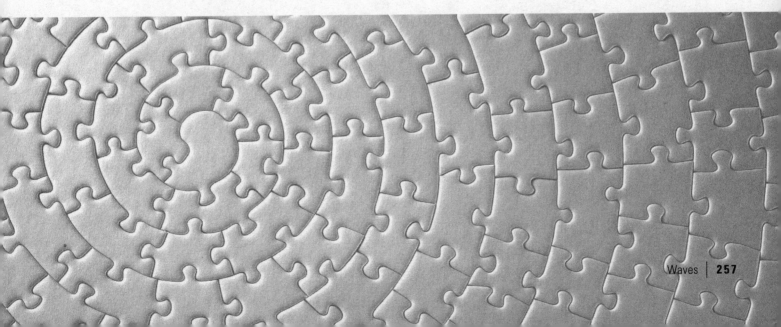

- Sound is a longitudinal wave. Its speed is given by $v = \sqrt{\dfrac{B}{\rho}}$, where B is the bulk modulus (this indicates how the medium responds to compression) and the density of air. At room temperature and at normal atmospheric pressure, the speed of sound is approximately 343 m/s.

- Two waves interfering at slightly different frequencies will form a beat frequency given by $f_{beat} = |f_1 - f_2|$.

- Resonance of sound in a tube depends on whether the tube is closed on one end or open on both ends. For a closed-ended tube resonance occurs for $f_n = n\dfrac{v}{4L}$ and $\lambda = \dfrac{4L}{n}$, where n must be an odd integer. For an open-ended tube, resonance occurs for $f_n = n\dfrac{v}{2L}$ and $\lambda = \dfrac{2L}{n}$, where n can be *any* integer.

- The Doppler effect tells you how the frequency of the sounds will vary if there is a relative motion between the source of the sound and the detector (observer). If the relative motion of the objects is toward each other, the frequency will be perceived as higher. If the relative motion is away from each other, the frequency will be perceived as lower. This idea is summed up in the equation $f_D = \dfrac{v \pm v_D}{v \mp v_S} \cdot f_S$.

Chapter 11
Electric Forces and Fields

"Invention is the most important product of man's creative brain. The ultimate purpose is the complete mastery of mind over the material world, the harnessing of human nature to human needs."

—Nikola Tesla

Long before the knowledge of electricity, many people were aware of the electric force. Descriptions of electricity were reported by Egyptians, Greeks, and Romans. Electricity would remain a mystery until 1600, when English scientist William Gilbert made a careful study of electricity and magnetism. Further work was continued by other notable scientists including Ben Franklin (his famous metal key and kite experiment in a storm). Real progress came in the nineteenth century, with regard to electricity and magnetism, from Hans Christian Orsted, Andre-Marie Ampere, Michael Faraday, and George Ohm. James Clerk Maxwell would definitively link them together into the concept electromagnetism.

ELECTRIC CHARGE

The basic components of atoms are protons, neutrons, and electrons. Protons and neutrons form the nucleus (and are referred to collectively as *nucleons*), while the electrons keep their distance, swarming around the nucleus. Most of an atom consists of empty space. In fact, if a nucleus were the size of the period at the end of this sentence, then the electrons would be five meters away. So what holds such an apparently tenuous structure together? One of the most powerful forces in nature: the *electromagnetic force*. Protons and electrons have a quality called **electric charge** that gives them an attractive force. Electric charge comes in two varieties: positive and negative. A positive particle always attracts a negative particle, and particles of the same charge always repel each other. Protons are positively charged, and electrons are negatively charged.

A Note About Electricity Transfer

The transfer of electricity is done by electrons (negative) not protons (positive). This is because the electrons are located in the outer shells of an atom, whereas, the protons and neutrons are tightly tucked away in the nucleus. Any transfer of electric charge is either a loss or gain of electrons.

Protons and electrons are intrinsically charged, but bulk matter is not. This is because the amount of charge on a proton exactly balances the charge on an electron, which is quite remarkable in light of the fact that protons and electrons are very different particles. Since most atoms contain an equal number of protons and electrons, their overall electric charge is 0, because the negative charges cancel out the positive charges. Therefore, in order for matter to be **charged**, an imbalance between the numbers of protons and electrons must exist. This can be accomplished by either the removal or addition of electrons (that is, by the **ionization** of some of the object's atoms). If you remove electrons, then the object becomes positively charged, while if you add electrons, then it becomes negatively charged. Furthermore, charge is **conserved**. For example, if you rub a glass rod with a piece of silk, then the silk will acquire a negative charge and the glass will be left with an *equal* positive charge. *Net charge cannot be created or destroyed.* (*Charge* can be created or destroyed—it happens all the time—but *net* charge cannot.)

The magnitude of charge on an electron (and therefore on a proton) is denoted e. This stands for **elementary charge** because it's the basic unit of electric charge. The charge of an ionized atom must be a whole number times e because charge can be added or subtracted only in lumps of size e. For this reason, we say that charge is **quantized**. To remind us of the quantized nature of electric charge, the charge of a particle (or object) is denoted by the letter q. In the SI system of units, charge is expressed in **coulombs** (abbreviated **C**). One coulomb is a tremendous amount of charge; the value of e is about 1.6×10^{-19} C.

> Charge is Quantized
>
> $$q = n(\pm e)$$
>
> where $n = 0, 1, 2...$

COULOMB'S LAW

The electric force between two charged particles obeys a law that is very similar to that describing the gravitational force between two masses; they are both inverse-square laws. The **electric force** between two particles with charges of q_1 and q_2, separated by a distance r, is given by the equation

$$F_E = k \frac{q_1 q_2}{r^2}$$

This is **Coulomb's Law.** We interpret a negative F_E as an attraction between the charges and a positive F_E as a repulsion. The value of the proportionality constant, k, depends on the material between the charged particles. In empty space (vacuum)—or air, for all practical purposes—it is called **Coulomb's constant** and has the approximate value $k_0 = 9 \times 10^9$ N $\cdot$ m^2/C^2. For reasons that will become clear later in this chapter, k_0 is usually written in terms of a fundamental constant known as the **permittivity of free space**, denoted ε_0, whose numerical value is approximately 8.85×10^{-12} C^2/(N·m^2). The equation that gives k_0 in terms of ε_0 is:

$$k_0 = \frac{1}{4\pi\varepsilon_0}$$

Coulomb's Law for the force between two point charges is then written as

$$F_E = \frac{1}{4\pi\varepsilon_0} \frac{q_1 q_2}{r^2}$$

Recall that the value of the universal gravitational constant, G, is 6.67×10^{-11} N $\cdot$ m^2/kg^2. The relative sizes of these fundamental constants show the relative strengths of the electric and gravitational forces. The value of k_0 is twenty orders of magnitude larger than G.

> **Example 1** Consider two small spheres, one carrying a charge of +1.5 nC and the other a charge of –2.0 nC, separated by a distance of 1.5 cm. Find the electric force between them. ("n" is the abbreviation for "nano," which means 10^{-9}.)

Solution. The electric force between the spheres is given by Coulomb's Law:

$$F_E = \frac{1}{4\pi\varepsilon_0} \frac{q_1 q_2}{r^2} = (9 \times 10^9 \text{ N} \cdot \text{m}^2/\text{C}^2) \frac{(1.5 \times 10^{-9} \text{ C})(-2.0 \times 10^{-9} \text{ C})}{(1.5 \times 10^{-2} \text{ m})^2} = -1.2 \times 10^{-4} \text{ N}$$

Electric Force = Strong Force
Electric force is a strong force in comparison with gravitational force. Electric force is what keeps matter together. If you were to compare the electric force (using their charges) between an electron and proton with the gravitational force (using their masses), the electric force is about 135,000 times stronger than the gravitational force!

The fact that F_E is negative means that the force is one of *attraction*, which we naturally expect, since one charge is positive and the other is negative. The force between the spheres is along the line that joins the charges, as we've illustrated below. The two forces shown form an action/reaction pair.

$$q_1 \oplus \xrightarrow{\ \mathbf{F}_E\ } \qquad \xleftarrow{\ \mathbf{F}_E\ } \ominus q_2$$

Superposition

Consider three point charges: q_1, q_2, and q_3. The total electric force acting on, say, q_2 is simply the sum of $\mathbf{F}_{1\text{-on-}2}$, the electric force on q_2 due to q_1, and $\mathbf{F}_{3\text{-on-}2}$, the electric force on q_2 due to q_3:

$$\mathbf{F}_{\text{on }2} = \mathbf{F}_{1\text{-on-}2} + \mathbf{F}_{3\text{-on-}2}$$

The fact that electric forces can be added in this way is known as **superposition**.

This is also true for more than two charges, for n charges:

$$\mathbf{F}_{\text{on }2} = \mathbf{F}_{1\text{-on-}2} + \mathbf{F}_{3\text{-on-}2} + \mathbf{F}_{4\text{-on-}2} + \dots + \mathbf{F}_{n\text{-on-}2}$$

Example 2 Consider four equal, positive point charges that are situated at the vertices of a square. Find the net electric force on a negative point charge placed at the square's center.

Solution. Refer to the diagram below. The attractive forces due to the two charges on each diagonal cancel out: $\mathbf{F}_1 + \mathbf{F}_3 = \mathbf{0}$, and $\mathbf{F}_2 + \mathbf{F}_4 = \mathbf{0}$, because the distances between the negative charge and the positive charges are all the same and the positive charges are all equivalent. Therefore, by symmetry, the net force on the center charge is zero.

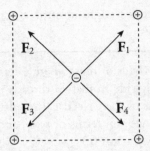

Example 3 If the two positive charges on the bottom side of the square in the previous example were removed, what would be the net electric force on the negative charge? Assume that each side of the square is 4.0 cm, each positive charge is 1.5 μC, and the negative charge is −6.2 nC. (μ is the symbol for "micro-," which equals 10^{-6}.)

Solution. If we break down $\mathbf{F}_1$ and $\mathbf{F}_2$ into horizontal and vertical components, then by symmetry the two horizontal components will cancel each other out, and the two vertical components will add:

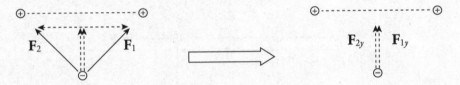

Since the diagram on the left shows the components of $\mathbf{F}_1$ and $\mathbf{F}_2$ making right triangles with legs each of length 2 cm, it must be that $F_{1y} = F_1 \sin 45°$ and $F_{2y} = F_2 \sin 45°$. Also, the magnitude of $\mathbf{F}_1$ equals that of $\mathbf{F}_2$. So the net electric force on the negative charge is $F_{1y} + F_{2y} = 2F \sin 45°$, where F is the strength of the force between the negative charge and each of the positive charges. If s is the length of each side of the square, then the distance r between each positive charge and the negative charge is $r = \frac{1}{2}s\sqrt{2}$ and

$$F_E = 2F \sin 45° = 2\frac{1}{4\pi\varepsilon_0}\frac{q_1 q_2}{r^2}\sin 45°$$

$$= 2(9\times10^9 \text{ N}\cdot\text{m}^2/\text{C}^2)\frac{(1.5\times10^{-6}\text{ C})(6.2\times10^{-9}\text{ C})}{(\frac{1}{2}\cdot 4.0\times10^{-2}\cdot\sqrt{2}\text{ m})^2}\sin 45°$$

$$= 0.15 \text{ N}$$

The direction of the net force is straight upward, toward the center of the line that joins the two positive charges.

> **Example 4** Two pith balls of mass m are each given a charge of $+q$. They are hung side-by-side from two threads each of length L, and move apart as a result of their electrical repulsion. Find the equilibrium separation distance x in terms of m, q, and L. (Use the fact that if θ is small, then $\tan \theta \approx \sin \theta$.)

Solution. Three forces act on each ball: weight, tension, and electrical repulsion.

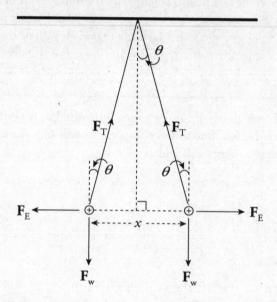

When the balls are in equilibrium, the net force each feels is zero. Therefore, the vertical component of $\mathbf{F}_T$ must cancel out $\mathbf{F}_w$ and the horizontal component of $\mathbf{F}_T$ must cancel out $\mathbf{F}_E$:

$$F_T \cos \theta = F_w \qquad \text{and} \qquad F_T \sin \theta = F_E$$

Dividing the second equation by the first, we get $\tan \theta = F_E / F_w$. Therefore,

$$\tan \theta = \frac{k \dfrac{q^2}{x^2}}{mg} = \frac{kq^2}{mgx^2}$$

Now, to approximate: If θ is small, then $\tan \theta \approx \sin \theta$ and, from the diagram, $\sin \theta = \frac{1}{2}x/L$. Therefore, the equation above becomes

$$\frac{\frac{1}{2}x}{L} = \frac{kq^2}{mgx^2} \quad \Rightarrow \quad \frac{1}{2}mgx^3 = kq^2 L \quad \Rightarrow \quad x = \sqrt[3]{\frac{2kq^2 L}{mg}}$$

THE ELECTRIC FIELD

The presence of a massive body such as the Earth causes objects to experience a gravitational force directed toward the Earth's center. For objects located outside the Earth, this force varies inversely with the square of the distance and directly with the mass of the gravitational source. A vector diagram of the gravitational field surrounding the Earth looks like this:

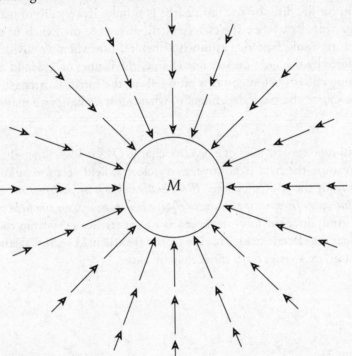

We can think of the space surrounding the Earth as permeated by a **gravitational field** that's created by the Earth. Any mass that's placed in this field then experiences a gravitational force due to this field.

The same process is used to describe the electric force. Rather than having two charges reach out across empty space to each other to produce a force, we can instead interpret the interaction in the following way: The presence of a charge creates an **electric field** in the space that surrounds it. Another charge placed in the field created by the first will experience a force due to the field.

Consider a point charge Q in a fixed position and assume that it is positive. Now imagine moving a tiny positive test charge q around to various locations near Q. At each location, measure the force that the test charge experiences, and call it $F_{on\ q}$. Divide this force by the test charge q; the resulting vector is the **electric field vector**, E, at that location:

$$E = \frac{\mathbf{F}_{\mathrm{on}\,q}}{q}$$

The reason for dividing by the test charge is simple. If we were to use a different test charge with, say, twice the charge of the first one, then each of the forces **F** we'd measure would be twice as much as before. But when we divided this new, stronger force by the new, greater test charge, the factors of 2 would cancel, leaving the same ratio as before. So this ratio tells us the intrinsic strength of the field due to the source charge, independent of whatever test charge we may use to measure it.

What would the electric field of a positive charge Q look like? Since the test charge used to measure the field is positive, every electric field vector would point radially away from the source charge. *If the source charge is positive, the electric field vectors point away from it; if the source charge is negative, then the field vectors point toward it.* And, since the force decreases as we get farther away from the charge (as $1/r^2$), so does the electric field. This is why the electric field vectors farther from the source charge are shorter than those that are closer.

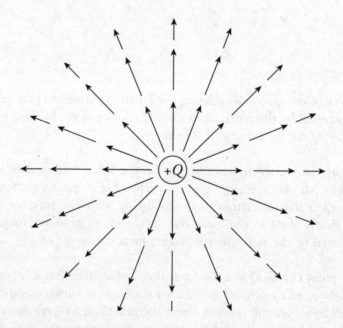

Since the force on the test charge q has a strength of $qQ/4\pi\varepsilon_0 r^2$, when we divide this by q, we get the expression for the strength of the electric field created by a point-charge source of magnitude Q:

$$E = \frac{1}{4\pi\varepsilon_0} \frac{Q}{r^2}$$

To make it easier to sketch an electric field, lines are drawn through the vectors such that the electric field vector is tangent to the line everywhere it's drawn.

Now, your first thought might be that obliterating the individual field vectors deprives us of information, since the length of the field vectors told us how strong the field was. Well, although the individual field vectors are gone, the strength of the field can be figured out by looking at the density of the field lines. Where the field lines are denser, the field is stronger.

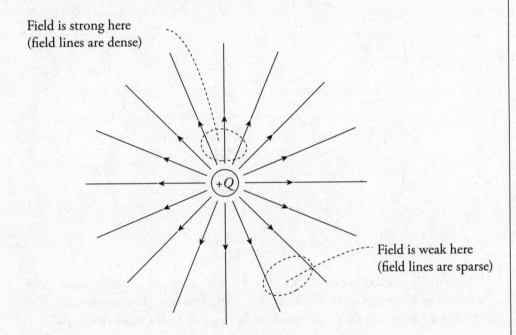

Field is strong here
(field lines are dense)

Field is weak here
(field lines are sparse)

Electric field vectors can be added like any other vectors. If we had two source charges, their fields would overlap and effectively add; a third charge wandering by would feel the effect of the combined field. At each position in space, add the electric field vector due to one of the charges to the electric field vector due to the other charge: $\mathbf{E}_{total} = \mathbf{E}_1 + \mathbf{E}_2$. (This is another instance of superposition.) In the diagram below, $\mathbf{E}_1$ is the electric field vector at a particular location due to the charge $+Q$, and $\mathbf{E}_2$ is the electric field vector at that same location due to the other charge, $-Q$. Adding these vectors gives the overall field vector $\mathbf{E}_{total}$ at that location.

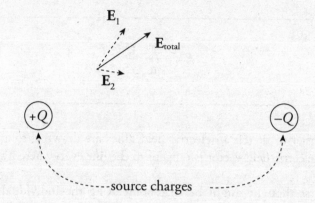

If this is done at enough locations, the electric field lines can be sketched.

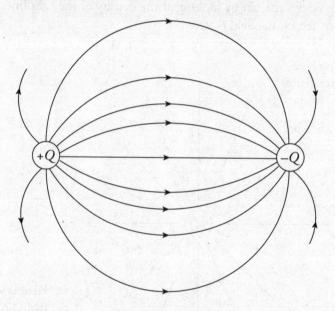

Note that, like electric field vectors, electric field lines always point away from positive source charges and toward negative ones. Two equal but opposite charges, like the ones shown in the diagram above, form a pair called an **electric dipole**.

If a positive charge $+q$ were placed in the electric field above, it would experience a force that is in the same direction as the field line passing through $+q$'s location. After all, electric fields are sketched from the point of view of what a positive test charge would do. On the other hand, if a negative charge $-q$ were placed in the electric field, it would experience a force that is in the direction opposite from the field line passing through $-q$'s location.

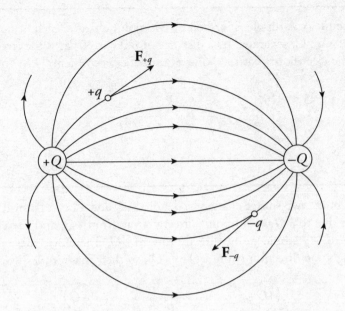

Finally, notice that electric field lines never cross.

The force generated on a charge that is in an electric field is given by the equation:

$$F = qE$$

Example 5 A charge $q = +3.0$ nC is placed at a location at which the electric field strength is 400 N/C. Find the force felt by the charge q.

Solution. From the definition of the electric field, we have the following equation:

$$\mathbf{F}_{\text{on } q} = q\mathbf{E}$$

Therefore, in this case, $F_{\text{on } q} = qE = (3 \times 10^{-9}\ \text{C})(400\ \text{N/C}) = 1.2 \times 10^{-6}\ \text{N}$.

Example 6 A dipole is formed by two point charges, each of magnitude 4.0 nC, separated by a distance of 6.0 cm. What is the strength of the electric field at the point midway between them?

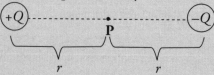

Solution. Let the two source charges be denoted $+Q$ and $-Q$. At Point P, the electric field vector due to $+Q$ would point directly away from $+Q$, and the electric field vector due to $-Q$ would point directly toward $-Q$. Therefore, these two vectors point in the same direction (from $+Q$ to $-Q$), so their magnitudes would add.

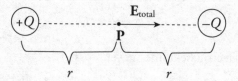

Using the equation for the electric field strength due to a single point charge, we find that

$$E_{total} = \frac{1}{4\pi\varepsilon_0}\frac{Q}{r^2} + \frac{1}{4\pi\varepsilon_0}\frac{Q}{r^2} = 2\frac{1}{4\pi\varepsilon_0}\frac{Q}{r^2}$$

$$= 2(9\times10^9 \text{ N}\cdot\text{m}^2/\text{C}^2)\frac{4.0\times10^{-9}\text{ C}}{\left[\frac{1}{2}(6.0\times10^{-2}\text{ m})\right]^2}$$

$$= 8.0\times10^4 \text{ N/C}$$

Example 7 If a charge $q = -5.0$ pC were placed at the midway point described in the previous example, describe the force it would feel. ("p" is the abbreviation for "pico-," which means 10^{-12}.)

Solution. Since the field E at this location is known, the force felt by q is easy to calculate:

$$\mathbf{F}_{on\,q} = q\mathbf{E} = (-5.0\times10^{-12}\text{ C})(8.0\times10^4\text{ N/C to the right}) = 4.0\times10^{-7}\text{ N to the } \textit{left}$$

Example 8 What can you say about the electric force that a charge would feel if it were placed at a location at which the electric field was zero?

Solution. Remember that $\mathbf{F}_{on\,q} = q\mathbf{E}$. So if $\mathbf{E} = \mathbf{0}$, then $\mathbf{F}_{on\,q} = \mathbf{0}$. (Zero field means zero force.)

An important subset of problems with electric fields deals with a uniform electric field. One method of creating this uniform field is to have two large conducting sheets, each storing charge, some distance apart. Near the edges of each sheet, the field may not be uniform, but near the middle, for all practical purposes the field is uniform. Having a uniform field means you can use kinematics equations and Newton's laws just as if you had a uniform gravitational field (as in previous chapters).

Example 9 Positive charge is distributed uniformly over a large, horizontal plate, which then acts as the source of a vertical electric field. An object of mass 5 g is placed at a distance of 2 cm above the plate. If the strength of the electric field at this location is 10^6 N/C, how much charge would the object need to have in order for the electrical repulsion to balance the gravitational pull?

Solution. Clearly, since the plate is positively charged, the object would also have to carry a positive charge so that the electric force would be repulsive.

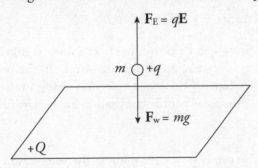

Let q be the charge on the object. Then, in order for F_E to balance mg, we must have

$$qE = mg \quad \Rightarrow \quad q = \frac{mg}{E} = \frac{(5 \times 10^{-3} \text{ kg})(10 \text{ N/kg})}{10^6 \text{ N/C}} = 5 \times 10^{-8} \text{ C} = 50 \text{ nC}$$

> **Example 10** A proton, neutron, and electron are in a uniform electric field of 20 N/C that is caused by two large charged plates that are 30 cm apart. The particles are far enough apart that they don't interact with each other. They are released from rest equidistant from each plate.
>
> (a) What is the magnitude of the net force acting on each particle?
>
> (b) What is the magnitude of the acceleration of each particle?
>
> (c) How much work will be done on the particle when it collides with one of the charged plates?
>
> (d) What is the speed of each particle when it strikes the plate?
>
> (e) How long does it take to reach the plate?

Solution.

(a) Since $E = \dfrac{F}{q}$, $F = qE$. Plugging in the values, we get

proton: $F = (1.6 \times 10^{-19}\,C)(20\,N/C) = 3.2 \times 10^{-18}\,N$

electron: $F = (1.6 \times 10^{-19}\,C)(20\,N/C) = 3.2 \times 10^{-18}\,N$

neutron: $F = (0\,C)(20\,N/C) = 0\,N$

Note: Because the proton and electron have the same magnitude, they will experience the same force. If you're asked for the direction, the proton travels in the same direction as the electric field and the electron travels in the opposite direction as the electric field. (This is as per the convention for electric fields.)

(b) Since $F = ma$, $a = \dfrac{F}{m}$. Plugging in the values, we get

proton: $a = \dfrac{3.2 \times 10^{-18}\,N}{1.67 \times 10^{-27}\,kg} = 1.9 \times 10^{-18}\,m/s^2$

electron: $a = \dfrac{3.2 \times 10^{-18}\,N}{9.11 \times 10^{-31}\,kg} = 3.5 \times 10^{12}\,m/s^2$

neutron: $a = \dfrac{0\,N}{1.67 \times 10^{-27}\,kg} = 0\,m/s^2$

Notice that although the charges have the same magnitude of force, the electron experiences an acceleration almost 2,000 times greater due to its mass being almost 2,000 times smaller than the proton's mass.

(c) Since $W = Fd$, we get $W = qEd$. Plugging in the values, we get

$$\text{proton: } W = (1.6 \times 10^{-19}\,\text{C})(20\,\text{N/C})(0.15\,\text{m}) = 4.8 \times 10^{-19}\,\text{N}$$

$$\text{electron: } W = (1.6 \times 10^{-19}\,\text{C})(20\,\text{N/C})(0.15\,\text{m}) = 4.8 \times 10^{-19}\,\text{N}$$

$$\text{neutron: } W = (0\,\text{C})(20\,\text{N/C})(0.15\,\text{m}) = 0\,\text{N}$$

(d) Recall one of the big five kinematics equations:

$$v_f^2 = v^2 + 2a(x - x_o) \rightarrow v_f = \sqrt{2a(x - x_o)}$$

If the particles are midway between the 30 cm plates, they will travel 0.15 m.

$$\text{proton: } v_f = \sqrt{2(1.9 \times 10^9\,\text{m/s}^2)(0.15\,\text{m})} \rightarrow v_f = 24{,}000\,\text{m/s}$$

$$\text{electron: } v_f = \sqrt{2(3.5 \times 10^{12}\,\text{m/s}^2)(0.15\,\text{m})} \rightarrow v_f = 1.0 \times 10^6\,\text{m/s}$$

$$\text{neutron: } v_f = \sqrt{2(0\,\text{m/s}^2)(0.15\,\text{m})} \rightarrow v_f = 0\,\text{m/s}$$

Notice that, even though the force is the same and the same work is done on both charges, there is a significant difference in final velocities due to the large mass difference. An alternative solution to this would be using $W = \Delta KE \rightarrow W = \frac{1}{2}mv^2$. You would have obtained the same answers.

(e) Recall one of the big five kinematics equations:

$$v_f = v_i + at \rightarrow t = \frac{v_f - v_i}{a} \rightarrow t = \frac{v_f}{a}$$

$$\text{proton: } t = \frac{24{,}000\,\text{m/s}}{1.9 \times 10^9\,\text{m/s}^2} \rightarrow 1.3 \times 10^{-5}\,\text{s}$$

$$\text{electron: } t = \frac{1 \times 10^6\,\text{m/s}}{-3.5 \times 10^{12}\,\text{m/s}^2} \rightarrow 2.9 \times 10^{-7}\,\text{s}$$

neutron: The neutron never accelerates, so it will never hit the plate

Chapter 11 Review Questions

Answers and Explanations can be found in Chapter 13.

Section I: Multiple-Choice

1. If the distance between two positive point charges is tripled, then the strength of the electrostatic repulsion between them will decrease by a factor of

   (A) 3
   (B) 6
   (C) 8
   (D) 9

2. Two 1 kg spheres each carry a charge of magnitude 1 C. How does F_E, the strength of the electric force between the spheres, compare to F_G, the strength of their gravitational attraction?

   (A) $F_E < F_G$
   (B) $F_E = F_G$
   (C) $F_E > F_G$
   (D) If the charges on the spheres are of the same sign, then $F_E > F_G$; but if the charges on the spheres are of the opposite sign, then $F_E < F_G$.

3. The figure below shows three point charges, all positive. If the net electric force on the center charge is zero, what is the value of y/x ?

   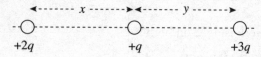

   $+2q$ $+q$ $+3q$

   (A) $\dfrac{4}{9}$

   (B) $\sqrt{\dfrac{2}{3}}$

   (C) $\sqrt{\dfrac{3}{2}}$

   (D) $\dfrac{3}{2}$

4.

   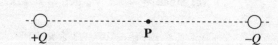

   $+Q$ P $-Q$

   The figure above shows two point charges, $+Q$ and $-Q$. If the negative charge were absent, the electric field at Point P due to $+Q$ would have strength E. With $-Q$ in place, what is the strength of the total electric field at P, which lies at the midpoint of the line segment joining the charges?

   (A) 0

   (B) $\dfrac{E}{2}$

   (C) E

   (D) $2E$

5. A sphere of charge $+Q$ is fixed in position. A smaller sphere of charge $+q$ is placed near the larger sphere and released from rest. The small sphere will move away from the large sphere with

   (A) decreasing velocity and decreasing acceleration
   (B) decreasing velocity and increasing acceleration
   (C) increasing velocity and decreasing acceleration
   (D) increasing velocity and increasing acceleration

6. An object of charge $+q$ feels an electric force $\mathbf{F}_E$ when placed at a particular location in an electric field, $\mathbf{E}$. Therefore, if an object of charge $-2q$ were placed at the same location where the first charge was, it would feel an electric force of

   (A) $\dfrac{-\mathbf{F}_E}{2}$

   (B) $-2\mathbf{F}_E$

   (C) $-2q\mathbf{F}_E$

   (D) $\dfrac{-2\mathbf{F}_E}{q}$

7. A charge of $-3Q$ is transferred to a solid metal sphere of radius r. Where will this excess charge reside?

(A) $-Q$ at the center, and $-2Q$ on the outer surface

(B) $-3Q$ at the center

(C) $-3Q$ on the outer surface

(D) $-Q$ at the center, $-Q$ in a ring of radius $\frac{1}{2} r$, and $-Q$ on the outer surface

8. How far apart are two charges ($q_1 = 8 \times 10^{-6}$ C and $q_2 = 6 \times 10^{-6}$ μC) if the electric force exerted by the charges on each other has a magnitude of 2.7×10^{-2} N ?

(A) 1 m

(B) 2 m

(C) 3 m

(D) 4 m

9. Two charges (q_1 and q_2) are separated by a distance r. If the ratio of F_G/F_E is equal to 9.0×10^{43}, what is the new ratio if the distance between the two charges is now $3r$?

(A) 1.0×10^{43}

(B) 3.0×10^{43}

(C) 9.0×10^{43}

(D) 27.0×10^{43}

Section II: Free-Response

1. In the figure shown, all four charges (+Q, +Q, −q, and −q) are situated at the corners of a square. The net electric force on each charge +Q is zero.

   (a) Express the magnitude of q in terms of Q.

   (b) Is the net electric force on each charge −q also equal to zero? Justify your answer.

   (c) Determine the electric field at the center of the square.

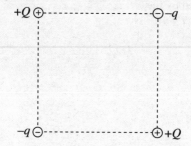

2. Two charges, +Q and +2Q, are fixed in place along the y-axis of an x-y coordinate system as shown in the figure below. Charge 1 is at the point (0, a), and Charge 2 is at the point (0, −2a).

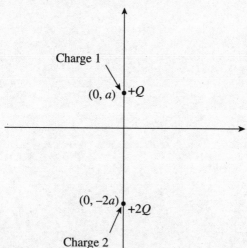

   (a) Find the electric force (magnitude and direction) felt by Charge 1 due to Charge 2.

   (b) Find the electric field (magnitude and direction) at the origin created by both Charges 1 and 2.

   (c) Is there a point on the x-axis where the total electric field is zero? If so, where? If not, explain briefly.

   (d) Is there a point on the y-axis where the total electric field is zero? If so, where? If not, explain briefly.

   (e) If a small negative charge, −q, of mass m were placed at the origin, determine its initial acceleration (magnitude and direction).

3. Two point charges ($q_1 = +4\ \mu C$ and $q_2 = +3\ \mu C$) are fixed in place according to the figure below.

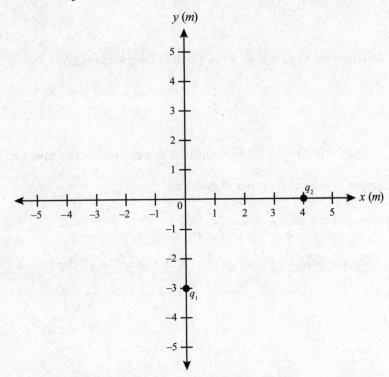

(a) What are the magnitudes of the horizontal and vertical components of the electric field at the point (0, 4)?

(b) What are the magnitudes of the horizontal and vertical components of the electric force on a charge $q_3 = -2\ \mu C$ placed at the point (0, 4)?

(c) In which quadrant would a positive charge q_4 have to be placed in order for the electric field at point (0, 4) to be equal to 0?

Summary

- Coulomb's Law describes the force acting on two point charges and is given by

$$F_e = \frac{1}{4\pi\varepsilon_0} \frac{q_1 q_2}{r^2}$$

  where $\dfrac{1}{4\pi\varepsilon_0}$ is a constant equal to $9.0 \times 10^9 \text{ N} \cdot \text{m}^2/\text{C}^2$. You can use all the strategies you used in the Newton's Laws chapter to solve these types of problems.

- The electric field is given by $E = \dfrac{F}{q}$ or $E = \dfrac{1}{4\pi\varepsilon_0} \dfrac{q}{r^2}$.

- Both the electric force and field are vector quantities and therefore all the rules for vector addition apply.

Chapter 12
Direct Current
Circuits

Today, electricity is well known and appears all around us from the lighting in our houses to the computing in our personal computers. All of these are powered by complex circuits. In this chapter, we will study the basics of simple direct current circuits.

ELECTRIC CURRENT

Picture a piece of metal wire. Within the metal, electrons are zooming around at speeds of about a million meters per second in random directions, colliding with other electrons and positive ions in the lattice. This constitutes charge in motion, but it doesn't constitute *net* movement of charge because the electrons move randomly. If there's no net movement of charge, there's no current. However, if we were to create a potential difference between the ends of the wire, meaning if we set up an electric field, the electrons would experience an electric force, and they would start to drift through the wire. This is current. Although the electric field would travel through the wire at nearly the speed of light, the electrons themselves would still have to make their way through a crowd of atoms and other free electrons, so their **drift speed**, v_d, would be relatively slow, on the order of a millimeter per second.

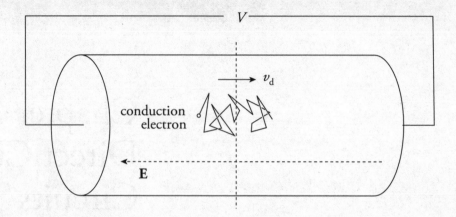

To measure the current, we have to measure how much charge crosses a plane per unit time. If an amount of charge of magnitude ΔQ crosses an imaginary plane in a time interval Δt, then the **average current** is

$$I_{avg} = \frac{\Delta Q}{\Delta t}$$

Because current is charge per unit time, it's expressed in coulombs per second. One coulomb per second is an **ampere** (abbreviated **A**), or amp. So 1 C/s = 1 A.

Although the charge carriers that constitute the current within a metal are electrons, the direction of the current is taken to be the direction that *positive* charge carriers would move. (This is explicitly stated on the AP Physics 1 Exam.) So, if the conduction electrons drift to the right, we'd say the current points toward the left.

RESISTANCE

Let's say we had a copper wire and a glass fiber that had the same length and cross-sectional area, and that we hooked up the ends of the metal wire to a source of potential difference and measured the resulting current. If we were to do the same thing with the glass fiber, the current would probably be too small to measure, but why? Well, the glass provided more resistance to the flow of charge. If the potential difference is V and the current is I, then the **resistance** is

$$R = \frac{V}{I} \quad \text{or} \quad V = IR$$

This is known as Ohm's Law. Notice for the same voltage if the current is large, the resistance is low, and if the current is small, then resistance is high. The Δ in the equation above is often omitted, but you should always assume that in this context, $V = \Delta V$ = potential difference, also called voltage.

Because resistance is voltage divided by current, it is expressed in volts per amp. One volt per amp is one **ohm** (Ω, *omega*). So, 1 V/A = 1 Ω.

The River: Resistance
In terms of the river, we can think of resistance as a portion of the river that zigzags (increases the length). This provides resistance to the flow of water.

VOLTAGE

Now that we know how to measure current, the next question is, what causes it? Why does an electron drift through the circuit? One answer is to say that there's an electric field inside the wire, and since negative charges move in the direction opposite to the electric field lines, electrons would drift opposite the electric field.

Another (equivalent) answer to the question is that there's a potential difference (a voltage) between the ends of the wire. Positive charges naturally move from regions of higher potential to lower potential.

> Voltage is what creates a current.

The River: Voltage
In terms of the river, a river will not flow if it is flat. A river will flow from higher ground to lower ground. We can think of voltage as the mountain that gives the river height to flow down.

It is not uncommon to see the voltage that creates a current referred to as electromotive force (emf), since it is the cause that sets the charges into motion into a preferred direction. The voltage provided in a circuit is generally supplied by the battery.

ELECTRIC CIRCUITS

An electric current is maintained when the terminals of a voltage source (a battery, for example) are connected by a conducting pathway, in what's called a **circuit**. If the current always travels in the same direction through the pathway, it's called a **direct current**.

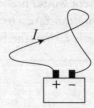

The job of the voltage source is to provide a potential difference called an **emf**, or **electromotive force**, which drives the flow of charge. The emf isn't really a force; it's the work done per unit charge, and it's measured in volts.

To try to imagine what's happening in a circuit in which a steady-state current is maintained, let's follow one of the charge carriers that's drifting through the pathway. (Remember we're pretending that the charge carriers are positive.) The charge is introduced by the positive terminal of the battery and enters the wire, where it's pushed by the electric field. It encounters resistance, bumping into the relatively stationary atoms that make up the metal's lattice and setting them into greater motion. So the electrical potential energy that the charge had when it left the battery is turning into heat. By the time the charge reaches the negative terminal, all of its original electrical potential energy is lost. In order to keep the current going, the voltage source must do positive work on the charge, forcing it to move from the negative terminal toward the positive terminal. The charge is now ready to make another journey around the circuit.

Energy and Power

When a carrier of positive charge q drops by an amount V in potential, it loses potential energy in the amount qV. If this happens in time t, then the rate at which this energy is transformed is equal to $(qV)/t = (q/t)V$. But q/t is equal to the current, I, so the rate at which electrical energy is transferred is given by the equation

$$P = IV$$

This equation works for the power delivered by a battery to the circuit as well as for resistors. The power dissipated in a resistor, as electrical potential energy is turned into heat, is given by $P = IV$, but, because of the relationship $V = IR$, we can express this in two other ways:

$$P = IV = I(IR) = I^2 R$$

or

$$P = IV = \frac{V}{R} \cdot V = \frac{V^2}{R}$$

Resistors become hot when current passes through them.

CIRCUIT ANALYSIS

We will now develop a way of specifying the current, voltage, and power associated with each element in a circuit. Our circuits will contain three basic elements: batteries, resistors, and connecting wires. As we've seen, the resistance of an ordinary metal wire is negligible; resistance is provided by devices that control the current: **resistors**. All the resistance of the system is concentrated in the resistors, which are symbolized in a circuit diagram by this symbol:

Batteries are denoted by the symbol:

where the longer line represents the **positive** (higher potential) terminal, and the shorter line is the **negative** (lower potential) terminal. Sometimes a battery is denoted by more than one pair of such lines:

Here's a simple circuit diagram:

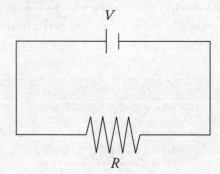

The electric potential (V) of the battery is indicated, as is the resistance (R) of the resistor. Determining the current in this case is straightforward because there's only one resistor. The equation $V = IR$ gives us

$$I = \frac{V}{R}$$

Combinations of Resistors

Two common ways of combining resistors within a circuit is to place them either in **series** (one after the other),

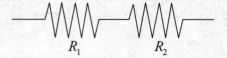

or in **parallel** (that is, side-by-side):

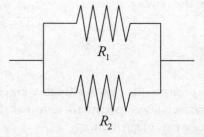

In order to simplify the circuit, our goal is to find the equivalent resistance of combinations. Resistors are said to be in series if they all share the same current and if the total voltage drop across them is equal to the sum of the individual voltage drops.

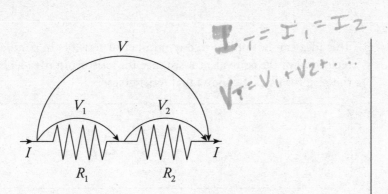

$I_T = I_1 = I_2$

$V_T = V_1 + V_2 + \ldots$

In this case, then, if V denotes the voltage drop across the combination, we have

$$R_{\text{equiv}} = \frac{V}{I} = \frac{V_1 + V_2}{I} = \frac{V_1}{I} + \frac{V_2}{I} = R_1 + R_2$$

This idea can be applied to any number of resistors in series (not just two):

$$R_S = \sum_i R_i \qquad R_T = R_1 + R_2 + R_3 \ldots$$

Resistors are said to be in parallel if they all share the same voltage drop, and the total current entering the combination is split among the resistors. Imagine that a current I enters the combination. It splits; some of the current, I_1, would go through R_1, and the remainder, I_2, would go through R_2.

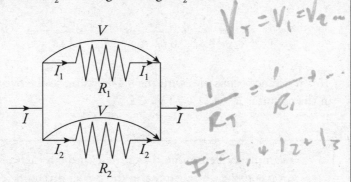

$V_T = V_1 = V_2 \ldots$

$\frac{1}{R_T} = \frac{1}{R_1} + \ldots$

$I = I_1 + I_2 + I_3$

So if V is the voltage drop across the combination, we have

$$I = I_1 + I_2 \quad \Rightarrow \quad \frac{V}{R_{\text{equiv}}} = \frac{V}{R_1} + \frac{V}{R_2} \quad \Rightarrow \quad \frac{1}{R_{\text{equiv}}} = \frac{1}{R_1} + \frac{1}{R_2}$$

Series Circuits and Parallel Circuits

In a series circuit, the current is the same through each resistor. In a parallel circuit, the voltage drop is the same.

Which Would You Rather Travel?

In a set of parallel resistors, current tends to travel through the path of least resistance. If two resistors are in parallel and the first resistor has the greater resistance, the second resistor will have more of the current running through it.

This idea can be applied to any number of resistors in parallel (not just two): The reciprocal of the equivalent resistance for resistors in parallel is equal to the sum of the reciprocals of the individual resistances:

$$\frac{1}{R_P} = \sum_i \frac{1}{R_i}$$

Example 1 Calculate the equivalent resistance for the following circuit:

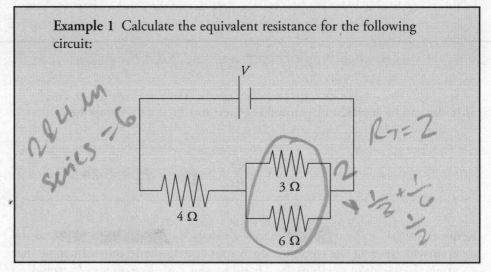

Solution. First find the equivalent resistance of the two parallel resistors:

$$\frac{1}{R_P} = \frac{1}{3\ \Omega} + \frac{1}{6\ \Omega} \quad \Rightarrow \quad \frac{1}{R_P} = \frac{1}{2\ \Omega} \quad \Rightarrow \quad R_P = 2\ \Omega$$

This resistance is in series with the 4 Ω resistor, so the overall equivalent resistance in the circuit is $R = 4\ \Omega + 2\ \Omega = 6\ \Omega$.

Example 2 Determine the current through each resistor, the voltage drop across each resistor, and the power given off (dissipated) as heat in each resistor of this circuit:

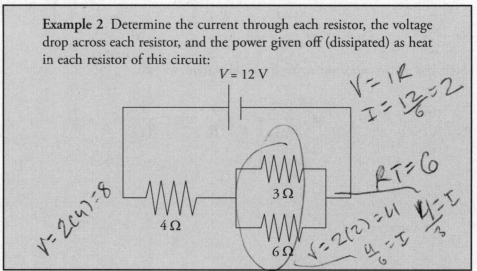

Solution. You might want to redraw the circuit each time we replace a combination of resistors by its equivalent resistance. From our work in the preceding example, we have

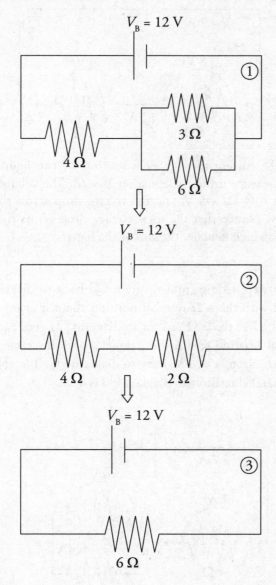

$V_B = 12$ V

$V_B = 12$ V

$V_B = 12$ V

Make It Simple, Then Work Backward

The typical method to working a circuit problem is to make the circuit provided into a simple circuit with a battery and an equivalent resistor. Then to solve for individual values on each resistor, we work backward and build the circuit back to its original form.

In the diagram to the left we moved backward and made the circuit simple from diagram 1 to 2 to 3. Then to solve for the resistors we go from 3 to 2 to 1.

From diagram ③, which has just one resistor, we can figure out the current:

$$I = \frac{V_B}{R} = \frac{12 \text{ V}}{6 \text{ }\Omega} = 2 \text{ A}$$

Now we can work our way back to the original circuit (diagram ①). In going from ③ to ②, we are going back to a series combination, and what do resistors in series share? That's right, the same current. So, we take the current, $I = 2$ A, back to diagram ②. The current through each resistor in diagram ② is 2 A.

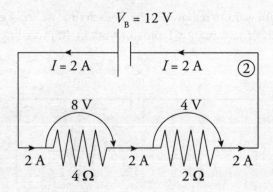

Since we know the current through each resistor, we can figure out the voltage drop across each resistor using the equation $V = IR$. The voltage drop across the 4 Ω resistor is (2 A)(4 Ω) = 8 V, and the voltage drop across the 2 Ω resistor is (2 A)(2 Ω) = 4 V. Notice that the total voltage drop across the two resistors is 8 V + 4 V = 12 V, which matches the emf of the battery.

Now for the last step! Going from diagram ② back to diagram ①. Nothing needs to be done with the 4 Ω resistor; nothing about it changes in going from diagram ② to ①, but the 2 Ω resistor in diagram ② goes back to the parallel combination. And what do resistors in parallel share? The same voltage drop. So we take the voltage drop, $V = 4$ V, back to diagram ①. The voltage drop across each of the two parallel resistors in diagram ① is 4 V.

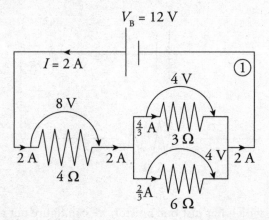

Since we know the voltage drop across each resistor, we can figure out the current through each resistor by using the equation $I = V/R$. The current through the 3 Ω resistor is (4 V)/(3 Ω) = $\frac{4}{3}$ A, and the current through the 6 Ω resistor is (4 V)/ (6 Ω) = $\frac{2}{3}$ A. Note that the current entering the parallel combination (2 A) equals the total current passing through the individual resistors $\left(\frac{4}{3} \text{ A} + \frac{2}{3} \text{ A} \right)$. Again, this was expected.

Finally, we will calculate the power dissipated as heat by each resistor. We can use any of the equivalent formulas: $P = IV$, $P = I^2R$, or $P = V^2/R$.

$$\text{For the 4 } \Omega \text{ resistor: } P = IV = (2 \text{ A})(8 \text{ V}) = 16 \text{ W}$$

$$\text{For the 3 } \Omega \text{ resistor: } P = IV = (\tfrac{4}{3}\text{A})(4 \text{ V}) = \tfrac{16}{3}\text{W}$$

$$\text{For the 6 } \Omega \text{ resistor: } P = IV = (\tfrac{2}{3}\text{A})(4 \text{ V}) = \tfrac{8}{3}\text{W}$$

So, the resistors are dissipating a total of

$$16 \text{ W} + \frac{16}{3}\text{ W} + \frac{8}{3}\text{ W} = 24 \text{ W}$$

If the resistors are dissipating a total of 24 J every second, then they must be provided with that much power. This is easy to check: $P = IV = (2 \text{ A})(12 \text{ V}) = 24 \text{ W}$.

Example 3 For the following circuit
  (a)   In which direction will current flow and why?
  (b)   What's the overall emf?
  (c)   What's the current in the circuit?
  (d)   At what rate is energy consumed by, and provided to, this circuit?

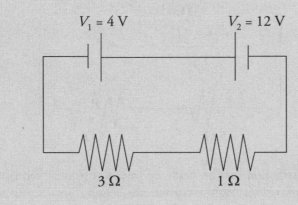

$V_1 = 4 \text{ V}$ $V_2 = 12 \text{ V}$

3Ω 1Ω

Solution.

(a)   The battery V_1 wants to send current clockwise, while the battery V_2 wants to send current counterclockwise. Since $V_2 > V_1$, the battery whose emf is V_2 is the more powerful battery, so the current will flow counterclockwise.

(b)   Charges forced through V_1 will lose, rather than gain, 4 V of potential, so the overall emf of this circuit is $V_2 - V_1 = 8 \text{ V}$.

(c)   Since the total resistance is $3 \Omega + 1 \Omega = 4 \Omega$, the current will be $I = (8 \text{ V})/(4 \Omega) = 2 \text{ A}$.

(d) V_2 will provide energy at a rate of $P_2 = IV_2 = (2 \text{ A})(12 \text{ V}) = 24 \text{ W}$, while V_1 will absorb at a rate of $P_1 = IV_1 = (2 \text{ A})(4 \text{ V}) = 8 \text{ W}$. Finally, energy will be dissipated in these resistors at a rate of $I^2R_1 + I^2R_2 = (2 \text{ A})^2(3 \Omega) + (2 \text{ A})^2(1 \Omega) = 16 \text{ W}$. Once again, energy is conserved; the power delivered (24 W) equals the power taken (8 W + 16 W = 24 W).

Example 4 All real batteries contain **internal resistance, r.** Determine the current in the following circuit when the switch S is closed:

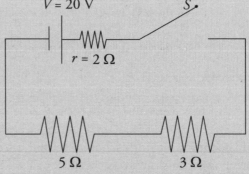

Solution. Before the switch is closed, there is no complete conducting pathway from the positive terminal of the battery to the negative terminal, so no current flows through the resistors. However, once the switch is closed, the resistance of the circuit is $2\ \Omega + 3\ \Omega + 5\ \Omega = 10\ \Omega$, so the current in the circuit is $I = (20\ \text{V})/(10\ \Omega) = 2\ \text{A}$. Often the battery and its internal resistance are enclosed in a dashed box:

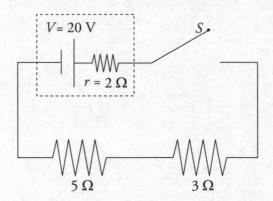

In this case, a distinction can be made between the emf of the battery and the actual voltage it provides once the current has begun. Since $I = 2$ A, the voltage drop across the internal resistance is $IR = (2\ \text{A})(2\ \Omega) = 4\ \text{V}$, so the effective voltage provided by the battery to the rest of the circuit—called the **terminal voltage**—is lower than the ideal voltage. It is $V = V_{\text{B}} - IR = 20\ \text{V} - 4\ \text{V} = 16\ \text{V}$.

Example 5 A student has three $30\ \Omega$ resistors and an ideal 90 V battery. (A battery is *ideal* if it has a negligible internal resistance.) Compare the current drawn from—and the power supplied by—the battery when the resistors are arranged in parallel versus in series.

Solution. Resistors in series always provide an equivalent resistance that's greater than any of the individual resistances, and resistors in parallel always provide an equivalent resistance that's smaller than their individual resistances. So, hooking up the resistors in parallel will create the smallest resistance and draw the greatest total current:

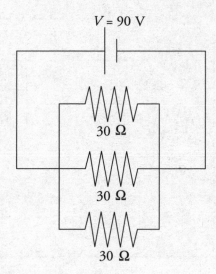

In this case, the equivalent resistance is

$$\frac{1}{R_P} = \frac{1}{30\ \Omega} + \frac{1}{30\ \Omega} + \frac{1}{30\ \Omega} \quad \Rightarrow \quad \frac{1}{R_P} = \frac{1}{10\ \Omega} \quad \Rightarrow \quad R_P = 10\ \Omega$$

and the total current is $I = V/R_P = (90\ \text{V})/(10\ \Omega) = 9$ A. (You could verify that 3 A of current would flow in each of the three branches of the combination.) The power supplied by the battery will be $P = IV = (9\ \text{A})(90\ \text{V}) = 810$ W.

If the resistors are in series, the equivalent resistance is $R_S = 30\ \Omega + 30\ \Omega + 30\ \Omega = 90\ \Omega$, and the current drawn is only $I = V/R_S = (90\ \text{V})/(90\ \Omega) = 1$ A. The power supplied by the battery in this case is just $P = IV = (1\ \text{A})(90\ \text{V}) = 90$ W.

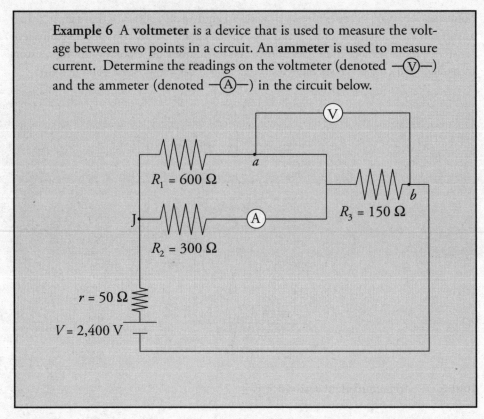

Example 6 A **voltmeter** is a device that is used to measure the voltage between two points in a circuit. An **ammeter** is used to measure current. Determine the readings on the voltmeter (denoted —Ⓥ—) and the ammeter (denoted —Ⓐ—) in the circuit below.

Solution. We consider the ammeter to be ideal; this means it has negligible resistance, so it doesn't alter the current that it's trying to measure. Similarly, we consider the voltmeter to have an extremely high resistance, so it draws negligible current away from the circuit.

Our first goal is to find the equivalent resistance in the circuit. The 600 Ω and 300 Ω resistors are in parallel; they're equivalent to a single 200 Ω resistor. This is in series with the battery's internal resistance, r, and R_3. The overall equivalent resistance is therefore $R = 50\ \Omega + 200\ \Omega + 150\ \Omega = 400\ \Omega$, so the current supplied by the battery is $I = V/R$; $I = (2{,}400\ \text{V})/(400\ \Omega) = 6$ A. At the junction marked J, this current splits. Since R_1 is twice R_2, half as much current will flow through R_1 as through R_2; the current through R_1 is $I_1 = 2$ A, and the current through R_2 is $I_2 = 4$ A. The voltage drop across each of these resistors is $I_1R_1 = I_2R_2 = 1{,}200$ V (matching voltages verify the values of currents I_1 and I_2). Since the ammeter is in the branch that contains R_2, it will read $I_2 = 4$ A.

The voltmeter will read the voltage drop across R_3, which is $V_3 = IR_3 = (6\ \text{A})$ $(150\ \Omega) = 900$ V. So the potential at point b is 900 V lower than at point a.

Example 7 The diagram below shows a point a at potential $V = 20$ V connected by a combination of resistors to a point (denoted G) that is **grounded.** *The ground is considered to be at potential zero.* If the potential at point a is maintained at 20 V, what is the current through R_3?

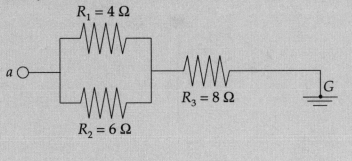

Solution. R_1 and R_2 are in parallel; their equivalent resistance is R_P, where

$$\frac{1}{R_P} = \frac{1}{4\ \Omega} + \frac{1}{6\ \Omega} \quad \Rightarrow \quad R_P = \frac{24}{10}\ \Omega = 2.4\ \Omega$$

R_P is in series with R_3, so the equivalent resistance is:

$$R = R_P + R_3 = (2.4\ \Omega) + (8\ \Omega) = 10.4\ \Omega$$

and the current that flows through R_3 is

$$I_3 = \frac{V}{R} = \frac{20\ \text{V}}{10.4\ \Omega} = 1.9\ \text{A}$$

Kirchhoff's Rules

When the resistors in a circuit cannot be classified as either in series or in parallel, we need another method for analyzing the circuit. The rules of Gustav Kirchhoff (pronounced "Keer koff") can be applied to any circuit:

Kirchhoff's First Law, also known as **The Junction Rule,** also known as the **Node Rule:**

> The total current that enters a junction must equal
> the total current that leaves the junction.

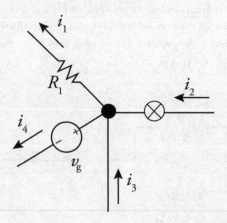

The current entering any junction is equal to the current leaving that junction: $i_2 + i_3 = i_1 + i_4$.

What Goes In Must Come Out
Knowing that the total drops must equal the total rise in potential is a good way to keep yourself in check when doing circuit problems. If 60 V came out of a battery, by the end, no more and no less must be used.

Kirchhoff's Second Law is also known as **The Loop Rule**.

> The sum of the potential differences (positive and negative) that traverse any closed loop in a circuit must be zero.

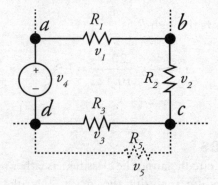

The sum of all voltages around the loop is equal to zero: $v_1 + v_2 + v_3 - v_4 = 0$.

The Loop Rule just says that, starting at any point, by the time we get back to that same point by following any closed loop, we have to be back to the same potential. Therefore, the total drop in potential must equal the total rise in potential. Put another way, the Loop Rule says that all the decreases in electrical potential energy (for example, caused by resistors in the direction of the current) must be balanced by all the increases in electrical potential energy (for example, caused by a source of emf from the negative to positive terminal). So the Loop Rule is basically a restatement of the Law of Conservation of Energy.

Similarly, the Junction Rule simply says that the charge (per unit time) that goes into a junction must equal the charge (per unit time) that comes out. This is basically a statement of the Law of Conservation of Charge.

In practice, the Junction Rule is straightforward to apply. The most important things to remember about the Loop Rule can be summarized as follows:

- When going across a resistor in the *same* direction as the current, the potential *drops* by *IR*.

- When going across a resistor in the *opposite* direction from the current, the potential *increases* by *IR*.

- When going from the negative to the positive terminal of a source of emf, the potential *increases* by *V*.

- When going from the positive to the negative terminal of a source of emf, the potential *decreases* by *V*.

Example 8 Use Kirchhoff's Rules to determine the current through R_2 in the following circuit:

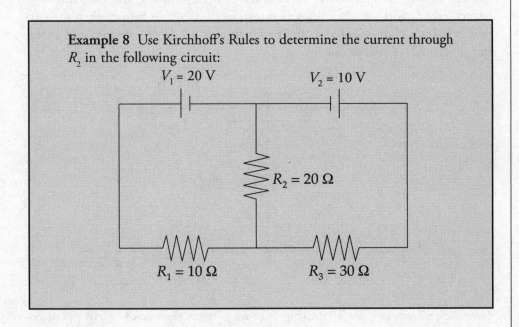

$V_1 = 20$ V $V_2 = 10$ V

$R_2 = 20\ \Omega$

$R_1 = 10\ \Omega$ $R_3 = 30\ \Omega$

Solution. First, let's label some points in the circuit.

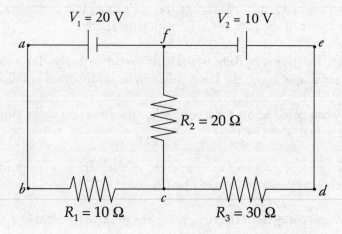

The points c and f are junctions (nodes). We have two nodes and three branches: one branch is *fabc*, another branch is *cdef*, and the third branch is *cf*. Each branch has one current throughout. If we label the current in *fabc* I_1 and the current in branch *cdef* I_2 (with the directions as shown in the diagram below), then the current in branch *cf* must be $I_1 - I_2$, by the Junction Rule: I_1 comes into c, and a total of $I_2 + (I_1 - I_2) = I_1$ comes out.

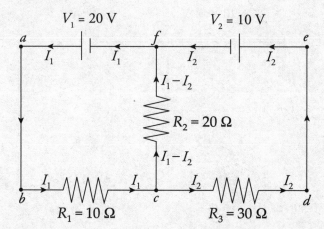

Now pick a loop; say, *abcfa*. Starting at a, we go to b, then across R_1 in the direction of the current, so the potential drops by I_1R_1. Then we move to c, then up through R_2 in the direction of the current, so the potential drops by $(I_1 - I_2)R_2$. Then we reach f, turn left and travel through V_1 from the negative to the positive terminal, so the potential increases by V_1. We now find ourselves back at a. By the Loop Rule, the total change in potential around this closed loop must be zero, and

$$-I_1R_1 - (I_1 - I_2)R_2 + V_1 = 0 \quad (1)$$

Since we have two unknowns (I_1 and I_2), we need two equations, so now pick another loop; let's choose *cdefc*. From *c* to *d*, we travel across the resistor in the direction of the current, so the potential drops by I_2R_3. From *e* to *f*, we travel through V_2 from the positive to the negative terminal, so the potential *drops* by V_2. Heading down from *f* to *c*, we travel across R_2 but in the direction opposite to the current, so the potential *increases* by $(I_1 - I_2)R_2$. At *c*, our loop is completed, so

$$-I_2R_3 - V_2 + (I_1 - I_2)R_2 = 0 \quad (2)$$

Substituting in the given numerical values for R_1, R_2, R_3, V_1, and V_2, and simplifying, these two equations become

$$3I_1 - 2I_2 = 2 \quad (1')$$
$$2I_1 - 5I_2 = 1 \quad (2')$$

Solving this pair of simultaneous equations, we get

$$I_1 = \frac{8}{11} \text{ A} = 0.73 \text{ A} \quad \text{and} \quad I_2 = \frac{1}{11} \text{ A} = 0.09 \text{ A}$$

So the current through R_2 is $I_1 - I_2 = \frac{7}{11}$ A $= 0.64$ A.

The choice of directions of the currents at the beginning of the solution was arbitrary. Don't worry about trying to guess the actual direction of the current in a particular branch. Just pick a direction, stick with it, and obey the Junction Rule. At the end, when you solve for the values of the branch current, a negative value will alert you that the direction of the current is actually opposite to the direction you originally chose for it in your diagram.

Chapter 12 Review Questions

Answers and Explanations can be found in Chapter 13.

Section I: Multiple-Choice

1. For an ohmic conductor, doubling the voltage without changing the resistance will cause the current to

   (A) decrease by a factor of 4
   (B) decrease by a factor of 2
   (C) increase by a factor of 2
   (D) increase by a factor of 4

2. If a 60-watt light bulb operates at a voltage of 120 V, what is the resistance of the bulb?

   (A) 2 Ω
   (B) 30 Ω
   (C) 240 Ω
   (D) 720 Ω

3. A battery whose emf is 40 V has an internal resistance of 5 Ω. If this battery is connected to a 15 Ω resistor R, what will the voltage drop across R be?

   (A) 10 V
   (B) 30 V
   (C) 40 V
   (D) 50 V

4.

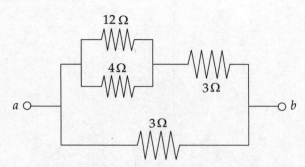

   Determine the equivalent resistance between points a and b.

   (A) 0.25 Ω
   (B) 0.333 Ω
   (C) 1.5 Ω
   (D) 2 Ω

5.

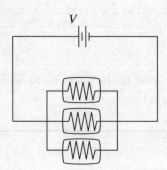

Three identical light bulbs are connected to a source of emf, as shown in the diagram above. What will happen if the middle bulb burns out?

   (A) The light intensity of the other two bulbs will decrease (but they won't go out).
   (B) The light intensity of the other two bulbs will increase.
   (C) The light intensity of the other two bulbs will remain the same.
   (D) More current will be drawn from the source of emf.

6.

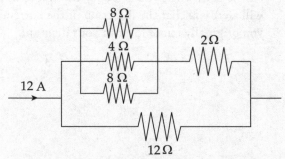

What is the voltage drop across the 12 Ω resistor in the portion of the circuit shown above?

   (A) 24 V
   (B) 36 V
   (C) 48 V
   (D 72 V

7.

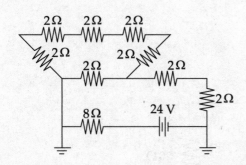

What is the current through the 8 Ω resistor in the circuit shown above?

(A) 0.5 A
(B) 1.0 A
(C) 1.5 A
(D) 3.0 A

8. How much energy is dissipated as heat in 20 s by a 100 Ω resistor that carries a current of 0.5 A ?

(A) 50 J
(B) 100 J
(C) 250 J
(D) 500 J

9. Three light bulbs are initially connected in series to a battery. If a fourth light bulb is then added and is also in series with the other three light bulbs, what happens to the current delivered by the battery?

(A) The current increases.
(B) The current remains the same.
(C) The current decreases.
(D) The current increases and then decreases.

Section II: Free-Response

1. Consider the following circuit:

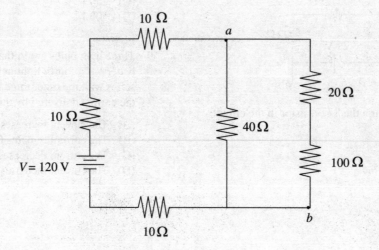

   (a)   At what rate does the battery deliver energy to the circuit?

   (b)   Find the current through the 40 Ω resistor.

   (c)   (i)   Determine the potential difference between points *a* and *b*.

   (ii)   At which of these two points is the potential higher?

   (d)   Find the energy dissipated by the 100 Ω resistor in 10 s.

2. Consider the following circuit:

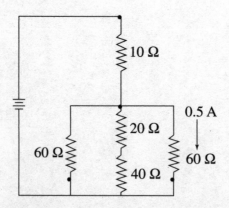

   (a)   What is the current through each resistor?

   (b)   What is the potential difference across each resistor?

   (c)   What is the equivalent resistance of the circuit?

Summary

- The current is the rate at which charge is transferred and given by $I_{avg} = \dfrac{\Delta Q}{\Delta t}$.

- Many objects obey Ohm's Law, which is given by $V = IR$.

- The electrical power in a circuit is given by

$$P = IV \text{ or } P = I^2R \text{ or } P = \dfrac{V^2}{R}$$

This is the same power we've encountered in our discussion of energy $P = \dfrac{W}{t}$.

- In a series circuit   In a parallel circuit

$$V_B = V_1 + V_2 + \ldots \qquad V_B = V_1 = V_2 = \ldots$$

$$I_B = I_1 = I_2 = \ldots \qquad I_B = I_1 + I_2 + \ldots$$

$$R_{eq} = R_1 + R_2 + \ldots \qquad 1/R_{eq} = 1/R_1 + 1/R_2 + \ldots$$

$$R_S = \sum_i R_i \qquad \dfrac{1}{R_P} = \sum_i \dfrac{1}{R_i}$$

- Kirchhoff's Loop Rule tells us that the sum of the potential differences in any closed loop in a circuit must be zero.

- Kirchhoff's Junction Rule (Node Rule) tells us the total current that enters a junction must equal the total current that leaves the junction.

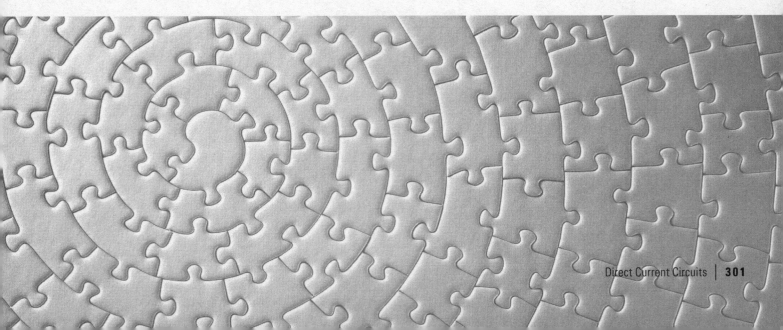

Chapter 13
Answers and Explanations to the Chapter Review Questions

CHAPTER 3 REVIEW QUESTIONS

Section I: Multiple-Choice

1. **B**   To add the vectors, draw the first vector and then from the end of the first draw the second. The resultant vector is from the beginning of the first vector to the end of the second:

The direction of the resultant vector is therefore northeast, eliminating (C) and (D). Since vectors **A** and **B** are perpendicular to each other and equal in magnitude, the magnitude of the resultant vector can be found using the Pythagorean Theorem:

$$a^2 + b^2 = c^2 \Rightarrow m^2 + m^2 = c^2 \Rightarrow 2m^2 = c^2 \Rightarrow m\sqrt{2} = c$$

This makes (B) correct.

2. **B**   To add the three vectors, add their components separately:

$$\mathbf{F}_1 + \mathbf{F}_2 + \mathbf{F}_3 = (0 - 10 + 5)\hat{\imath} + (-20 + 0 + 10)\hat{\jmath} = -5\hat{\imath} - 10\hat{\jmath}$$

3. **C**   The magnitude of a vector is given by:

$$A = \sqrt{(Ax)^2 + (Ay)^2}$$

If both components of the vector are doubled, the new magnitude, A', will be:

$$A' = \sqrt{(2Ax)^2 + (2Ay)^2} = \sqrt{4(Ax)^2 + 4(Ay)^2} = \sqrt{4[(Ax)^2 + (Ay)^2]} = 2\sqrt{(Ax)^2 + (Ay)^2} = 2A$$

So the magnitude of the vector will also be doubled.

The angle/direction of a vector is given by:

$$\theta = \tan^{-1}(A_y/A_x)$$

As can be observed in this equation, doubling the magnitude of both components of the vector will have no effect on the direction of the vector.

4. **D**   Subtracting the vector v_0 is equivalent to adding the negative of v_0 to v_f. Since v_f and $-v_0$ both point south, adding the two vectors results in a vector that is the sum of their two magnitudes and also points south:

$$v_f = 5\text{ms} \downarrow$$

$$-v_0 = 15\text{ms} \downarrow$$

This corresponds to (D).

5. **B** The components of the vector must satisfy the equation:

$$A = 10 = \sqrt{(Ax)^2 + (Ay)^2}$$

The only answer choice with components that satisfy this equation is (B): $10 = \sqrt{6^2 + 8^2}$.

6. **B** The vector sum $\mathbf{A} + \mathbf{B}$ can be found by adding the individual components:

$$\mathbf{A} + \mathbf{B} = (1 + 4)\hat{\mathbf{i}} + (-2 - 5)\hat{\mathbf{j}} = 5\hat{\mathbf{i}} - 7\hat{\mathbf{j}}$$

The angle that this vector makes with the x-axis is then found with:

$$\theta = \tan^{-1} \frac{(\mathbf{A} + \mathbf{B})_y}{(\mathbf{A} + \mathbf{B})_x} = \tan^{-1} \frac{7}{5}$$

This matches (B).

7. **A** If the object travels along the vectors $\boldsymbol{d}_1$ then $\boldsymbol{d}_2$, then the total distance travelled with be the sum of the two vectors $\boldsymbol{d}_1 + \boldsymbol{d}_2$:

$$\boldsymbol{d}_1 + \boldsymbol{d}_2 = (4 + 2)\hat{\mathbf{i}} + (5 + (-3))\hat{\mathbf{j}}$$
$$= 6\hat{\mathbf{i}} + 2\hat{\mathbf{j}}$$

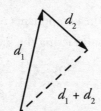

The total distance of the object from the starting position is therefore the magnitude $\boldsymbol{d}_1 + \boldsymbol{d}_2$:

$$|\boldsymbol{d}_1 + \boldsymbol{d}_2| = \sqrt{((\boldsymbol{d}_1 + \boldsymbol{d}_2)_x)^2 + ((\boldsymbol{d}_1 + \boldsymbol{d}_2)_y)^2} = \sqrt{6^2 + 2^2} = \sqrt{40} \approx 6.3 \text{ m}$$

That's (A).

8. **A** Subtracting vector B from A is equivalent to adding the negative of the second vector to the first:

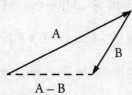

This results in a vector pointing the direction of (A).

9.　**C**　The x-component of vector A can be used to calculate the magnitude of A:

$$A_x = A \cos \theta$$
$$-42 = A \cos 130°$$

$$A = \frac{-42}{\cos 130°}$$

The y-component can then be calculated:

$$A_y = A \sin \theta = \left(\frac{-42}{\cos 130°}\right) \sin 130° = -42 \tan 130° \approx 50.1$$

which corresponds to (C).

10.　**B**　If the two vectors were pointed in the same direction, their magnitudes would add, and since magnitudes are positive, the result could not be zero, eliminating (A). If the vectors point in opposite directions, then the magnitude of the resultant vector would be obtained by subtracting the individual magnitudes of the two vectors. The only way for the difference to equal zero is if these two vectors have the same magnitude, eliminating (C) and (D). Choice (B) is therefore correct.

Section II: Free-Response

1.　(a)　The magnitude of A is:

$$|A| = \sqrt{(A_x)^2 + (A_y)^2} = \sqrt{3^2 + 6^2} = \sqrt{9+36} = \sqrt{45} \approx 6.7$$

(b)　Vector **C** is subtracted from **B** by adding –**C** to **B**:

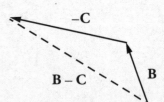

The components of **B** – **C** are found by subtracting the individual components:

$$\mathbf{B} - \mathbf{C} = (-1 - 5)\mathbf{\hat{i}} + (4 - (-2))\mathbf{\hat{j}}$$
$$= -6\mathbf{\hat{i}} + -6\mathbf{\hat{j}}$$

(c)　The components of 2**B** are found by multiplying each component of **B** by 2:

$$2\mathbf{B} = (2 \times -1)\mathbf{\hat{i}} + (2 \times 4)\mathbf{\hat{j}} = -2\mathbf{\hat{i}} + 8\mathbf{\hat{j}}$$

The components of **A** + 2**B** are found by adding the individual components:

$$\mathbf{A} + 2\mathbf{B} = (3 + (-2))\mathbf{\hat{i}} + (6 + 8)\mathbf{\hat{j}} = \mathbf{\hat{i}} + 14\mathbf{\hat{j}}$$

(d)　The components of **A** – **B** – **C** are found by working with the individual components:

$$\mathbf{A} - \mathbf{B} - \mathbf{C} = (3 - (-1) - 5)\mathbf{\hat{i}} + (6 - 4 - (-2))\mathbf{\hat{j}} = -1\mathbf{\hat{i}} + 4\mathbf{\hat{j}}$$

Let $\mathbf{D} = \mathbf{A} - \mathbf{B} - \mathbf{C}$ The magnitude of $\mathbf{D}$ is found using

$$D = \sqrt{(D_x)^2 + (D_y)^2} = \sqrt{(-1)^2 + 4^2} = \sqrt{17} \approx 4.1$$

and the direction of $\mathbf{D}$ relative to the horizontal is found using

$$\theta = \tan^{-1}\frac{4}{-1} \approx -76°$$

As the resultant vector has components $-1\hat{\imath} + 4\hat{\jmath}$, it is located in Quadrant II. To obtain an angle in Quadrant II, add 180° to get the correct answer of $\theta \approx 104°$ *or* 76° north of west.

2. (a)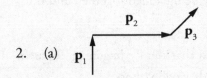

(b) The distance the ant has travelled, *d*, is the magnitude of the sum of the individual vector's in the ant's path: $\mathbf{d} = \mathbf{p_1} + \mathbf{p_2} + \mathbf{p_3}$. To do this addition, the components of $\mathbf{p_3}$ must be calculated. If due east is the *x*-direction and due north is the *y*-direction, then northeast is 45° about the horizontal. Thus,

$$p_{3,x} = p_3\cos\theta = (14 \text{ cm}) \cos 45° \approx 10 \text{ cm}$$
$$p_{3,y} = p_3\sin\theta = (14 \text{ cm}) \sin 45° \approx 10 \text{ cm}$$

Adding the individual components,

$$\mathbf{d} = \mathbf{p_1} + \mathbf{p_2} + \mathbf{p_3} = (20 \text{ cm} + 0 + 10 \text{ cm})\hat{\imath} + (0 + 30 \text{ cm} + 10 \text{ cm})\hat{\jmath}$$

$$= 30 \text{ cm } \hat{\imath} + 40 \text{ cm } \hat{\jmath}$$

The magnitude of this vector is:

$$d = \sqrt{(30 \text{ cm}^2) + (40 \text{ cm}^2)} = \sqrt{2500 \text{ cm}^2} = 50 \text{ cm}$$

(c) The ant's position is $\mathbf{d} = 40 \text{ cm } \hat{\imath} + 30 \text{ cm } \hat{\jmath}$. If the ant walks due north, the *x*-component of its position will not change, but the *y*-component will increase. Since its final position vector will have a magnitude of 80 cm, the *y*-component of the final position vector, $\mathbf{f}$, can be calculated:

$$f = \sqrt{(f_x)^2 + (f_y)^2} \Rightarrow 80 \text{ cm} = \sqrt{(40 \text{ cm})^2 + (f_y)^2} \Rightarrow (80 \text{ cm})^2 - (40 \text{ cm})^2 = (f_y)^2 \Rightarrow f_y \approx 69.3 \text{ cm}$$

Since the ant was already 30 cm north of his original position, he can walk another 69.3 cm – 30 cm = 39.3 cm due north.

3. (a) The components of $\mathbf{p} = 140 \text{ m}$ directed 10° above the horizontal are given by

$$p_x = p\cos\theta = (140 \text{ m}) \cos 10° = 138 \text{ cm}$$
$$p_y = p\sin\theta = (140 \text{ m}) \sin 10° = 24.3 \text{ cm}$$

thus, $\mathbf{p} = 138\hat{\imath} + 24.3\hat{\jmath}$.

(b)   The vector addition **p'** = **p** + **w** is accomplished by starting to draw the tail of the second vector at the end of the first:

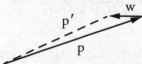

(c)   Adding the components of **p** + **w** yields p' = (138 m − 30 m)î + 24.3 m ĵ = 108 m î + 24.3 m ĵ. The new angle with the horizontal is therefore $\theta = \tan^{-1}\frac{24.3}{108} = 12.7°$.

4.   (a)   Since **v** depends on **c** and **r**, it makes sense to use v as the resultant vector. The diagram shows that r and c are connected tip-to-tail, indicating that these vectors are being added together, and thus, **v** = **r** + **c**.

(b)   If the angle between **v** and **c** is 90°, a right triangle is formed, so that the Pythagorean Theorem can be used to solve for the magnitude of **v** as a function of the other two vectors:

$$v^2 + c^2 = r^2 \Rightarrow v^2 = r^2 - c^2 \Rightarrow v = \sqrt{r^2 - c^2}$$

(c)   Let due east be the î direction and due north be the ĵ direction. Then the vectors **c** and **v** are defined as: **c** = $5\frac{m}{s}$î and **v** = $-10\frac{m}{s}$ĵ. Solving the vector equation from part (a) for **r** yields

$$\mathbf{r} = \mathbf{v} - \mathbf{c} = (0 - 5\tfrac{m}{s})\hat{\imath} + (-10\tfrac{m}{s} - 0)\hat{\jmath} = -5\tfrac{m}{s}\hat{\imath} - 10\tfrac{m}{s}\hat{\jmath}$$

Using these components, the magnitude and direction of **r** are given by:

$$|r| = \sqrt{(r_x)^2 + (r_y)^2} = \sqrt{(-5)^2 + (-10)^2} = \sqrt{125} = 5\sqrt{5} \approx 11.2\tfrac{m}{s}$$

$$\theta = \tan^{-1}\frac{-10}{-5} = -116.6° = 26.6° \text{ west of south or } 63.4° \text{ south of west}$$

CHAPTER 4 REVIEW QUESTIONS

Section I: Multiple-Choice

1. **A, D** Traveling once around a circular path means that the final position is the same as the initial position. Therefore, the displacement is zero. The average speed, which is *total* distance traveled divided by elapsed time, cannot be zero. Remember that velocity has direction and magnitude. When the object travels in a circular path the direction is always changing so the velocity is always changing. Because the velocity is changing, this means there is an acceleration that is delivered on the object.

2.   **D**   Section 1 represents a constant positive speed. Section 2 shows an object slowing down, moving in the positive direction. Section 3 represents an object speeding up in the negative direction. Section 4 demonstrates a constant negative speed, and section 5 represents the correct answer: slowing down moving in the negative direction. Though the slope is positive, this corresponds to acceleration, indicating that the direction of acceleration is opposite to the direction of velocity and thus is slowing down. However, the section remains in the negative quadrant, and the velocity becomes slower but is still negative.

3.   **B, D**   In parabolic motion, a projectile experiences only the constant direction due to gravity but the velocity does not point in the same direction. Eliminate (A). Zero acceleration means no change in speed (or direction), making (B) and (D) true. An object whose speed remains constant but whose velocity vector is changing direction is accelerating. Eliminate (C).

4.   **B**   The baseball is still under the influence of Earth's gravity. Its acceleration throughout the *entire* flight is constant, equal to g downward.

5.   **A**   Use Big Five #3 with $v_0 = 0$:

$$x = x_0 + v_0 t + \frac{1}{2}at^2 = \frac{1}{2}at^2 \quad \Rightarrow \quad t = \sqrt{\frac{2\Delta x}{a}} = \sqrt{\frac{2(200 \text{ m})}{5 \text{ m/s}^2}} = 9 \text{ s}$$

6.   **C**   Use Big Five #5 with $v_0 = 0$ (calling *down* the positive direction):

$$v^2 = v_0^2 + 2a(x - x_0) = 2a(x - x_0) \quad \Rightarrow \quad (x - x_0) = \frac{v^2}{2a} = \frac{v^2}{2g} = \frac{(30 \text{ m/s})^2}{2(10 \text{ m/s}^2)} = 45 \text{ m}$$

7.   **C**   Apply Big Five #3 to the vertical motion, calling *down* the positive direction:

$$\Delta y = v_{0y}t + \frac{1}{2}a_y t^2 = \frac{1}{2}a_y t^2 = \frac{1}{2}gt^2 \quad \Rightarrow \quad t = \sqrt{\frac{2\Delta y}{g}} = \sqrt{\frac{2(80 \text{ m})}{10 \text{ m/s}^2}} = 4 \text{ s}$$

Note that the stone's initial horizontal speed ($v_{0x} = 10$ m/s) is irrelevant.

8.   **B**   First, determine the time required for the ball to reach the top of its parabolic trajectory (which is the time required for the vertical velocity to drop to zero).

$$v_y \overset{\text{set}}{=} 0 \quad \Rightarrow \quad v_{0y} - gt = 0 \quad \Rightarrow \quad t = \frac{v_{0y}}{g}$$

The total flight time is equal to twice this value:

$$t_t = 2t = 2\frac{v_{0y}}{g} = 2\frac{v_0 \sin\theta_0}{g} = \frac{2(10 \text{ m/s})\sin 30°}{10 \text{ m/s}^2} = 1 \text{ s}$$

9. **C** After 4 seconds, the stone's vertical speed has changed by $\Delta v_y = a_y t = (10 \text{ m/s}^2)(4 \text{ s}) = 40 \text{ m/s}$. Since $v_{0y} = 0$, the value of v_y at $t = 4$ is 40 m/s. The horizontal speed does not change. Therefore, when the rock hits the water, its velocity has a horizontal component of 30 m/s and a vertical component of 40 m/s.

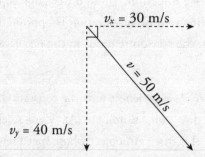

By the Pythagorean Theorem, the magnitude of the total velocity, v, is 50 m/s.

10. **D** Since the acceleration of the projectile is always downward (because of its gravitational acceleration), the vertical speed decreases as the projectile rises and increases as the projectile falls. Choices (A), (B), and (C) are all false.

11. **B** Use Big Five #2:

$$v = v_0 + at = 5 \text{ m/s} + (-10 \text{ m/s}^2)(3 \text{ s}) = -25 \text{ m/s}$$

Because we called up the positive direction, the negative sign for the velocity indicates that the stone is traveling downward.

12. **C** The variables involved in this question are the initial velocity (given in both cases), the acceleration (constant), final velocity (0 in both cases), and displacement (the skidding distance). As the missing variable is time, use Big Five #5 to solve for the displacement:

$$v^2 = v^2 + 2a(x - x_0)$$

$$x - x_0 = \frac{v^2 - v_0^2}{2a}$$

As the initial position, x_0, and the final velocity, v^2, are equal to 0, this equation simplifies to:

$$x = \frac{-v_0^2}{2a}$$

When the initial velocity is doubled, the final position quadruples. (Note that the acceleration in this problem is negative, as the brakes cause the car to decelerate).

Section II: Free-Response

1. (a) At time $t = 1$ s, the car's velocity starts to decrease as the acceleration (which is the slope of the given v versus t graph) changes from positive to negative.

(b)   The average velocity between $t = 0$ and $t = 1$ s is $\frac{1}{2}(v_{t=0} + v_{t=1}) = \frac{1}{2}(0 + 20 \text{ m/s}) = 10$ m/s, and the average velocity between $t = 1$ and $t = 5$ is $\frac{1}{2}(v_{t=1} + v_{t=5}) = \frac{1}{2}(20 \text{ m/s} + 0) = 10$ m/s. The two average velocities are the same.

(c)   The displacement is equal to the area bounded by the graph and the t-axis, taking areas above the t-axis as positive and those below as negative. In this case, the displacement from $t = 0$ to $t = 5$ s is equal to the area of the triangular region whose base is the segment along the t-axis from $t = 0$ to $t = 5$ s:

$$\Delta x \ (t = 0 \text{ to } t = 5 \text{ s}) = \frac{1}{2} \times \text{base} \times \text{height} = \frac{1}{2}(5 \text{ s})(20 \text{ m/s}) = 50 \text{ m}$$

The displacement from $t = 5$ s to $t = 7$ s is equal to the negative of the area of the triangular region whose base is the segment along the t-axis from $t = 5$ s to $t = 7$ s:

$$\Delta x \ (t = 5 \text{ s to } t = 7 \text{ s}) = -\frac{1}{2} \times \text{base} \times \text{height} = -\frac{1}{2}(2 \text{ s})(10 \text{ m/s}) = -10 \text{ m}$$

Therefore, the displacement from $t = 0$ to $t = 7$ s is

$$\Delta x \ (t = 0 \text{ to } t = 5 \text{ s}) + \Delta s \ (t = 5 \text{ s to } t = 7 \text{ s}) = 50 \text{ m} + (-10 \text{ m}) = 40 \text{ m}$$

(d)   The acceleration is the slope of the v versus t graph. The segment of the graph from $t = 0$ to $t = 1$ s has a slope of $a = \Delta v/\Delta t = (20 \text{ m/s} - 0)/(1 \text{ s} - 0) = 20 \text{ m/s}^2$, and the segment of the graph from $t = 1$ s to $t = 7$ s has a slope of $a = \Delta v/\Delta t = (-10 \text{ m/s} - 20 \text{ s})/(7 \text{ s} - 1 \text{ s}) = -5 \text{ m/s}^2$. Therefore, the graph of a versus t is

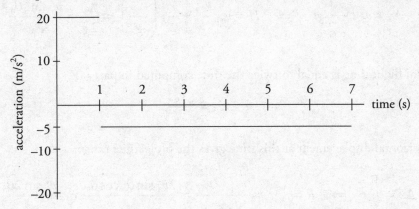

(e)

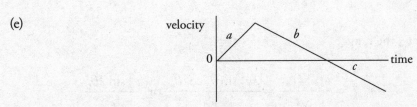

Section a shows that the object is speeding up in the positive direction. Section b shows that the object is slowing down, yet still moving in the positive direction. At five seconds, the object has stopped for an instant. Section *c* shows that the object is moving in the negative direction and speeding up. The corresponding position-versus-time graph for each section would look like this:

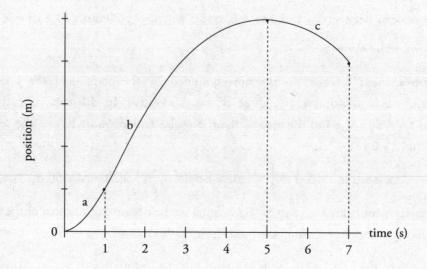

2. (a) The maximum height of the projectile occurs at the time at which its vertical velocity drops to zero:

$$v_y \overset{set}{=} 0 \quad \Rightarrow \quad v_{0y} - gt = 0 \quad \Rightarrow \quad t = \frac{v_{0y}}{g}$$

The vertical displacement of the projectile at this time is computed as follows:

$$\Delta y = v_{0y}t - \frac{1}{2}gt^2 \quad \Rightarrow \quad H = v_{0y}\frac{v_{0y}}{g} - \frac{1}{2}g\left(\frac{v_{0y}}{g}\right)^2 = \frac{v_{0y}^2}{2g} = \frac{v_0^2 \sin^2\theta_0}{2g}$$

(b) The total flight time is equal to twice the time computed in part (a):

$$t_t = 2t = 2\frac{v_{0y}}{g}$$

The horizontal displacement at this time gives the projectile's range:

$$\Delta x = v_{0x}t \quad \Rightarrow \quad R = v_{0x}t_t = \frac{v_{0x} \cdot 2v_{0y}}{g} = \frac{2v_0^2 \sin\theta_0 \cos\theta_0}{g} \quad \text{or} \quad \frac{v_0^2 \sin 2\theta_0}{g}$$

(c) For any given value of v_0, the range,

$$\Delta x = v_{0x}t \quad \Rightarrow \quad R = v_{0x}t_t = \frac{v_{0x} \cdot 2v_{0y}}{g} = \frac{2v_0^2 \sin\theta_0 \cos\theta_0}{g} \quad \text{or} \quad \frac{v_0^2 \sin 2\theta_0}{g}$$

will be maximized when $\sin 2\theta_0$ is maximized. This occurs when $2\theta_0 = 90°$, that is, when $\theta_0 = 45°$.

(d) Set the general expression for the projectile's vertical displacement equal to h and solve for the two values of t (assuming that $g = +10 \text{ m/s}^2$):

$$v_{0y}t - \frac{1}{2}gt^2 \overset{\text{set}}{=} h \quad \Rightarrow \quad \frac{1}{2}gt^2 - v_{0y}t + h = 0$$

Applying the quadratic formula, find that

$$t = \frac{v_{0y} \pm \sqrt{(-v_{0y})^2 - 4(\frac{1}{2}g)(h)}}{2(\frac{1}{2}g)} = \frac{v_{0y} \pm \sqrt{v_{0y}^2 - 2gh}}{g}$$

Therefore, the two times at which the projectile crosses the horizontal line at height h are

$$t_1 = \frac{v_{0y} - \sqrt{v_{0y}^2 - 2gh}}{g} \quad \text{and} \quad t_2 = \frac{v_{0y} + \sqrt{v_{0y}^2 - 2gh}}{g}$$

so the amount of time that elapses between these events is

$$\Delta t = t_2 - t_1 = \frac{2\sqrt{v_{0y}^2 - 2gh}}{g}$$

3. (a) The cannonball will certainly reach the wall (which is only 220 m away) since the ball's range is

$$R = \frac{v_0^2 \sin 2\theta_0}{g} = \frac{(50 \text{ m/s})^2 \sin 2(40°)}{9.8 \text{ m/s}^2} = 251 \text{ m}$$

You simply need to make sure that the cannonball's height is less than 30 m at the point where its horizontal displacement is 220 m (so that the ball actually hits the wall rather than flying over it). To do this, find the time at which $x = 220$ m by first writing

$$x = v_{0x}t \quad \Rightarrow \quad t = \frac{x}{v_{0x}} = \frac{x}{v_0 \cos\theta_0} \qquad (1)$$

Thus, the cannonball's vertical position can be written in terms of its horizontal position as follows:

$$y = v_{0y}t - \tfrac{1}{2}gt^2 = v_0 \sin\theta_0 \frac{x}{v_0 \cos\theta_0} - \tfrac{1}{2}g\left(\frac{x}{v_0 \cos\theta_0}\right)^2$$

$$= x \tan\theta_0 - \frac{gx^2}{2v_0^2 \cos^2\theta_0} \qquad (2)$$

Substituting the known values for x, θ_0, g, and v_0, you get

$$y(\text{at } x = 220 \text{ m}) = (220 \text{ m})\tan 40° - \frac{(9.8 \text{ m/s}^2)(220 \text{ m})^2}{2(50 \text{ m/s})^2 \cos^2 40°}$$

$$= 23 \text{ m}$$

This is indeed less than 30 m, as desired.

(b)   From Equation (1) derived in part (a),

$$t = \frac{x}{v_0 \cos\theta_0} = \frac{220 \text{ m}}{(50 \text{ m/s})\cos 40°} = 5.7 \text{ s}$$

(c)   The height at which the cannonball strikes the wall was determined in part (a) to be 23 m.

4.   (a)   For parabolic trajectories, the total flight time can be determined by doubling the amount of time it takes for the projectile to reach its apex. However, this trajectory is NOT parabolic. As the initial position is 25 m, the final position is 0 m, the initial velocity is $v_0 = v_0 \sin\theta = 40 \sin 30° = 40 \left(\frac{1}{2}\right) = 20 \text{ m/s}$, and the acceleration is -10 m/s^2, the missing variable is the final velocity so the flight time of the cannonball can be computed using Big Five #3:

$$x_y = x_0 + v_0 t + \frac{1}{2}at^2$$

$$0 = 25 \text{ m} + 20 \text{ m/s} \cdot t + \frac{1}{2}(-10 \text{ m/s}^2)\cdot t^2$$

$$0 = 25 \text{ m} + 20 \text{ m/s} \cdot t - 5 \text{ m/s}^2 \cdot t^2$$

$$0 = -5 \text{ m/s}^2 \cdot t^2 + 20 \text{ m/s} \cdot t + 25 \text{ m}$$

Applying the quadratic formula, find that:

$$t = \frac{-20 \pm \sqrt{20^2 - 4(25)(-5)}}{2(-5)} = \frac{-20 \pm \sqrt{400 + 500}}{-10} = \frac{-20 \pm \sqrt{900}}{-10} = \frac{-20 \pm 30}{-10}$$

$$t = -1 \text{ s or } t = 5 \text{ s}$$

As time cannot be a negative value, the flight time of the cannonball is 5 seconds.

(b)   As the horizontal speed of a projectile is constant, the range is given by:

$$\Delta x = v_{0x} t = 40 \text{ m/s} \cdot \cos(30°) \cdot 5 \text{ s} = 173 \text{ m}$$

(c)   The cannonball is fired with an initial horizontal velocity of:

$$v_{0x} = 40 \text{ m/s} \cdot \cos(30°) = 34.6 \text{ m/s}$$

As there is no horizontal acceleration the horizontal speed of the projectile remains constant throughout the entire flight, leading to a flat line with a y-value of 34.6 m/s over the flight time of 5 s. The horizontal speed vs. time graph can then be plotted

Horizontal Speed vs. Time

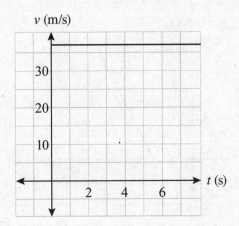

The initial vertical velocity of the cannonball is:

$$v_{0x} = 40 \text{ m/s} \cdot \sin(30°) = 20 \text{ m/s}$$

In the vertical direction, there is a constant acceleration due to gravity ($a = 10$ m/s^2). As the slope of the velocity time graph is equal to the acceleration, the resulting velocity time graph is a linear line with a negative slope since acceleration due to gravity points downwards. As the magnitude of the acceleration of gravity is 10 m/s^2, the vertical velocity of the cannonball thus decreases by 10 m/s every second. With an initial velocity of 20 m/s, this means that after 2 s, the vertical velocity of the projectile is 0 m/s. After a total of 5 s, the vertical velocity of the projectile is –30 m/s. This can be visualized in the vertical velocity vs. time graph below:

Vertical Velocity vs. Time

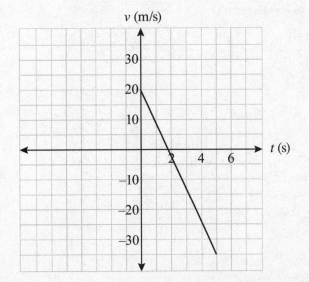

As the problem asks for the graph of the vertical speed vs. time, the absolute value of the graph must be drawn to get the correct plot of

Vertical Speed vs. Time

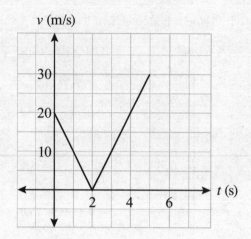

CHAPTER 5 REVIEW QUESTIONS

Section I: Multiple-Choice

1. **B**   Because the person is not accelerating, the net force he feels must be zero. Therefore, the magnitude of the upward normal force from the floor must balance that of the downward gravitational force. Although these two forces have equal magnitudes, they do not form an action/reaction pair because they both act on the same object (namely, the person). The forces in an action/reaction pair always act on different objects. The correct action-reaction pair in this situation is the Earth's pull on the person (the weight) and the person's pull on the Earth.

2. **D**   First, draw a free-body diagram:

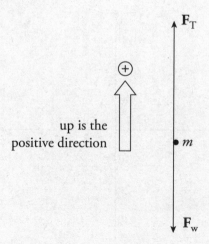

The person exerts a downward force on the scale, and the scale pushes up on the person with an equal (but opposite) force, F_N. Thus, the scale reading is F_N, the magnitude of the normal force. Since $F_N - F_w = ma$, you get $F_N = F_w + ma = (800 \text{ N}) + [800 \text{ N}/(10 \text{ m/s}^2)](5 \text{ m/s}^2) = 1{,}200 \text{ N}$.

3. **A** The net force that the object feels on the inclined plane is $mg \sin \theta$, the component of the gravitational force that is parallel to the ramp. Since $\sin \theta = (5 \text{ m})/(20 \text{ m}) = \dfrac{1}{4}$, you get $F_{net} = (2 \text{ kg})(10 \text{ N/kg})(\dfrac{1}{4})$ $= 5 \text{ N}$.

4. **C** The net force on the block is $F - F_f = F - \mu_k F_N = F - \mu_k F_w = (18 \text{ N}) - (0.4)(20 \text{ N}) = 10 \text{ N}$. Since $F_{net} = ma = (F_w/g)a$, you get $10 \text{ N} = [(20 \text{ N})/(10 \text{ m/s}^2)]a$, which gives $a = 5 \text{ m/s}^2$.

5. **A** The force pulling the block down the ramp is $mg \sin \theta$, and the maximum force of static friction is $\mu_s F_N = \mu_s mg \cos \theta$. If $mg \sin \theta$ is greater than $\mu_s mg \cos \theta$, then there is a net force down the ramp, and the block will accelerate down. So, the question becomes, "Is $\sin \theta$ greater than $\mu_s \cos \theta$?" Since $\theta = 30°$ and $\mu_s = 0.5$, the answer is "yes."

6. **D** One way to attack this question is to notice that if the two masses happen to be equal, that is, if $M = m$, then the blocks won't accelerate (because their weights balance). The only expression given that becomes zero when $M = m$ is the one given in (D). Draw a free-body diagram:

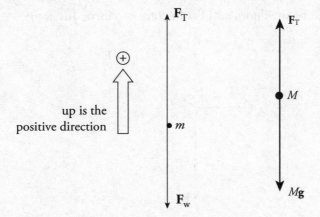

Newton's Second Law gives the following two equations:

$F_T - mg = ma$ (1)

$F_T - Mg = M(-a)$ (2)

Subtracting these equations yields $Mg - mg = ma + Ma = (M + m)a$, so

$$a = \frac{Mg - mg}{M + m} = \frac{M - m}{M + m} g$$

7. **D** If $F_{net} = 0$, then $a = 0$. No acceleration means constant speed (possibly, but not necessarily zero) with no change in direction. Therefore, (B) and (C) are false, and (A) is not necessarily true.

8. **C** The horizontal motion across the frictionless tables is unaffected by (vertical) gravitational acceleration. It would take as much force to accelerate the block across the table on Earth as it would on the Moon. (If friction *were* taken into account, then the smaller weight of the block on the Moon would imply a smaller normal force by the table and hence a smaller frictional force. Less force would be needed on the Moon in this case.)

9. **D** The maximum force which static friction can exert on the crate is $\mu_s F_N = \mu_s F_w = \mu_s mg = (0.4)(100 \text{ kg})(10 \text{ N/kg}) = 400 \text{ N}$. Since the force applied to the crate is only 344 N, static friction is able to apply that same magnitude of force on the crate, keeping it stationary. Choice (B) is incorrect because the static friction force is *not* the reaction force to F; both F and $F_{f(static)}$ act on the same object (the crate) and therefore cannot form an action/reaction pair.

10. **A** With Crate #2 on top of Crate #1, the force pushing downward on the floor is greater, so the normal force exerted by the floor on Crate #1 is greater, which increases the friction force. Choices (B), (C), and (D) are all false.

Section II: Free-Response

1. (a) The forces acting on the crate are F_T (the tension in the rope), F_w (the weight of the block), F_N (the normal force exerted by the floor), and F_f (the force of kinetic friction):

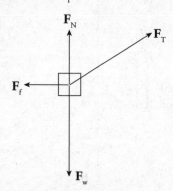

    (b) First, break F_T into its horizontal and vertical components:

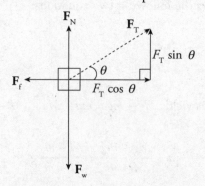

    Since the net vertical force on the crate is zero, you get $F_N + F_T \sin \theta = F_w$, so $F_N = F_w - F_T \sin \theta = mg - F_T \sin \theta$.

(c) From part (b), notice that the net horizontal force acting on the crate is

$$F_T \cos\theta - F_f = F_T \cos\theta - \mu F_N = F_T \cos\theta - \mu(mg - F_T \sin\theta)$$

so the crate's horizontal acceleration across the floor is

$$a = \frac{F_{net}}{m} = \frac{F_T \cos\theta - \mu(mg - F_T \sin\theta)}{m}$$

2. (a) The forces acting on Block #1 are $\mathbf{F}_T$ (the tension in the string connecting it to Block #2), $\mathbf{F}_{w1}$ (the weight of the block), and $\mathbf{F}_{N1}$ (the normal force exerted by the tabletop):

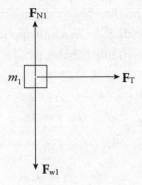

(b) The forces acting on Block #2 are $\mathbf{F}$ (the pulling force), $\mathbf{F}_T$ (the tension in the string connecting it to Block #1), $\mathbf{F}_{w2}$ (the weight of the block), and $\mathbf{F}_{N2}$ (the normal force exerted by the tabletop):

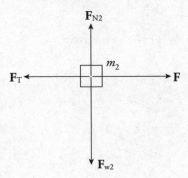

(c) Newton's Second Law applied to Block #2 yields $F - F_T = m_2 a$, and applied to Block #1 yields $F_T = m_1 a$. Adding these equations, you get $F = (m_1 + m_2)a$, so

$$a = \frac{F}{m_1 + m_2}$$

(d) Substituting the result of part (c) into the equation $F_T = m_1 a$, you get

$$F_T = m_1 a = \frac{m_1}{m_1 + m_2} F$$

(e) (i) Since the force **F** must accelerate all three masses—m_1, m, and m_2—the common acceleration of all parts of the system is

$$a = \frac{F}{m_1 + m + m_2}$$

(ii) Let $\mathbf{F}_{T1}$ denote the tension force in the connecting string acting on Block #1, and let $\mathbf{F}_{T2}$ denote the tension force in the connecting string acting on Block #2. Then, Newton's Second Law applied to Block #1 yields $F_{T1} = m_1 a$ and applied to Block #2 yields $F - F_{T2} = m_2 a$. Therefore, using the value for a computed above, you get

$$
\begin{aligned}
F_{T2} - F_{T1} &= (F - m_2 a) - m_1 a \\
&= F - (m_1 + m_2)a \\
&= F - (m_1 + m_2)\frac{F}{m_1 + m + m_2} \\
&= F\left(1 - \frac{m_1 + m_2}{m_1 + m + m_2}\right) \\
&= F\frac{m}{m_1 + m + m_2}
\end{aligned}
$$

3. (a) First, draw free-body diagrams for the two boxes:

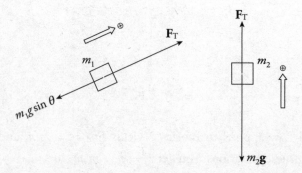

Applying Newton's Second Law to the boxes yields the following two equations:

$$F_T - m_1 g \sin\theta = m_1 a \qquad (1)$$

$$F_T - m_2 g = m_2(-a) \qquad (2)$$

Subtract the equations and solve for a:

$$m_2 g - m_1 g \sin\theta = (m_1 + m_2)a$$

$$a = \frac{m_2 - m_1 \sin\theta}{m_1 + m_2}g$$

(i) For a to be positive, you must have $m_2 - m_1 \sin\theta > 0$, which implies that $\sin\theta < m_2/m_1$, or, equivalently, $\theta < \sin^{-1}(m_2/m_1)$.

(ii) For a to be zero, you must have $m_2 - m_1 \sin\theta = 0$, which implies that $\sin\theta = m_2/m_1$, or, equivalently, $\theta = \sin^{-1}(m_2/m_1)$.

(b) Including the force of kinetic friction, the force diagram for m_1 is

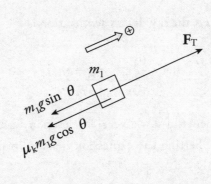

Since $F_f = \mu_k F_N = \mu_k m_1 g \cos\theta$, applying Newton's Second Law to the boxes yields these two equations:

$$F_T - m_1 g \sin\theta - \mu_k mg \cos\theta = m_1 a \qquad (1)$$

$$m_2 g - F_T = m_2 a \qquad (2)$$

Add the equations and solve for a:

$$m_2 g - m_1 g \sin\theta - \mu_k mg \cos\theta = (m_1 + m_2)a$$

$$a = \left(\frac{m_2 - m_1 (\sin\theta + \mu_k \cos\theta)}{m_1 + m_2} \right) g$$

In order for a to be equal to zero (so that the box of mass m_1 slides up the ramp with constant velocity),

$$m_2 - m_1 (\sin\theta + \mu_k \cos\theta) = 0$$

$$\sin\theta + \mu_k \cos\theta = \frac{m_2}{m_1}$$

4. (a) The forces acting on the sky diver are $\mathbf{F}_r$, the force of air resistance (upward), and $\mathbf{F}_w$, the weight of the sky diver (downward):

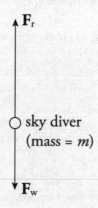

(b) Since $F_{net} = F_w - F_r = mg - kv$, the sky diver's acceleration is

$$a = \frac{F_{net}}{m} = \frac{mg - kv}{m}$$

(c) Terminal speed occurs when the sky diver's acceleration becomes zero, since then the descent velocity becomes constant. Setting the expression derived in part (b) equal to 0, find the speed $v = v_t$ at which this occurs:

$$v = v_t \text{ when } a = 0 \quad \Rightarrow \quad \frac{mg - kv_t}{m} = 0 \quad \Rightarrow \quad v_t = \frac{mg}{k}$$

(d) The sky diver's descent speed is initially v_0 and the acceleration is (close to) g. However, once the parachute opens, the force of air resistance provides a large (speed-dependent) upward acceleration, causing her descent velocity to decrease. The slope of the v versus t graph (the acceleration) is not constant but instead decreases to zero as her descent speed decreases from v_0 to v_t. Therefore, the graph is not linear.

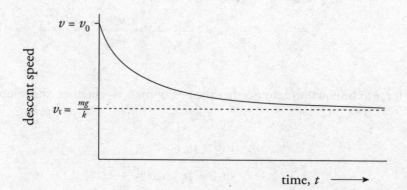

CHAPTER 6 REVIEW QUESTIONS

Section I: Multiple-Choice

1. **A** Since the force **F** is perpendicular to the displacement, the work it does is zero.

2. **B** By the Work–Energy Theorem,

$$W = \Delta K = \frac{1}{2}m(v^2 - v_0^2) = \frac{1}{2}(4 \text{ kg})[(6 \text{ m/s})^2 - (3 \text{ m/s})^2] = 54 \text{ J}$$

3. **B** Since the box (mass m) falls through a vertical distance of h, its gravitational potential energy decreases by mgh. The length of the ramp is irrelevant here. If friction were involved then the length of the plane would matter.

4. **A** The gravitational force points downward while the book's displacement is upward. Therefore, the work done by gravity is $-mgh = -(2 \text{ kg})(10 \text{ N/kg})(1.5 \text{ m}) = -30 \text{ J}$.

5. **C** The only force doing work on the block as it slides down the inclined plane is the force of gravity, so the Work–Energy Theorem can be applied with the total work equal to the work done by gravity. The work done by gravity as the block slides down the inclined plane is equal to the negative change in potential energy.

$$W = -\Delta U = U_\text{f} - U_\text{i} = mgh = \Delta K = \frac{1}{2}m(v^2 - v_0^2) = \frac{1}{2}mv^2 \Rightarrow v = \sqrt{2gh} = \sqrt{2(10)(6.4)(\sin 30°)} = 8 \text{ m/s}$$

6. **C** Since a nonconservative force (namely, friction) is acting during the motion, use the modified Conservation of Mechanical Energy equation.

$$K_\text{i} + U_\text{i} + W_\text{friction} = K_\text{i} + U_\text{f}$$

$$0 + mgh - Fs = K_\text{f} + 0$$

$$mgh - Fs = K_\text{f}$$

7.   **D**   Apply Conservation of Mechanical Energy (including the negative work done by $\mathbf{F}_r$, the force of air resistance):

$$K_i + U_i + W_r = K_f + U_f$$

$$0 + mgh - F_r h = \frac{1}{2}mv^2 + 0$$

$$v = \sqrt{\frac{2h(mg - F_r)}{m}}$$

$$= \sqrt{\frac{2(40 \text{ m})[(4 \text{ kg})(10 \text{ N/kg}) - 20 \text{ N}]}{4 \text{ kg}}}$$

$$= 20 \text{ m/s}$$

8.   **D**   Because the rock has lost half of its gravitational potential energy, its kinetic energy at the halfway point is half of its kinetic energy at impact. Since K is proportional to v^2, if $K_{\text{at halfway point}}$ is equal to $\frac{1}{2}K_{\text{at impact}}$, then the rock's speed at the halfway point is $\sqrt{1/2} = 1/\sqrt{2}$ its speed at impact.

9.   **D**   Using the equation $P = Fv$, find that $P = (200 \text{ N})(2 \text{ m/s}) = 400 \text{ W}$.

Section II: Free-Response

1.   (a)   Applying Conservation of Energy,

$$K_A + U_A = K_{\text{at } H/2} + U_{\text{at } H/2}$$

$$0 + mgH = \frac{1}{2}mv^2 + mg(\frac{1}{2}H)$$

$$\frac{1}{2}mgH = \frac{1}{2}mv^2$$

$$v = \sqrt{gH}$$

(b)   Applying Conservation of Energy again,

$$K_A + U_A = K_B + U_B$$

$$0 + mgH = \frac{1}{2}mv_B^2 + 0$$

$$v_B = \sqrt{2gH}$$

(c)   By the Work–Energy Theorem, we want the work done by friction to be equal (but opposite) to the kinetic energy of the box at Point B:

$$W = \Delta K = \frac{1}{2}m(v_C^2 - v_B^2) = -\frac{1}{2}mv_B^2 = -\frac{1}{2}m(\sqrt{2gH})^2 = -mgH$$

Therefore,

$$W = -mgH \quad \Rightarrow \quad -F_f x = -mgH \quad \Rightarrow \quad -\mu_k mgx = -mgH \quad \Rightarrow \quad \mu_k = H/x$$

(d) Apply Conservation of Energy (including the negative work done by friction as the box slides up the ramp from B to C):

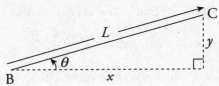

$$K_B + U_B + W_f = K_C + U_C$$

$$\frac{1}{2}m\left(\sqrt{2gH}\right)^2 + 0 - F_f L = 0 + mgy$$

$$mgH + 0 - F_f L = 0 + mgy$$

$$mg(H - y) - (\mu_k mg\cos\theta)(L) = 0$$

$$\mu_k = \frac{H - y}{L\cos\theta} = \frac{H - y}{x}$$

(e) The result of part (b) reads $v_B = \sqrt{2gH}$. Therefore, by Conservation of Mechanical Energy (with the work done by the frictional force on the slide included), you get

$$K_A + U_A + W_f = K_B' + U_B$$

$$0 + mgH + W_f = \frac{1}{2}m\left(\frac{1}{2}v_B\right)^2 + 0$$

$$mgH + W_f = \frac{1}{2}m\left(\frac{1}{2}\sqrt{2gH}\right)^2$$

$$mgH + W_f = \frac{1}{4}mgH$$

$$W_f = -\frac{3}{4}mgH$$

2. (a) Using Conservation of Energy $K_i + U_i = K_f + U_f$ and $v_i = 0$, this becomes $U_i = K_f + U_f$ or $K_f = U_i - U_f$. This is equivalent to $\frac{1}{2}mv^2 = mgh_i - mgh_f$, which simplifies to $\frac{1}{2}v^2 - gh_i = -gh_f$ or $h_f = h_i - \frac{v^2}{2g}$. Now fill in the table.

Time (s)	Velocity (m/s)	Height (m)
0.00	0.00	1.5
0.05	1.41	1.4
0.10	2.45	1.2
0.15	3.74	0.8
0.20	3.74	0.8
0.25	3.46	0.9
0.30	3.16	1.0
0.35	2.83	1.1
0.40	3.46	0.9
0.45	4.24	0.6
0.50	4.47	0.5

(b) The greatest acceleration would occur where there is the greatest change in velocity. This occurs between 0.00 and 0.05 seconds. The acceleration during that time interval is given by

$$a = \frac{v_f - v_i}{t_f - t_i} \Rightarrow \frac{1.41 - 0}{0.05 - 0.00} \text{ or } a = 28 \text{ m/s}^2.$$

(c) Changing the mass does not affect the time spent falling or the velocity of the object. Thus, a change in mass will not affect the results.

3. (a) Use the Work–Energy Theorem:

$$W_{total} = \Delta K$$

The force doing work during the motion is provided by the force of friction:

$$W = F_f \cdot d \cdot \cos \theta = \Delta K$$

$$\mu_k F_N \cdot d \cdot \cos \theta = \frac{1}{2} m(v^2 - v_0^2)$$

As the force of friction is antiparallel to the direction of the displacement, $\theta = 180°$.

$$\mu_k F_N \cdot d \cdot \cos \theta = \frac{1}{2} m(v^2 - v_0^2)$$

$$\mu_k mg \cdot d \cdot \cos \theta = \frac{1}{2} m(v^2 - v_0^2)$$

$$\mu_k g \cdot d \cdot \cos \theta = \frac{(v^2 - v_0^2)}{2}$$

$$d = \frac{(v^2 - v_0^2)}{2\mu_k \cdot g \cdot \cos \theta} = \frac{(0 - 10^2)}{2(0.2) \cdot 10 \cdot \cos 180°} = \frac{-100}{-4} = 25 \text{ m}$$

The skidding distance is 25 m.

(b) The final equation for the skidding distance is

$$d = \frac{(v^2 - v_0^2)}{2\mu_k \cdot g \cdot \cos \theta}$$

Since the final velocity is 0, the stopping distance is proportional to the initial speed. So if the initial speed is doubled, the skidding distance is quadrupled. This can also be determined by plugging in an initial velocity of 20 m/s:

$$d = \frac{(v^2 - v_0^2)}{2\mu_k \cdot g \cdot \cos \theta} = \frac{(0 - 20^2)}{2(0.2) \cdot 10 \cdot \cos 180°} = \frac{-400}{-4} = 100 \; m$$

(c) Look back at the equation for the skidding distance:

$$d = \frac{(v^2 - v_0^2)}{2\mu_k \cdot g \cdot \cos \theta}$$

This equation does not include mass, so mass does not affect the skidding distance. (Note: While doubling the mass doubles the initial kinetic energy of the car, it also doubles the normal force and thus the frictional force acting on the car.)

CHAPTER 7 REVIEW QUESTIONS

Section I: Multiple-Choice

1. **C** The magnitude of the object's linear momentum is $p = mv$. If $p = 6$ kg · m/s and $m = 2$ kg, then $v = 3$ m/s. Therefore, the object's kinetic energy is $K = \frac{1}{2}mv^2 = \frac{1}{2}(2 \text{ kg})(3 \text{ m/s})^2 = 9$ J.

2. **B** The impulse delivered to the ball, $J = F\Delta t$, equals its change in momentum. Since the ball started from rest, you get

    $$F\Delta t = mv \quad \Rightarrow \quad \Delta t = \frac{mv}{F} = \frac{(0.5 \text{ kg})(4 \text{ m/s})}{20 \text{ N}} = 0.1 \text{ s}$$

3. **D** The impulse delivered to the ball, $J = \bar{F}\Delta t$, equals its change in momentum. Thus,

    $$\bar{F}\Delta t = \Delta p = p_{fi} - p = m(v_{fi} - v) \quad \Rightarrow \quad \bar{F} = \frac{m(v_{fi} - v)}{\Delta t} = \frac{(2 \text{ kg})(8 \text{ m/s} - 4 \text{ m/s})}{0.5 \text{ s}} = 16 \text{ N}$$

4. **C** The impulse delivered to the ball is equal to its change in momentum. The momentum of the ball was $m\mathbf{v}$ before hitting the wall and $m(-\mathbf{v})$ after. Therefore, the change in momentum is $m(-\mathbf{v}) - m\mathbf{v} = -2m\mathbf{v}$, so the magnitude of the momentum change (and the impulse) is $2mv$.

5. **B** By definition of *perfectly inelastic*, the objects move off together with one common velocity, $\mathbf{v}'$, after the collision. By Conservation of Linear Momentum,

    $$m_1\mathbf{v}_1 + m_2\mathbf{v}_2 = (m_1 + m_2)\mathbf{v}'$$
    $$\mathbf{v}' = \frac{m_1v_1 + m_2v_2}{m_1 + m_2}$$
    $$= \frac{(3 \text{ kg})(2 \text{ m/s}) + (5 \text{ kg})(-2 \text{ m/s})}{3 \text{ kg} + 5 \text{ kg}}$$
    $$= 0.5 \text{ m/s}$$

6. **D** First, apply Conservation of Linear Momentum to calculate the speed of the combined object after the (perfectly inelastic) collision:

    $$m_1v_1 + m_2v_2 = (m_1 + m_2)v'$$
    $$v' = \frac{m_1v_1 + m_2v_2}{m_1 + m_2}$$
    $$= \frac{m_1v_1 + (2m_1)(0)}{m_1 + 2m_1}$$
    $$= \frac{1}{3}v_1$$

Therefore, the ratio of the kinetic energy after the collision to the kinetic energy before the collision is

$$\frac{K'}{K} = \frac{\frac{1}{2}m'v'^2}{\frac{1}{2}m_1v_1^2} = \frac{\frac{1}{2}(m_1 + 2m_1)(\frac{1}{3}v_1)^2}{\frac{1}{2}m_1v_1^2} = \frac{1}{3}$$

7. **C** Total linear momentum is conserved in a collision during which the net external force is zero. If kinetic energy is lost, then by definition, the collision is not elastic.

8. **D** Because the two carts are initially at rest, the initial momentum is zero. Therefore, the final total momentum must be zero.

9. **C** The linear momentum of the bullet must have the same magnitude as the linear momentum of the block in order for their combined momentum after impact to be zero. The block has momentum MV to the left, so the bullet must have momentum MV to the right. Since the bullet's mass is m, its speed must be $v = MV/m$.

10. **C** In a perfectly inelastic collision, kinetic energy is never conserved; some of the initial kinetic energy is always lost to heat and some is converted to potential energy in the deformed shapes of the objects as they lock together.

11. **B** Total linear momentum is conserved in the absence of external forces. If the final speed of both objects is 0, that means the total linear momentum after the collision is 0, which then implies that the total linear momentum before the collision is also 0. As Object 2 has half the mass of Object 1, Object 2 must have twice the initial speed of Object 1 (and must be travelling in the opposite direction) in order for the total linear momentum to equal 0.

Section II: Free-Response

1. (a) First, draw a free-body diagram:

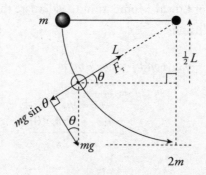

The net force toward the center of the steel ball's circular path provides the centripetal force. From the geometry of the diagram, you get

$$F_T - mg\sin\theta = \frac{mv^2}{L} \qquad (*)$$

In order to determine the value of mv^2, use Conservation of Mechanical Energy:

$$K_i + U_i = K_f + U_f$$
$$0 + mgL = \frac{1}{2}mv^2 + mg(\frac{1}{2}L)$$
$$\frac{1}{2}mgL = \frac{1}{2}mv^2$$
$$mgL = mv^2$$

Substituting this result into Equation (*), you get

$$F_T - mg\sin\theta = \frac{mgL}{L}$$
$$F_T = mg(1+\sin\theta)$$

Now, from the free-body diagram, $\sin\theta = \frac{1}{2}L/L = \frac{1}{2}$, so

$$F_T = mg(1+\frac{1}{2}) = \frac{3}{2}mg$$

(b) Apply Conservation of Energy to find the speed of the ball just before impact:

$$K_i + U_i = K_f + U_f$$
$$0 + mgL = \frac{1}{2}mv^2 + 0$$
$$v = \sqrt{2gL}$$

Now use Conservation of Linear Momentum and conservation of kinetic energy for the elastic collision to derive the expressions for the speeds of the ball, v_1, and the block, v_2, immediately after the collision. Applying Conservation of Linear Momentum yields:

$$m\sqrt{2gL} = mv_1 + 4mv_2$$
$$\sqrt{2gL} = v_1 + 4v_2$$
$$v_1 = \sqrt{2gL} - 4v_2$$

Applying the conservation of kinetic energy for the elastic collision yields:

$$\frac{1}{2}m\left(\sqrt{2gL}\right)^2 = \frac{1}{2}mv_1^2 + \frac{1}{2}(4m)v_2^2$$

$$2gL = v_1^2 + (4m)v_2^2$$

Plugging in the expression for v_1 from the Conservation of Linear Momentum yields:

$$2gL = \left(\sqrt{2gL} - 4v_2\right)^2 + 4v_2^2$$

$$2gL = 2gL - 8\sqrt{2gL}v_2 + 16v_2^2 + 4v_2^2$$

$$0 = 20v_2^2 - 8\sqrt{2gL}v_2$$

$$0 = 2v_2\left(10v_2 - 4\sqrt{2gL}\right)$$

$$0 = 10v_2 - 4\sqrt{2gL}$$

$$10v_2 = 4\sqrt{2gL}$$

$$v_2 = \frac{2\sqrt{2}}{5}\sqrt{gL}$$

(c) Using the velocity of the block immediately after the collision, v_2, solve for the velocity of the ball immediately after the collision:

$$v_1 = \sqrt{2gL} - 4v_2$$

$$v_1 = \sqrt{2gL} - 4\left(\frac{2\sqrt{2}}{5}\sqrt{gL}\right)$$

$$v_1 = \sqrt{2gL} - \frac{4 \cdot 2}{5}\sqrt{2gL}$$

Factor out $\sqrt{2gL}$ to get:

$$v_1 = \sqrt{2gL}\left(1 - \frac{8}{5}\right)$$

$$v_1 = -\frac{3}{5}\sqrt{2gL}$$

Now, apply Conservation of Mechanical Energy to find

$$K_i + U_i = K_f + U_f$$

$$\frac{1}{2}mv_1'^2 + 0 = 0 + mgh$$

$$h = \frac{v_1'^2}{2g} = \frac{\left(\frac{3}{5}\sqrt{2gL}\right)^2}{2g} = \frac{9}{25}L$$

2. (a) By Conservation of Linear Momentum, $mv = (m + M)v'$, so $v' = \dfrac{mv}{m + M}$

Now, by Conservation of Mechanical Energy,

$$K_i + U_i = K_f + U_f$$

$$\frac{1}{2}(m + M)v'^2 + 0 = 0 + (m + M)gy$$

$$\frac{1}{2}v'^2 = gy$$

$$\frac{1}{2}\left(\frac{mv}{m + M}\right)^2 = gy$$

$$v = \frac{m + M}{m}\sqrt{2gy}$$

(b) Use the result derived in part (a) to compute the kinetic energy of the block and bullet immediately after the collision:

$$K' = \frac{1}{2}(m + M)v'^2 = \frac{1}{2}(m + M)\left(\frac{mv}{m + M}\right)^2 = \frac{1}{2}\frac{m^2v^2}{m + M}$$

Since $K = \dfrac{1}{2}mv^2$, the difference is

$$\Delta K = K' - K = \frac{1}{2}\frac{m^2v^2}{m + M} - \frac{1}{2}mv^2$$

$$= \frac{1}{2}mv^2\left(\frac{m}{m + M} - 1\right)$$

$$= K\left(\frac{-M}{m + M}\right)$$

Therefore, the fraction of the bullet's original kinetic energy that was lost is $M/(m + M)$. This energy is manifested as heat (the bullet and block are warmer after the collision than before), and some was used to break the intermolecular bonds within the wooden block to allow the bullet to penetrate.

(c) From the geometry of the diagram,

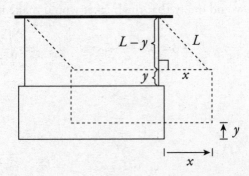

the Pythagorean Theorem implies that $(L-y)^2 + x^2 = L^2$. Therefore,

$$L^2 - 2Ly + y^2 + x^2 = L^2 \quad \Rightarrow \quad y = \frac{x^2}{2L}$$

(where we have used the fact that y^2 is small enough to be neglected). Substituting this into the result of part (a), derive the following equation for the speed of the bullet in terms of x and L instead of y:

$$v = \frac{m+M}{m}\sqrt{2gy} = \frac{m+M}{m}\sqrt{2g\frac{x^2}{2L}} = \frac{m+M}{m}x\sqrt{\frac{g}{L}}$$

(d) No; momentum is conserved only when the net external force on the system is zero (or at least negligible). In this case, the block and bullet feel a net nonzero force that causes it to slow down as it swings upward. Since its speed is decreasing as it swings upward, its linear momentum cannot remain constant.

CHAPTER 8 REVIEW QUESTIONS

Section I: Multiple-Choice

1. **A, C** In uniform circular motion the speed is constant but the velocity is constantly changing because the direction is changing. Because the direction is changing, there is an acceleration. The magnitude of acceleration is constant; however, the acceleration is directed toward the center and perpendicular to the velocity. As a result, since the velocity is always changing, so is the direction of acceleration.

2. **B** When the bucket is at the lowest point in its vertical circle, it feels a tension force $\mathbf{F}_T$ upward and the gravitational force $\mathbf{F}_w$ downward. The net force toward the center of the circle, which is the centripetal force, is $F_T - F_w$. Thus,

$$F_T - F_w = m\frac{v^2}{r} \quad \Rightarrow \quad v = \sqrt{\frac{r(F_T - mg)}{m}} = \sqrt{\frac{(0.60 \text{ m})[50 \text{ N} - (3 \text{ kg})(10 \text{ N/kg})]}{3 \text{ kg}}} = 2 \text{ m/s}$$

3. **C** When the bucket reaches the topmost point in its vertical circle, the forces acting on the bucket are its weight, $\mathbf{F}_w$, and the downward tension force, $\mathbf{F}_T$. The net force, $\mathbf{F}_w + \mathbf{F}_T$, provides the centripetal force. In order for the rope to avoid becoming slack, $\mathbf{F}_T$ must not vanish. Therefore, the cut-off speed for ensuring that the bucket makes it around the circle is the speed at which $\mathbf{F}_T$ just becomes zero; any greater speed would imply that the bucket would make it around. Thus,

$$F_w + F_T = m\frac{v^2}{r} \quad \Rightarrow \quad F_w + 0 = m\frac{v^2_{\text{cut-off}}}{r} \quad \Rightarrow \quad v_{\text{cut-off}} = \sqrt{\frac{rF_w}{m}} = \sqrt{gr}$$

$$= \sqrt{(10 \text{ m/s}^2)(0.60 \text{ m})}$$
$$= 2.4 \text{ m/s}$$

4. **D** Centripetal acceleration is given by the equation $a_c = v^2/r$. Since the object covers a distance of $2\pi r$ in 1 revolution, its speed is $2\pi r$. Therefore,

$$a_c = \frac{v^2}{r} = \frac{(2\pi r)^2}{r} = 4\pi^2 r$$

5. **C** The torque is $\tau = rF = (0.20 \text{ m})(20 \text{ N}) = 4 \text{ N} \cdot \text{m}$.

6. **D** From the diagram,

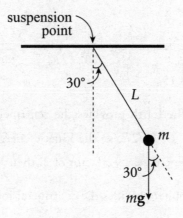

calculate that

$$\tau = rF \sin\theta = Lmg \sin\theta$$
$$= (0.80 \text{ m})(0.50 \text{ kg})(10 \text{ N/kg})(\sin 30°)$$
$$= 2.0 \text{ N} \cdot \text{m}$$

7. **B** The stick will remain at rest in the horizontal position if the torques about the suspension point are balanced:

$$\tau_{CCW} = \tau_{CW}$$
$$r_1 F_1 = r_2 F_2$$
$$r_1 m_1 g = r_2 m_2 g$$
$$m_2 = \frac{r_1 m_1}{r_2} = \frac{(50 \text{ cm})(3 \text{ kg})}{30 \text{ cm}} = 5 \text{ kg}$$

8. **A** Gravitational force obeys an inverse-square law: $F_{grav} \propto 1/r^2$. Therefore, if r increases by a factor of 2, then F_{grav} decreases by a factor of $2^2 = 4$.

9. **D** Mass is an intrinsic property of an object and does not change with location. This eliminates (B). If an object's height above the surface of the Earth is equal to $2R_E$, then its distance from the

center of the Earth is $3R_E$. Thus, the object's distance from the Earth's center increases by a factor of 3, so its weight decreases by a factor of $3^2 = 9$.

10. **C** The gravitational force that the Moon exerts on the planet is equal in magnitude to the gravitational force that the planet exerts on the Moon (Newton's Third Law).

11. **D** The gravitational acceleration at the surface of a planet of mass M and radius R is given by the equation $g = GM/R^2$. Therefore, for the dwarf planet Pluto:

$$g_{\text{Pluto}} = G \frac{M_{\text{Pluto}}}{R_{\text{Pluto}}^2} = G \frac{\dfrac{1}{500} M_{\text{Earth}}}{\left(\dfrac{1}{15} R_{\text{Earth}}\right)^2} = \frac{15^2}{500} \cdot G \frac{M_{\text{Earth}}}{R_{\text{Earth}}^2} = \frac{225}{500}(10g \text{ m/s}^2) = \frac{225}{50} \text{ m/s}^2$$

12. **B** The gravitational pull by the Earth provides the centripetal force on the satellite, so $GMm/R^2 = mv^2/R$. This gives $\frac{1}{2}mv^2 = GMm/2R$, so the kinetic energy K of the satellite is inversely proportional to R. Therefore, if R increases by a factor of 2, then K decreases by a factor of 2.

13. **D** The gravitational pull by Jupiter provides the centripetal force on its moon:

$$G\frac{Mm}{R^2} = \frac{mv^2}{R}$$
$$G\frac{M}{R} = v^2$$
$$G\frac{M}{R} = \left(\frac{2\pi R}{T}\right)^2$$
$$G\frac{M}{R} = \frac{4\pi^2 R^2}{T^2}$$
$$M = \frac{4\pi^2 R^3}{GT^2}$$

14. **D** Let the object's distance from Body A be x; then its distance from Body B is $R - x$. In order for the object to feel no net gravitational force, the gravitational pull by A must balance the gravitational pull by B. Therefore, if you let M denote the mass of the object, then

$$G\frac{m_A M}{x^2} = G\frac{m_B M}{(R-x)^2}$$
$$\frac{m}{x^2} = \frac{4m}{(R-x)^2}$$
$$\frac{(R-x)^2}{x^2} = \frac{4m}{m}$$
$$\left(\frac{R-x}{x}\right)^2 = 4$$
$$\left(\frac{R}{x} - 1\right)^2 = 4$$
$$\frac{R}{x} - 1 = 2$$
$$x = R/3$$

15. **B** Because the planet is spinning clockwise and the velocity is tangent to the circle, the velocity must point down. The acceleration and force point toward the center of the circle.

16. **B, D** If there were no forces or balanced forces, in and out, the satellite would have a net force of zero. If the net force were zero, the satellite would continue in a straight line and not orbit the planet.

The force of gravity produces the centripetal force that keeps the object orbiting (where r is the distance between the center of the planet and the orbiting satellite).

$$F_g \rightarrow F_C$$

$$G\frac{m_{planet}m_{satellite}}{r^2} = m_{satellite}\frac{v^2_{orbit}}{r}$$

$$v_{orbit} = \sqrt{G\frac{m_{planet}}{r}}$$

The mass of the satellite does not determine the orbital speed. Only the mass of the planet and its distance from the center of the planet determine orbital speed.

17. **B** Since the centripetal force always points along a radius toward the center of the circle, and the velocity of the object is always tangent to the circle (and thus perpendicular to the radius), the work done by the centripetal force is zero. Alternatively, since the object's speed remains constant, the Work–Energy Theorem tells you that no work is being performed.

Section II: Free-Response

1. (a) Given a mass of 1 kg, weight F_g = 5 N, and diameter = 8×10^6 m (this gives you $r = 4 \times 10^6$ m), you can fill in this equation:

$$F_G = \frac{Gm_1m_2}{r^2} \Rightarrow m_1 = \frac{F_G r^2}{Gm_2}$$

This becomes:

$$m_1 = \frac{(5\text{ N})(4\times10^6\text{ m})^2}{\left(6.67\times10^{-11}\ \dfrac{\text{N}\cdot\text{m}^2}{\text{kg}^2}\right)(1\text{ kg})} = 1.2 \times 10^{24}\text{ kg}$$

(b)
$$g = \frac{Gm_1}{r^2} \Rightarrow g = \frac{\left(6.67\times10^{-11}\ \dfrac{\text{N}\cdot\text{m}^2}{\text{kg}^2}\right)1.2\times10^{24}\text{ kg}}{(4\times10^6\text{ m})^2} = 5\text{ m/s}^2$$

Note that we could have also observed that, because a 1 kg mass (which normally weighs 10 N on the surface of the Earth) only weighed 5 N, gravity on this planet must be half the Earth's gravity.

If you want to look at it in terms of g, g is 10 m/s^2 on Earth, so you can simply convert.

$$5 \text{ m/s}^2 \left(\frac{1g}{10 \text{ m/s}^2} \right) = 0.5g$$

(c) Density is given by mass per unit volume ($\rho = \frac{m}{V}$). In addition, use the equation for the volume of a sphere as $V = \frac{4}{3}\pi r^3$ to get

$$\rho = \frac{m}{\frac{4}{3}\pi r^3} \Rightarrow \rho = \frac{3m}{4\pi r^3} \Rightarrow \rho = \frac{3(1.2 \times 10^{24} \text{ kg})}{4(3.14)(4 \times 10^6 \text{ m})^3} \text{ or}$$

$$\rho = 4{,}480 \frac{\text{kg}}{\text{m}^3}$$

2. (a) The centripetal acceleration is given by the equation $a_c = \frac{v^2}{R}$. You also know that for objects traveling in circles (Earth's orbit can be considered a circle), $v = \frac{2\pi R}{T}$. Substituting this v into the previous equation for centripetal acceleration, you get $a_c = \frac{4\pi^2 R}{T^2}$. This becomes

$$a_c = \frac{4\pi^2 1.5 \times 10^{11} \text{ m}}{(3.15 \times 10^7 \text{ s})^2} = 6.0 \times 10^{-3} \text{ m/s}^2.$$

(b) The gravitational force is the force that keeps the Earth traveling in a circle around the Sun. More specifically, $ma_c = (6 \times 10^{24} \text{ kg})(6.0 \times 10^{-3} \text{ m/s}^2) = 3.6 \times 10^{22}$ N.

(c) Use the Universal Law of Gravitation: $F = GmM/r^2$. You know F from the previous question, G is constant, m is the mass of the Earth (given in question), and r is the radius from Earth to Sun (also given in the question). So you can rearrange this equation to solve for M (mass of the Sun) as $M = F \cdot r^2/(G \cdot m)$ and then just plug in the appropriate values.

$F = GmM/r^2 \rightarrow M = Fr^2/(Gm) = (3.6 \times 10^{22} \text{ N})(1.5 \times 10^{11} \text{ m})^2/\{[6.67 \times 10^{-11} \text{ Nm}^2/(\text{kg}^2)]$ $[6.0 \times 10^{24} \text{ kg}]\} = 2.0 \times 10^{30}$

3. (a) The forces acting on a person standing against the cylinder wall are gravity ($\mathbf{F}_w$, downward), the normal force from the wall ($\mathbf{F}_N$, directed toward the center of the cylinder), and the force of static friction ($\mathbf{F}_f$, directed upward):

(b) In order to keep the passenger from sliding down the wall, the maximum force of static friction must be at least as great as the passenger's weight: $F_{f\,(max)} \geq mg$. Since $F_{f\,(max)} = \mu_s F_N$, this condition becomes $\mu_s F_N \geq mg$.

Now, consider the circular motion of the passenger. Neither $\mathbf{F}_f$ nor $\mathbf{F}_w$ has a component toward the center of the path, so the centripetal force is provided entirely by the normal force:

$$F_N = \frac{mv^2}{r}$$

Substituting this expression for F_N into the previous equation, you get

$$\mu_s \frac{mv^2}{r} \geq mg$$
$$\mu_s \geq \frac{gr}{v^2}$$

Therefore, the coefficient of static friction between the passenger and the wall of the cylinder must satisfy this condition in order to keep the passenger from sliding down.

(c) Since the mass m canceled out in deriving the expression for μ_s, the conditions are independent of mass. Thus, the inequality $\mu_s \geq gr/v^2$ holds for both the adult passenger of mass m and the child of mass $m/2$.

4. (a) The forces acting on the car are gravity ($\mathbf{F}_w$, downward), the normal force from the road ($\mathbf{F}_N$, upward), and the force of static friction ($\mathbf{F}_f$, directed toward the center of curvature of the road):

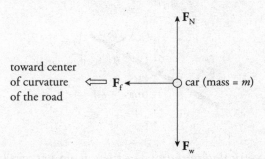

(b) The force of static friction (assume static friction because you *don't* want the car to slide) provides the necessary centripetal force:

$$F_f = \frac{mv^2}{r}$$

Therefore, to find the maximum speed at which static friction can continue to provide the necessary force, write

$$F_{f\,(max)} = \frac{mv_{max}^2}{r}$$

$$\mu_s F_N = \frac{mv_{max}^2}{r}$$

$$\mu_s mg = \frac{mv_{max}^2}{r}$$

$$v_{max} = \sqrt{\mu_s gr}$$

(c) Ignoring friction, the forces acting on the car are gravity ($\mathbf{F}_w$, downward) and the normal force from the road ($\mathbf{F}_N$, which is now tilted toward the center of curvature of the road):

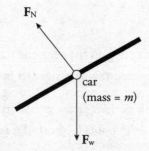

(d) Because of the banking of the turn, the normal force is tilted toward the center of curvature of the road. The component of $\mathbf{F}_N$ toward the center can provide the centripetal force, making reliance on friction unnecessary.

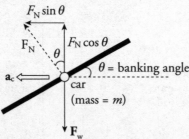

However, there's no vertical acceleration, so there is no net vertical force. Therefore, $F_N \cos\theta = F_w = mg$, so $F_N = mg/\cos\theta$. The component of F_N toward the center of curvature of the turn, $F_N \sin\theta$, provides the centripetal force:

$$F_N \sin\theta = \frac{mv^2}{r}$$

$$\frac{mg}{\cos\theta} \sin\theta = \frac{mv^2}{r}$$

$$g \tan\theta = \frac{v^2}{r}$$

$$\theta = \tan^{-1}\frac{v^2}{gr}$$

CHAPTER 9 REVIEW QUESTIONS

Section I: Multiple-Choice

1. **B, C** The acceleration of a simple harmonic oscillator is not constant, since the restoring force—and consequently, the acceleration—depends on the position. Eliminate (A). Choices (B) and (C) are defining characteristics of simple harmonic motion; therefore, (D) is also false. Period must also be independent of amplitude, as period and frequency are proportional to one another.

2. **B** The acceleration of the block has its maximum magnitude at the points where its displacement from equilibrium has the maximum magnitude (since $a = F/m = kx/m$). At the endpoints of the oscillation region, the potential energy is maximized and the kinetic energy (and hence the speed) is zero.

3. **D** By Conservation of Mechanical Energy, $K + U_S$ is a constant for the motion of the block. At the endpoints of the oscillation region, the block's displacement, x, is equal to $\pm A$. Since $K = 0$ here, all the energy is in the form of potential energy of the spring, $\frac{1}{2}kA^2$. Because $\frac{1}{2}kA^2$ gives the total energy at these positions, it also gives the total energy at any other position.

    Using the equation $U_S(x) = \frac{1}{2}kx^2$, find that, at $x = \frac{1}{2}A$,

    $$K + U_S = \frac{1}{2}kA^2$$
    $$K + \frac{1}{2}k(\frac{1}{2}A)^2 = \frac{1}{2}kA^2$$
    $$K = \frac{3}{8}kA^2$$

    Therefore,

    $$K/E = \frac{3}{8}kA^2 \bigg/ \frac{1}{2}kA^2 = \frac{3}{4}$$

4. **C** As we derived in Chapter 9, Example 2, the maximum speed of the block is given by the equation $v_{max} = A\sqrt{k/m}$. Therefore, v_{max} is inversely proportional to $\sqrt{m}$. If m is increased by a factor of 2, then v_{max} will decrease by a factor of $\sqrt{2}$.

5. **D** The period of a spring–block simple harmonic oscillator is independent of the value of g. (Recall that $T = 2\pi\sqrt{m/k}$.) Therefore, the period will remain the same.

6. **D** The frequency of a spring–block simple harmonic oscillator is given by the equation $f = (1/2\pi)\sqrt{k/m}$. Squaring both sides of this equation, you get $f^2 = (k/4\pi^2)(1/m)$. Therefore, if f^2 is plotted versus $(1/m)$, then the graph will be a straight line with slope $k/4\pi^2$. (Note: The slope of the line whose equation is $y = ax$ is a.)

7. **C** For small angular displacements, the period of a simple pendulum is essentially independent of amplitude.

8. **D** Combining Hooke's Law with Newton's Second Law, you get

$$F = kx = ma \Rightarrow a = \frac{kx}{m} = \frac{50 \text{ N/m} \cdot 4 \text{ m}}{20 \text{ kg}} = 10 \text{ m/s}^2$$

9. **B** By Conservation of Mechanical Energy, the energy of the block is the same throughout the motion. At the amplitude, the block has potential energy $U = \frac{1}{2}kA^2$ and zero kinetic energy. At the equilibrium position, the block has kinetic energy $K = \frac{1}{2}mv^2$ and zero potential energy. Applying the Conservation of Mechanical Energy to these two points in the motion yields

$$\frac{1}{2}kA^2 + 0 = 0 + \frac{1}{2}mv^2$$

$$kA^2 = mv^2$$

$$k = \frac{mv^2}{A^2} = \frac{10 \text{ kg} \cdot (4 \text{ m/s})^2}{(2 \text{ m})^2} = 40 \text{ kg/s}_2$$

The period of the block can then be calculated using the equation,

$$T = 2\pi\sqrt{\frac{m}{k}} = 2\pi\sqrt{\frac{10 \text{ kg}}{40 \text{ kg/s}^2}} = \pi \text{ s} \approx 3 \text{ s}$$

10. **B** The frequency of a spring-block simple harmonic oscillator is independent of the amplitude. The equation for the frequency of a spring-block simple harmonic oscillator is $f = \frac{1}{2\pi}\sqrt{\frac{k}{m}}$. The frequency is inversely proportional to the square root of the mass, so decreasing the mass of the block by a factor of 4 would increase the frequency by a factor of 2.

Section II: Free-Response

1. (a) Since the spring is compressed to 3/4 of its natural length, the block's position relative to equilibrium is $x = -\frac{1}{4}L$. Therefore, from $F_s = -kx$, find

$$a = \frac{F_s}{m} = \frac{-k(-\frac{1}{4}L)}{m} = \frac{kL}{4m}$$

(b) Let v_1 denote the velocity of Block 1 just before impact, and let v_1' and v_2' denote, respectively, the velocities of Block 1 and Block 2 immediately after impact. By Conservation of Linear Momentum, write $mv_1 = mv_1' + mv_2'$, or

$$v_1 = v_1' + v_2' \qquad (1)$$

The initial kinetic energy of Block 1 is $\frac{1}{2}mv_1^2$. If half is lost to heat, then $\frac{1}{4}mv_1^2$ is left to be shared

by Block 1 and Block 2 after impact: $\frac{1}{4}mv_1^2 = \frac{1}{2}mv_1'^2 + \frac{1}{2}mv_2'^2$, or

$$v_1^2 = 2v_1'^2 + 2v_2'^2 \qquad (2)$$

Square Equation (1) and multiply by 2 to give

$$2v_1^2 = 2v_1'^2 + 4v_1'v_2' + 2v_2'^2 \qquad (1')$$

then subtract Equation (2) from Equation (1'):

$$v_1^2 = 4v_1'v_2' \qquad (3)$$

Square Equation (1) again,

$$v_1^2 = v_1'^2 + 2v_1'v_2' + v_2'^2$$

and substitute into this the result of Equation (3):

$$4v_1'v_2' = v_1'^2 + 2v_1'v_2' + v_2'^2$$
$$0 = v_1'^2 - 2v_1'v_2' + v_2'^2$$
$$0 = (v_1' - v_2')^2$$
$$v_1' = v_2' \qquad (4)$$

Thus, combining Equations (1) and (4), find that

$$v_1' = v_2' = \frac{1}{2}v_1$$

(c) When Block 1 reaches its new amplitude position, A', all of its kinetic energy is converted to elastic potential energy of the spring. That is,

$$K_1' \rightarrow U_s' \quad \Rightarrow \quad \frac{1}{2}mv_1'^2 = \frac{1}{2}kA'^2$$
$$A'^2 = \frac{m}{k}v_1'^2$$
$$A'^2 = \frac{m}{k}\left(\frac{1}{2}v_1\right)^2$$
$$A'^2 = \frac{mv_1^2}{4k} \qquad (1)$$

But the original potential energy of the spring, $U_S = \frac{1}{2}k(-\frac{1}{4}L)^2$, gave K_1:

$$U_S \rightarrow K_1 \quad \Rightarrow \quad \frac{1}{2}k(-\frac{1}{4}L)^2 = \frac{1}{2}mv_1^2 \quad \Rightarrow \quad mv_1^2 = \frac{1}{16}kL^2 \quad (2)$$

Substituting this result into Equation (1) gives

$$A'^2 = \frac{\frac{1}{16}kL^2}{4k} = \frac{L^2}{64} \quad \Rightarrow \quad A' = \frac{1}{8}L$$

(d) The period of a spring–block simple harmonic oscillator depends only on the spring constant k and the mass of the block. Since neither of these changes, the period will remain the same; that is, $T' = T_0$.

(e) As shown in part (b), Block 2's velocity as it slides off the table is $\frac{1}{2}v_1$ (horizontally). The time required to drop the vertical distance H is found as follows (calling *down* the positive direction):

$$\Delta y = v_{0y}t + \frac{1}{2}gt^2 \quad \Rightarrow \quad H = \frac{1}{2}gt^2 \quad \Rightarrow \quad t = \sqrt{\frac{2H}{g}}$$

Therefore,

$$R = (\frac{1}{2}v_1)t = \frac{1}{2}v_1\sqrt{\frac{2H}{g}}$$

Now, from Equation (2) of part (c), $v_1 = \sqrt{\frac{kL^2}{16m}}$, so

$$R = \frac{1}{2}\sqrt{\frac{kL^2}{16m}}\sqrt{\frac{2H}{g}} = \frac{L}{8}\sqrt{\frac{2kH}{mg}}$$

2. (a) By Conservation of Linear Momentum,

$$mv = (m+M)v' \quad \Rightarrow \quad v' = \frac{mv}{m+M}$$

(b) When the block is at its amplitude position (maximum compression of spring), the kinetic energy it (and the embedded bullet) had just after impact will become the potential energy of the spring:

$$K' \rightarrow U_S$$

$$\frac{1}{2}(m+M)\left(\frac{mv}{m+M}\right)^2 = \frac{1}{2}kA^2$$

$$A = \frac{mv}{\sqrt{k(m+M)}}$$

(c) Since the mass on the spring is $m + M$, $f = \frac{1}{2\pi}\sqrt{\frac{k}{m+M}}$.

3. (a) By Conservation of Mechanical Energy, $K + U = E$, so

$$\frac{1}{2}Mv^2 + \frac{1}{2}k(\frac{1}{2}A)^2 = \frac{1}{2}kA^2$$
$$\frac{1}{2}Mv^2 = \frac{3}{8}kA^2$$
$$v = A\sqrt{\frac{3k}{4M}}$$

(b) Since the clay ball delivers no horizontal linear momentum to the block, horizontal linear momentum is conserved. Thus,

$$Mv = (M+m)v'$$
$$v' = \frac{Mv}{M+m} = \frac{MA}{M+m}\sqrt{\frac{3k}{4M}} = \frac{A}{M+m}\sqrt{\frac{3kM}{4}}$$

(c) Applying the general equation for the period of a spring–block simple harmonic oscillator,

$$T = 2\pi\sqrt{\frac{M+m}{k}}$$

(d) The total energy of the oscillator after the clay hits is $\frac{1}{2}kA'^2$, where A' is the new amplitude. Just after the clay hits the block, the total energy is

$$K' + U_S = \frac{1}{2}(M+m)v'^2 + \frac{1}{2}k(\frac{1}{2}A)^2$$

Substitute for v' from part (b), set the resulting sum equal to $\frac{1}{2}kA'^2$, and solve for A'.

$$\frac{1}{2}(M+m)\left(\frac{A}{M+m}\sqrt{\frac{3kM}{4}}\right)^2 + \frac{1}{2}k(\frac{1}{2}A)^2 = \frac{1}{2}kA'^2$$
$$\frac{A^2 \cdot 3kM}{8(M+m)} + \frac{1}{8}kA^2 = \frac{1}{2}kA'^2$$
$$A^2\left(\frac{3M}{M+m}+1\right) = 4A'^2$$
$$A' = \frac{1}{2}A\sqrt{\frac{3M}{M+m}+1}$$

(e) No, because the period depends only on the mass and the spring constant k.

(f) Yes. For example, if the clay had landed when the block was at $x = A$, the speed of the block would have been zero immediately before the collision and immediately after. No change in the block's speed would have meant no change in K, so no change in E, so no change in $A = \sqrt{2E/k}$.

CHAPTER 10 REVIEW QUESTIONS

Section I: Multiple-Choice

1. **C** From the equation $\lambda f = v$, find that

$$\lambda = \frac{v}{f} = \frac{10 \text{ m/s}}{5 \text{ Hz}} = 2 \text{ m}$$

2. **C** The speed of a transverse traveling wave on a stretched rope is given by the equation $v = \sqrt{F_T/\mu}$. Therefore,

$$v = \sqrt{\frac{F_T}{m/L}} = \sqrt{\frac{80 \text{ N}}{(1 \text{ kg})/(5 \text{ m})}} = \sqrt{400 \text{ m}^2/\text{s}^2} = 20 \text{ m/s}$$

3. **C** The time interval from a point moving from its maximum displacement above $y = 0$ (equilibrium) to its maximum displacement below equilibrium is equal to one-half the period of the wave. In this case,

$$T = \frac{1}{f} = \frac{\lambda}{v} = \frac{8 \text{ m}}{2 \text{ m/s}} = 4 \text{ s}$$

so the desired time is $\frac{1}{2}(4 \text{ s}) = 2 \text{ s}$.

4. **D** The standard equation for a transverse traveling wave has the form $y = A \sin (\omega t \pm kx)$, where ω is the angular frequency and k is the angular wave number. Here, $k = \frac{1}{2}\pi$ cm^{-1}. Since, by definition, $k = 2\pi/\lambda$, you get

$$\lambda = \frac{2\pi}{k} = \frac{2\pi}{\frac{1}{2}\pi \text{ cm}^{-1}} = 4 \text{ cm}$$

5. **D** The distance between successive nodes is always equal to $\frac{1}{2}\lambda$. If a standing wave on a string fixed at both ends has a total of 4 nodes, the string must have a length L equal to $3(\frac{1}{2}\lambda)$. If $L = 6$ m, then λ must equal 4 m.

6. **B** The previous question showed that $\lambda = 4$ m. Since $v = 40$ m/s, the frequency of this standing wave must be

$$f = \frac{v}{\lambda} = \frac{40 \text{ m/s}}{4 \text{ m}} = 10 \text{ Hz}$$

7. **A** In general, sound travels faster through solids than through gases. Therefore, when the wave enters the air from the metal rod, its speed will decrease. The frequency, however, will not change. Since $v = \lambda f$ must always be satisfied, a decrease in v implies a decrease in λ.

8.   A   The distance from S_2 to P is 5 m (it's the hypotenuse of 3-4-5 triangle), and the distance from S_1 to P is 4 m. The difference between the path lengths to Point P is 1 m, which is half the wavelength. Therefore, the sound waves are always exactly out of phase when they reach Point P from the two speakers, causing destructive interference there. By contrast, since Point Q is equidistant from the two speakers, the sound waves will always arrive in phase at Q, interfering constructively. Since there is destructive interference at P and constructive interference at Q, the amplitude at P will be less than at Q.

9.   B   An air column (such as an organ pipe) with one closed end resonates at frequencies given by the equation $f_n = nv/(4L)$ for odd integers n. The fundamental frequency corresponds, by definition, to $n = 1$. Therefore,

$$f_1 = \frac{v}{4L} = \frac{340 \text{ m/s}}{4(0.17 \text{ m})} = 500 \text{ Hz}$$

10. B   The speed of the chirp is

$$v = \lambda f = (8.75 \times 10^{-3} \text{ m})(40 \times 10^3 \text{ Hz}) = 350 \text{ m/s}$$

If the distance from the bat to the tree is d, then the wave travels a total distance of $d + d = 2d$ (round-trip distance). If T is the time for this round-trip, then

$$2d = vT \quad \Rightarrow \quad d = \frac{vT}{2} = \frac{(350 \text{ m/s})(0.4 \text{ s})}{2} = 70 \text{ m}$$

11. A   Since the car is traveling away from the stationary detector, the observed frequency will be lower than the source frequency. This eliminates (B) and (C). Using the Doppler effect equation, find that

$$f_D = \frac{v}{v + v_S} \cdot f_S = \frac{340 \text{ m/s}}{(340 + 20) \text{ m/s}} \cdot (600 \text{ Hz}) = \frac{34}{36}(600 \text{ Hz})$$

Section II: Free-Response

1.   (a)   The speed of a transverse traveling wave on a stretched rope is given by the equation $v = \sqrt{F_T/\mu}$. Therefore,

$$v^2 = \frac{F_T}{\mu} \quad \Rightarrow \quad F_T = \mu v^2 = (0.4 \text{ kg/m})(12 \text{ m/s})^2 = 58 \text{ N}$$

(b)   Use the fundamental equation $\lambda f = v$:

$$f = \frac{v}{\lambda} = \frac{12 \text{ m/s}}{2 \text{ m}} = 6 \text{ Hz}$$

(c) (i) Because higher harmonic numbers correspond to shorter wavelengths, the harmonic number of the 3.2 m standing wave must be higher than that of the 4 m standing wave. You're told that these harmonic numbers are consecutive integers, so if n is the harmonic number of the 4 m standing wave, then $n + 1$ is the harmonic number of the 3.2 m wave. Therefore,

$$\frac{2L}{n} = 4 \text{ m} \qquad \text{and} \qquad \frac{2L}{n+1} = 3.2 \text{ m}$$

The first equation says that $2L = 4n$, and the second one says that $2L = 3.2(n + 1)$. Therefore, $4n$ must equal $3.2(n + 1)$; solving this equation gives $n = 4$. Substituting this into either one of the displayed equations then gives $L = 8$ m.

(ii) Because $\mu = 0.4$ kg/m, the mass of the rope must be

$$m = \mu L = (0.4 \text{ kg/m})(8 \text{ m}) = 3.2 \text{ kg}$$

(d) You determined this in the solution to part (c) (i). The 4 m standing wave has harmonic number $n = 4$.

(e)

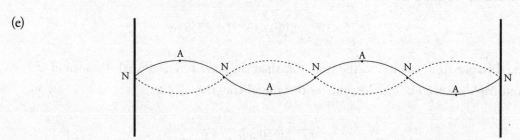

2. (a) For a stationary observer and a source moving away, the Doppler effect equation predicts that

$$f_D = \frac{v}{v + v_C} \cdot f_C = \frac{345 \text{ m/s}}{(345 + 50) \text{ m/s}}(500 \text{ Hz}) = 437 \text{ Hz}$$

(b) Yes. If the source is stationary and the observer moves away, the Doppler effect equation now gives

$$f_D = \frac{v - v_D}{v} \cdot f_C = \frac{(345 - 50) \text{ m/s}}{345 \text{ m/s}}(500 \text{ Hz}) = 428 \text{ Hz}$$

(c) Since $\theta = 60°$, the given equation becomes

$$f' = \frac{v}{v - v_C \cos\theta} \cdot f = \frac{345 \text{ m/s}}{345 \text{ m/s} - (50 \text{ m/s})\cos 60°}(500 \text{ Hz}) = 539 \text{ Hz}$$

Note that the observed frequency is higher than the source frequency, which you can expect since the car is traveling toward the students.

(d) With $\theta = 120°$, the given equation becomes

$$f' = \frac{v}{v - v_C \cos\theta} \cdot f = \frac{345 \text{ m/s}}{345 \text{ m/s} - (50 \text{ m/s})\cos 120°}(500 \text{ Hz}) = 466 \text{ Hz}$$

Note that the observed frequency is now lower than the source frequency, which you can expect since the car is traveling away from the students.

(e) When the car is far to the left of the students' position, θ is very small (approaches 0°). Therefore, for large negative x, the observed frequency should approach

$$f' = \frac{v}{v - v_C \cos\theta} \cdot f = \frac{345 \text{ m/s}}{345 \text{ m/s} - (50 \text{ m/s})\cos 0°}(500 \text{ Hz}) = 585 \text{ Hz}$$

When the car is far to the right of the students' position, θ approaches 180°. Therefore, for large positive x, the observed frequency should approach

$$f' = \frac{v}{v - v_C \cos\theta} \cdot f = \frac{345 \text{ m/s}}{345 \text{ m/s} - (50 \text{ m/s})\cos 180°}(500 \text{ Hz}) = 437 \text{ Hz}$$

When the car is directly in front of the students—that is, when $x = 0$ (and $\theta = 90°$)—the equation given predicts that f' will equal 500 Hz. The graph of f' versus x should therefore have the following general shape:

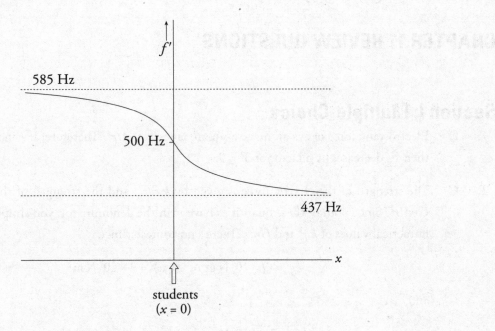

3. (a) Use the equation for the beat frequency,

$$f_{\text{beat}} = |f_1 - f_2| = |400 \text{ Hz} - 440 \text{ Hz}| = 40 \text{ Hz}$$

(b) As sound travels from air into water, the speed of sound will increase as sound travels faster in liquids than gases (recall the equation $v = \sqrt{\frac{B}{\rho}}$). By Wave Rule #2, when a wave passes into another medium, its speed changes but its frequency does not. By the equation $v = f\lambda$, it can be observed that if speed increases and frequency does not change then the wavelength must increase.

(c) As the sound source (tuning fork) is traveling toward the stationary detector (the student on the second floor), the observed frequency will be higher than the source frequency. However, the question is asking what happens to the observed frequency as the tuning fork travels upward. Due to acceleration from gravity, the speed of the tuning fork as it moves up will decrease. As the relative speed between the source and detector is decreasing, the impact of the Doppler effect will decrease. This will lead to a decrease in the observed frequency as the tuning fork travels upward (although the observed frequency will still be higher than the source frequency).

CHAPTER 11 REVIEW QUESTIONS

Section I: Multiple-Choice

1. **D** Electrostatic force obeys an inverse-square law: $F_E \propto 1/r^2$. Therefore, if r increases by a factor of 3, then F_E decreases by a factor of $3^2 = 9$.

2. **C** The strength of the electric force is given by kq^2/r^2, and the strength of the gravitational force is Gm^2/r^2. Since both of these quantities have r^2 in the denominator, you simply need to compare the numerical values of kq^2 and Gm^2. There's no contest: Since

$$kq^2 = (9 \times 10^9 \text{ N·m}^2/\text{C}^2)(1 \text{ C})^2 = 9 \times 10^9 \text{ N·m}^2$$

and

$$Gm^2 = (6.7 \times 10^{-11} \text{ N·m}^2/\text{kg}^2)(1 \text{ kg})^2 = 6.7 \times 10^{-11} \text{ N·m}^2$$

you can see that $kq^2 > Gm^2$, so F_E is much stronger than F_G.

3. **C** If the net electric force on the center charge is zero, the electrical repulsion by the $+2q$ charge must balance the electrical repulsion by the $+3q$ charge:

$$\frac{1}{4\pi\varepsilon_0}\frac{(2q)(q)}{x^2} = \frac{1}{4\pi\varepsilon_0}\frac{(3q)(q)}{y^2} \implies \frac{2}{x^2} = \frac{3}{y^2} \implies \frac{y^2}{x^2} = \frac{3}{2} \implies \frac{y}{x} = \sqrt{\frac{3}{2}} \, a = \frac{F_E}{m} = \frac{1}{4\pi\varepsilon_0}\frac{Qq}{mr^2}$$

4.   **D**   Since P is equidistant from the two charges, and the magnitudes of the charges are identical, the strength of the electric field at P due to $+Q$ is the same as the strength of the electric field at P due to $-Q$. The electric field vector at P due to $+Q$ points away from $+Q$, and the electric field vector at P due to $-Q$ points toward $-Q$. Since these vectors point in the same direction, the net electric field at P is (E to the right) + (E to the right) = ($2E$ to the right).

5.   **C**   The acceleration of the small sphere is

$$a = \frac{F_E}{m} = \frac{1}{4\pi\varepsilon_0} \frac{Qq}{mr^2}$$

As r increases (that is, as the small sphere is pushed away), a decreases. However, since a is always positive, the small sphere's speed, v, is always increasing.

6.   **B**   Since $\mathbf{F}_E$ (on q) = $q\mathbf{E}$, it must be true that $\mathbf{F}_E$ (on $-2q$) = $-2q\mathbf{E}$ = $-2\mathbf{F}_E$.

7.   **C**   All excess electric charge on a conductor resides on the outer surface.

8.   **D**   Use Coulomb's Law to solve for r,

$$F_E = k\frac{q_1 q_2}{r^2}$$

$$r^2 = \frac{kq_1 q_2}{F_E}$$

$$r^2 = \sqrt{\frac{kq_1 q_2}{F_E}} = \sqrt{\frac{(9 \times 10^9 \text{ N} \cdot \text{m}^2/\text{C}^2)(8 \times 10^{-6} \text{ C})(6 \times 10^{-6} \text{ C})}{2.7 \times 10^{-2} \text{ N}}} = \sqrt{\frac{432 \times 10^{-3} \text{ m}^2}{2.7 \times 10^{-2} \text{ N}}} = 4 \text{ m}$$

9.   **C** F_G and F_E are both proportional to $1/r^2$. Increasing the distance between the charges by a factor of 3 will reduce the magnitude of both forces by a factor of 9, but the ratio of F_G/F_E will remain the same.

Section II: Free-Response

1.   (a) From the figure below, you have $F_{1\text{-}2} = F_1/\cos 45°$.

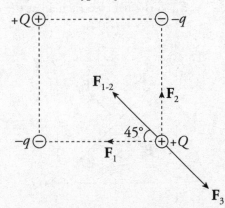

Since the net force on $+Q$ is zero, you want $F_{1\text{-}2} = F_3$. If s is the length of each side of the square, then:

$$F_{1\text{-}2} = F_3 \quad \Rightarrow \quad \frac{F_1}{\cos 45°} = F_3 \quad \Rightarrow \quad \frac{1}{\cos 45°}\frac{1}{4\pi\varepsilon_0}\frac{Qq}{s^2} = \frac{1}{4\pi\varepsilon_0}\frac{Q^2}{(s\sqrt{2})^2}$$

$$\sqrt{2}\cdot q = \frac{Q}{2}$$

$$q = \frac{Q}{2\sqrt{2}}$$

(b)   No. If $q = Q/2\sqrt{2}$, as found in part (a), then the net force on $-q$ is not zero.

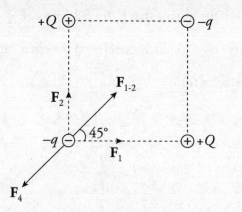

This is because $F_{1\text{-}2} \neq F_4$, as the following calculations show:

$$F_{1\text{-}2} = \frac{F_1}{\cos 45°} = \sqrt{2}\,\frac{1}{4\pi\varepsilon_0}\frac{Qq}{s^2} = \sqrt{2}\,\frac{1}{4\pi\varepsilon_0}\frac{Q\frac{Q}{2\sqrt{2}}}{s^2} = \frac{1}{8\pi\varepsilon_0}\frac{Q^2}{s^2}$$

but

$$F_4 = \frac{1}{4\pi\varepsilon_0}\frac{q^2}{(s\sqrt{2})^2} = \frac{1}{4\pi\varepsilon_0}\frac{\left(\frac{Q}{2\sqrt{2}}\right)^2}{(s\sqrt{2})^2} = \frac{1}{64\pi\varepsilon_0}\frac{Q^2}{s^2}$$

(c)   By symmetry, $E_1 = E_2$ and $E_3 = E_4$, so the net electric field at the center of the square is zero:

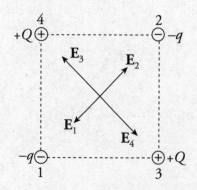

2. (a) The magnitude of the electric force on Charge 1 is

$$F_1 = \frac{1}{4\pi\varepsilon_0} \frac{(Q)(2Q)}{(a+2a)^2} = \frac{1}{18\pi\varepsilon_0} \frac{Q^2}{a^2}$$

The direction of $\mathbf{F}_1$ is directly away from Charge 2; that is, in the $+y$ direction, so

$$\mathbf{F}_1 = \frac{1}{18\pi\varepsilon_0} \frac{Q^2}{a^2} \hat{\mathbf{j}}$$

(b) The electric field vectors at the origin due to Charge 1 and due to Charge 2 are

$$\mathbf{E}_1 = \frac{1}{4\pi\varepsilon_0} \frac{Q}{a^2} (-\hat{\mathbf{j}}) \qquad \text{and} \qquad \mathbf{E}_2 = \frac{1}{4\pi\varepsilon_0} \frac{2Q}{(2a)^2} (+\hat{\mathbf{j}}) = \frac{1}{8\pi\varepsilon_0} \frac{Q}{a^2} (+\hat{\mathbf{j}})$$

Therefore, the net electric field at the origin is

$$\mathbf{E} = \mathbf{E}_1 + \mathbf{E}_2 = \frac{1}{4\pi\varepsilon_0} \frac{Q}{a^2} (-\hat{\mathbf{j}}) + \frac{1}{8\pi\varepsilon_0} \frac{Q}{a^2} (+\hat{\mathbf{j}}) = \frac{1}{8\pi\varepsilon_0} \frac{Q}{a^2} (-\hat{\mathbf{j}})$$

(c) No. The only point on the x-axis where the individual electric field vectors due to each of the two charges point in exactly opposite directions is the origin $(0, 0)$. But at that point, the two vectors are not equal and thus do not cancel.

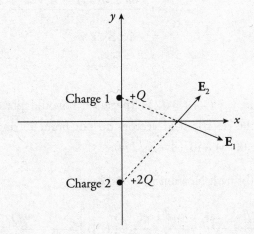

Therefore, at no point on the x-axis could the total electric field be zero.

(d)   Yes. There will be a Point P on the y-axis between the two charges,

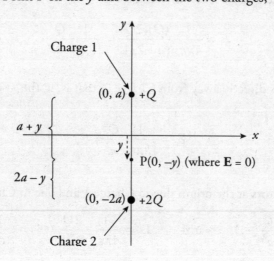

where the electric fields due to the individual charges will cancel each other out.

$$E_1 = E_2$$
$$\frac{1}{4\pi\varepsilon_0}\frac{Q}{(a+y)^2} = \frac{1}{4\pi\varepsilon_0}\frac{2Q}{(2a-y)^2}$$
$$\frac{1}{(a+y)^2} = \frac{2}{(2a-y)^2}$$
$$(2a-y)^2 = 2(a+y)^2$$
$$y^2 + 8ay - 2a^2 = 0$$
$$y = \frac{-8a \pm \sqrt{(8a)^2 - 4(-2a^2)}}{2}$$
$$= (-4 \pm 3\sqrt{2})a$$

Disregarding the value $y = (-4-3\sqrt{2})a$ (because it would place the Point P below Charge 2 on the y-axis, where the electric field vectors do not point in opposite directions), you find that $\mathbf{E} = \mathbf{0}$ at the Point P = $(0, -y) = (0, (4-3\sqrt{2})a)$.

(e)   Use the result of part (b) with Newton's Second Law:

$$\mathbf{a} = \frac{\mathbf{F}}{m} = \frac{-q\mathbf{E}}{m} = \frac{-q}{m}\left[\frac{1}{8\pi\varepsilon_0}\frac{Q}{a^2}(-\hat{\mathbf{j}})\right] = \frac{1}{8\pi\varepsilon_0}\frac{qQ}{ma^2}(+\hat{\mathbf{j}})$$

3.   (a)   Using the superposition principle, the electric field at the point (0, 4) is equal to the sum of the electric fields from q_1 and q_2.

$$\mathbf{E} = \mathbf{E}_{q_1} + \mathbf{E}_{q_2}$$

The magnitude of the horizontal and vertical components of the electric field at this point is then the sum of the individual horizontal and vertical components from q_1 and q_2

$$E_x = E_{q_1,x} + E_{q_2,x}$$
$$E_y = E_{q_1,y} + E_{q_2,y}$$

As electric fields point away from the positive charges, the electric field due to q_1 points upward in the vertical direction and has no horizontal component. The distance between q_1 and the point $(0, 4)$ is $r_1 = 7$ m. The electric field from q_2 points northwest, making a 45 degrees angle with the horizontal from the point $(0, 4)$, and has both a horizontal and vertical component. Using the Pythagorean Theorem, the distance between q_2 and the point $(0, 4)$ is $r^2 = \sqrt{(4 \text{ m})^2 + (4 \text{ m})^2} = 4\sqrt{2}$ m.

$$E_x = E_{q_1} + E_{q_2,x} = 0 + E_{q_2} \cos(45°) = \frac{kq_2}{r_2^2} \cos(45°) = \frac{(9 \times 10^9 \text{ N} \cdot \text{m}^2/\text{C}^2)(3 \times 10^{-6} \text{ C})}{(4\sqrt{2} \text{ m})^2} \times \frac{\sqrt{2}}{2} = 597 \text{ N/C}$$

$$E_y = E_{q_1,y} + E_{q_2,y} = \frac{kq_2}{r_1^2} + \frac{kq_2}{r_2^2} \sin(45°) = \frac{(9 \times 10^9 \text{ N} \cdot \text{m}^2/\text{C}^2)(4 \times 10^{-6} \text{ C})}{(7 \text{ m})^2} + \frac{(9 \times 10^9 \text{ N} \cdot \text{m}^2/\text{C}^2)(3 \times 10^{-6} \text{ C})}{(4\sqrt{2} \text{ m})^2} = 1331 \text{ N/C}$$

(b) When a charge is placed at a point with an electric field, the electric force felt by the charge is equal to the product of the charge and the electric field at that point.

$$F_E = qE$$

To determine the magnitude of the horizontal and vertical components of the electric field, plug in in the horizontal and vertical components of the electric field at the point solved in (a).

$$F_{E,x} = q_3 E_x = 2 \cdot 10^{-3} \text{ C} (597 \text{ N/C}) = 1.2 \text{ N}$$

(Note that the sign of q_3 is excluded, as it does not affect the magnitude of the electric force)

$$F_{E,y} = qE_y = -2 \cdot 10^{-3} \text{ C} (1331 \text{ N/C}) = 2.7 \text{ N}$$

(c) In order for the electric field at the point $(0, 4)$ to equal 0, the electric field produced by q_4 must be equal in magnitude but opposite in sign to the electric field produced from q_1 and q_2. As the electric field produced by q_1 points north and the electric field produced by q_2 points northwest, the electric field produced by q_4 must have a vertical component that points south and a horizontal component that points east. As electric fields point away from positive charges, this is possible only if q_4 is to the north and west of the point $(0, 4)$. The only points that fit this criterion are in Quadrant II.

CHAPTER 12 REVIEW QUESTIONS

Section I: Multiple-Choice

1. **C**   The equation $I = V/R$ implies that increasing V by a factor of 2 will cause I to increase by a factor of 2.

2. **C**   Use the equation $P = V^2/R$:

$$P = \frac{V^2}{R} \quad \Rightarrow \quad R = \frac{V^2}{P} = \frac{(120 \text{ V})^2}{60 \text{ W}} = 240 \text{ }\Omega$$

3. **B**   The current through the circuit is

$$I = \frac{\mathcal{E}}{r + R} = \frac{40 \text{ V}}{(5 \text{ }\Omega) + (15 \text{ }\Omega)} = 2 \text{ A}$$

Therefore, the voltage drop across R is $V = IR = (2 \text{ A})(15 \text{ }\Omega) = 30 \text{ V}$.

4. **D**   The 12 Ω and 4 Ω resistors are in parallel and are equivalent to a single 3 Ω resistor, because $\frac{1}{12 \text{ }\Omega} + \frac{1}{4 \text{ }\Omega} = \frac{1}{3 \text{ }\Omega}$. This 3 Ω resistor is in series with the top 3 Ω resistor, giving an equivalent resistance in the top branch of $3 + 3 = 6 \text{ }\Omega$. Finally, this 6 Ω resistor is in parallel with the bottom 3 Ω resistor, giving an overall equivalent resistance of 2 Ω, because $\frac{1}{6 \text{ }\Omega} + \frac{1}{3 \text{ }\Omega} = \frac{1}{2 \text{ }\Omega}$.

5. **C**   If each of the identical bulbs has resistance R, then the current through each bulb is ε/R. This is unchanged if the middle branch is taken out of the parallel circuit. (What *will* change is the total amount of current provided by the battery.)

6. **B**   The three parallel resistors are equivalent to a single 2 Ω resistor, because $\frac{1}{8 \text{ }\Omega} + \frac{1}{4 \text{ }\Omega} + \frac{1}{8 \text{ }\Omega} = \frac{1}{2 \text{ }\Omega}$. This 2 Ω resistance is in series with the given 2 Ω resistor, so their equivalent resistance is $2 + 2 = 4 \text{ }\Omega$. Therefore, three times as much current will flow through this equivalent 4 Ω resistance in the top branch as through the parallel 12 Ω resistor in the bottom branch, which implies that the current through the bottom branch is 3 A, and the current through the top branch is 9 A. The voltage drop across the 12 Ω resistor is therefore $V = IR = (3 \text{ A})(12 \text{ }\Omega) = 36 \text{ V}$.

7. **D** Since points *a* and *b* are grounded, they're at the same potential (call it zero).

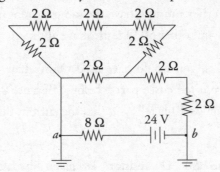

Traveling from *b* to *a* across the battery, the potential increases by 24 V, so it must decrease by 24 V across the 8 Ω resistor as we reach point *a*. Thus, $I = V/R = (24\ \text{V})/(8\ \Omega) = 3\ \text{A}$.

8. **D** The equation $P = I^2R$ gives

$$P = (0.5\ \text{A})^2(100\ \Omega) = 25\ \text{W} = 25\ \text{J/s}$$

Therefore, in 20 s, the energy dissipated as heat is

$$E = Pt = (25\ \text{J/s})(20\ \text{s}) = 500\ \text{J}$$

9. **C** As the resistance of resistors in series is additive, the addition of a fourth resistor (light bulb) in series will increase the total resistance of the circuit. As the voltage source (battery) is unchanged, this will lead to a decrease in the current supplied by the battery according to the equation $V = IR$.

Section II: Free-Response

1. (a) The two parallel branches, the one containing the 40 Ω resistor and the other a total of 120 Ω, is equivalent to a single 30 Ω resistance. This 30 Ω resistance is in series with the three 10 Ω resistors, giving an overall equivalent circuit resistance of $10 + 10 + 30 + 10 = 60\ \Omega$. Therefore, the current supplied by the battery is $I = V/R = (120\ \text{V})/(60\ \Omega) = 2\ \text{A}$, so it must supply energy at a rate of $P = IV = (2\ \text{A})(120\ \text{V}) = 240\ \text{W}$.

   (b) Since three times as much current will flow through the 40 Ω resistor as through the branch containing 120 Ω of resistance, the current through the 40 Ω resistor must be 1.5 A.

   (c) (i) $V_a - V_b = IR_{20} + IR_{100} = (0.5\ \text{A})(20\ \Omega) + (0.5\ \text{A})(100\ \Omega) = 60\ \text{V}$.

   (ii) Point *a* is at the higher potential (current flows from high to low potential).

   (d) Because energy is equal to power multiplied by time, you get

$$E = Pt = I^2Rt = (0.5\ \text{A})^2(100\ \Omega)(10\ \text{s}) = 250\ \text{J}$$

2. (a) There are many ways to solve this problem. If you notice that each of the three branches of the parallel section is 60 Ω, then they must all have the same current flowing through them. The currents through the 20 Ω, 40 Ω, and other 60 Ω resistor are all 0.5 A.

    If you had not noticed this, you would have used Ohm's Law to determine the voltage across the resistors and proceeded from there (see part b below). Because 0.5 A goes through each of the three pathways, Kirchhoff's Junction Rule tells you that the current that must have come though the 10 Ω resistor is 1.5 A.

    (b) The voltage across the 60 Ω resistor is given by Ohm's Law. Because $V = IR$, $V = (0.5 \text{ A})(60 \text{ }\Omega) = 30$ V. All three parallel branches must have the same voltage across them, so the other 60 Ω resistor also has 30 V across it and the combination of the 20 Ω and 40 Ω resistor must also be 30 V. To determine the voltage across the 20 Ω and 40 Ω resistor you can rely on the previously solved currents of 0.5 A and Ohm's Law to yield $V = IR = (0.5 \text{ A})(20 \text{ }\Omega) = 10$ V and $V = (0.5\text{A})(40 \text{ }\Omega) = 20$ V.

    You also could have used the ratio of the resistors. That is, we know the two voltages must sum to 30 V and the voltage drop across the 40 Ω must be twice the amount across the 20 Ω. The voltage across the 10 Ω resistor can be found using Ohm's Law: $V = (1.5 \text{ A})(10 \text{ }\Omega) = 15$ V.

    (c) The equivalent resistance of the circuit can be solved either by adding the resistances or by using Ohm's Law.

    If you want to add resistances, start by summing the 20 Ω and 40 Ω resistors to get 60 Ω. Then add the three parallel branches using $\dfrac{1}{R_P} = \sum_i \dfrac{1}{R_i}$ or $\dfrac{1}{R_P} = \sum \dfrac{1}{60 \text{ }\Omega} + \dfrac{1}{60 \text{ }\Omega} + \dfrac{1}{60 \text{ }\Omega}$, which becomes 20 Ω. Then adding this section in series to the 10 Ω resistor to get $R_P = R_1 + R_2 = 10 \text{ }\Omega + 20 \text{ }\Omega = 30 \text{ }\Omega$.

    You could have also realized that the total voltage drop across the battery is 45 V (15 V across the 10 Ω resistor and 30 V across the parallel branch). Using Ohm's Law again $R_{eq} = \dfrac{V_B}{I_B}$ or $R_{eq} = \dfrac{45 \text{ V}}{1.5 \text{ A}} = 30 \text{ }\Omega$.

Part VI
Practice Test 2

Practice Test 2

AP® Physics 1 Exam

SECTION I: Multiple-Choice Questions

DO NOT OPEN THIS BOOKLET UNTIL YOU ARE TOLD TO DO SO.

At a Glance

Total Time
90 minutes
Number of Questions
50
Percent of Total Grade
50%
Writing Instrument
Pen required

Instructions

Section I of this examination contains 50 multiple-choice questions. Fill in only the ovals for numbers 1 through 50 on your answer sheet.

CALCULATORS MAY BE USED ON BOTH SECTIONS OF THE AP PHYSICS 1 EXAM.

Indicate all of your answers to the multiple-choice questions on the answer sheet. No credit will be given for anything written in this exam booklet, but you may use the booklet for notes or scratch work. Please note that there are two types of multiple-choice questions: single-select and multi-select questions. After you have decided which of the suggested answers is best, completely fill in the corresponding oval(s) on the answer sheet. For single-select, you must give only one answer; for multi-select you must give BOTH answers in order to earn credit. If you change an answer, be sure that the previous mark is erased completely. Here is a sample question and answer.

Sample Question   Sample Answer

Chicago is a
(A) state
(B) city
(C) country
(D) continent

Use your time effectively, working as quickly as you can without losing accuracy. Do not spend too much time on any one question. Go on to other questions and come back to the ones you have not answered if you have time. It is not expected that everyone will know the answers to all the multiple-choice questions.

About Guessing

Many candidates wonder whether or not to guess the answers to questions about which they are not certain. Multiple-choice scores are based on the number of questions answered correctly. Points are not deducted for incorrect answers, and no points are awarded for unanswered questions. Because points are not deducted for incorrect answers, you are encouraged to answer all multiple-choice questions. On any questions you do not know the answer to, you should eliminate as many choices as you can, and then select the best answer among the remaining choices.

GO ON TO THE NEXT PAGE.

ADVANCED PLACEMENT PHYSICS 1 TABLE OF INFORMATION

CONSTANTS AND CONVERSION FACTORS	
Proton mass, $m_p = 1.67 \times 10^{-27}$ kg	Electron charge magnitude, $e = 1.60 \times 10^{-19}$ C
Neutron mass, $m_n = 1.67 \times 10^{-27}$ kg	Coulomb's law constant, $k = 1/4\pi\varepsilon_0 = 9.0 \times 10^9$ N·m^2/C^2
Electron mass, $m_e = 9.11 \times 10^{-31}$ kg	Universal gravitational constant, $G = 6.67 \times 10^{-11}$ m^3/kg·s^2
Speed of light, $c = 3.00 \times 10^8$ m/s	Acceleration due to gravity at Earth's surface, $g = 9.8$ m/s^2

UNIT SYMBOLS	meter,	m	kelvin,	K	watt,	W	degree Celsius,	°C
	kilogram,	kg	hertz,	Hz	coulomb,	C		
	second,	s	newton,	N	volt,	V		
	ampere,	A	joule,	J	ohm,	Ω		

PREFIXES		
Factor	Prefix	Symbol
10^{12}	tera	T
10^9	giga	G
10^6	mega	M
10^3	kilo	k
10^{-2}	centi	c
10^{-3}	milli	m
10^{-6}	micro	μ
10^{-9}	nano	n
10^{-12}	pico	p

VALUES OF TRIGONOMETRIC FUNCTIONS FOR COMMON ANGLES							
θ	$0°$	$30°$	$37°$	$45°$	$53°$	$60°$	$90°$
$\sin\theta$	0	1/2	3/5	$\sqrt{2}/2$	4/5	$\sqrt{3}/2$	1
$\cos\theta$	1	$\sqrt{3}/2$	4/5	$\sqrt{2}/2$	3/5	1/2	0
$\tan\theta$	0	$\sqrt{3}/3$	3/4	1	4/3	$\sqrt{3}$	∞

The following conventions are used in this exam.
  I. The frame of reference of any problem is assumed to be inertial unless otherwise stated.
  II. Assume air resistance is negligible unless otherwise stated.
  III. In all situations, positive work is defined as work done on a system.
  IV. The direction of current is conventional current: the direction in which positive charge would drift.
  V. Assume all batteries and meters are ideal unless otherwise stated.

GO ON TO THE NEXT PAGE.

ADVANCED PLACEMENT PHYSICS 1 EQUATIONS, EFFECTIVE 2015

MECHANICS

$$v_x = v_{x0} + a_x t$$

$$x = x_0 + v_{x0}t + \frac{1}{2}a_x t^2$$

$$v_x^2 = v_{x0}^2 + 2a_x(x - x_0)$$

$$\vec{a} = \frac{\sum \vec{F}}{m} = \frac{\vec{F}_{net}}{m}$$

$$\left|\vec{F}_f\right| \leq \mu \left|\vec{F}_n\right|$$

$$a_c = \frac{v^2}{r}$$

$$\vec{p} = m\vec{v}$$

$$\Delta \vec{p} = \vec{F}\Delta t$$

$$K = \frac{1}{2}mv^2$$

$$\Delta E = W = F_{\parallel}d = Fd\cos\theta$$

$$P = \frac{\Delta E}{\Delta t}$$

$$\theta = \theta_0 + \omega_0 t + \frac{1}{2}\alpha t^2$$

$$\omega = \omega_0 + \alpha t$$

$$x = A\cos(2\pi ft)$$

$$\vec{\alpha} = \frac{\sum \vec{\tau}}{I} = \frac{\vec{\tau}_{net}}{I}$$

$$\tau = r_{\perp}F = rF\sin\theta$$

$$L = I\omega$$

$$\Delta L = \tau \Delta t$$

$$K = \frac{1}{2}I\omega^2$$

$$\left|\vec{F}_s\right| = k\left|\vec{x}\right|$$

$$U_s = \frac{1}{2}kx^2$$

$$\rho = \frac{m}{V}$$

$$\Delta U_g = mg\Delta y$$

$$T = \frac{2\pi}{\omega} = \frac{1}{f}$$

$$T_s = 2\pi\sqrt{\frac{m}{k}}$$

$$T_p = 2\pi\sqrt{\frac{\ell}{g}}$$

$$\left|\vec{F}_g\right| = G\frac{m_1 m_2}{r^2}$$

$$\vec{g} = \frac{\vec{F}_g}{m}$$

$$U_G = -\frac{Gm_1 m_2}{r}$$

a = acceleration
A = amplitude
d = distance
E = energy
f = frequency
F = force
I = rotational inertia
K = kinetic energy
k = spring constant
L = angular momentum
ℓ = length
m = mass
P = power
p = momentum
r = radius or separation
T = period
t = time
U = potential energy
V = volume
v = speed
W = work done on a system
x = position
y = height
α = angular acceleration
μ = coefficient of friction
θ = angle
ρ = density
τ = torque
ω = angular speed

ELECTRICITY

$$\left|\vec{F}_E\right| = k\left|\frac{q_1 q_2}{r^2}\right|$$

$$I = \frac{\Delta q}{\Delta t}$$

$$R = \frac{\rho\ell}{A}$$

$$I = \frac{\Delta V}{R}$$

$$P = I\Delta V$$

$$R_s = \sum_i R_i$$

$$\frac{1}{R_p} = \sum_i \frac{1}{R_i}$$

A = area
F = force
I = current
ℓ = length
P = power
q = charge
R = resistance
r = separation
t = time
V = electric potential
ρ = resistivity

WAVES

$$\lambda = \frac{v}{f}$$

f = frequency
v = speed
λ = wavelength

GEOMETRY AND TRIGONOMETRY

Rectangle
$$A = bh$$

Triangle
$$A = \frac{1}{2}bh$$

Circle
$$A = \pi r^2$$
$$C = 2\pi r$$

Rectangular solid
$$V = \ell wh$$

Cylinder
$$V = \pi r^2 \ell$$
$$S = 2\pi r\ell + 2\pi r^2$$

Sphere
$$V = \frac{4}{3}\pi r^3$$
$$S = 4\pi r^2$$

A = area
C = circumference
V = volume
S = surface area
b = base
h = height
ℓ = length
w = width
r = radius

Right triangle
$$c^2 = a^2 + b^2$$
$$\sin\theta = \frac{a}{c}$$
$$\cos\theta = \frac{b}{c}$$
$$\tan\theta = \frac{a}{b}$$

GO ON TO THE NEXT PAGE.

THIS PAGE IS LEFT INTENTIONALLY BLANK.

GO ON TO THE NEXT PAGE.

AP PHYSICS 1

SECTION I

Note: To simplify calculations, you may use $g = 10$ m/s² in all problems.

Directions: Each of the questions or incomplete statements is followed by four suggested answers or completions. Select the one that is best in each case and then fill in the corresponding circle on the answer sheet.

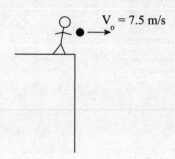

$V_o = 7.5$ m/s

1. An object is thrown horizontally to the right off a high cliff with an initial speed of 7.5 m/s. Which arrow best represents the direction of the object's velocity after 2 seconds? (Assume air resistance is negligible.)

   (A)

   (B) ⟍

   (C) ↓

   (D) ↓

Questions 2-4 refer to the following figure:

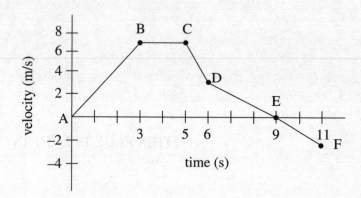

2. Which of the following correctly ranks the change in kinetic energy for each segment from least to greatest?

   (A) BC, EF, DE, CD, AB
   (B) AB, CD, DE, EF, BC
   (C) BC, EF, DE, CD, AB
   (D) CD, DE, EF, BC, AB

3. During which segment is the magnitude of average acceleration greatest?

   (A) AB
   (B) BC
   (C) CD
   (D) DE

4. What is the total distance traveled by the object?

   (A) 32 m
   (B) 34 m
   (C) 36 m
   (D) 38 m

GO ON TO THE NEXT PAGE.

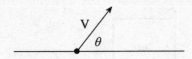

5. A ball is thrown in a projectile motion trajectory with an initial velocity v at an angle θ above the ground. If the acceleration due to gravity is $-g$, which of the following is the correct expression of the time it takes for the ball to reach its highest point, y, from the ground?

(A) $v^2 \sin t/g$

(B) $-v \cos \theta/g$

(C) $v \sin \theta/g$

(D) $v^2 \cos \theta/g$

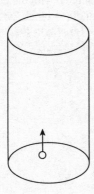

6. A bubble in a glass of water releases from rest at the bottom of the glass and rises at acceleration a to the surface in t seconds. How much farther does the bubble travel in its last second than in its first second?

(A) at

(B) $(t-1)a$

(C) $(t+1)a$

(D) $\dfrac{1}{2}at$

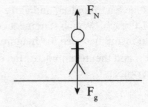

7. A person standing on a horizontal floor is acted upon by two forces: the downward pull of gravity and the upward normal force of the floor. These two forces

(A) have equal magnitudes and form an action-reaction pair

(B) have equal magnitudes and do not form an action-reaction pair

(C) have unequal magnitudes and form an action-reaction pair

(D) have unequal magnitudes and do not form an action-reaction pair

8. Which of the following graphs best represents the force of friction on an object starting at rest that eventually starts sliding across a level surface due to a gradually increasing horizontal force?

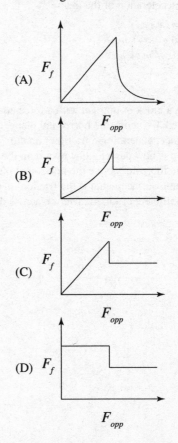

GO ON TO THE NEXT PAGE.

9. Two objects have a mass 1 kg and carry a charge of magnitude 1 C each. Which statement correctly identifies the relationship between the magnitude of the force of gravity, F_g, and the magnitude of the electric force, F_E, between the objects?

    (A) $F_g > F_E$
    (B) $F_g < F_E$
    (C) $F_g = F_E$
    (D) Cannot be determined without knowing the sign of the charges.

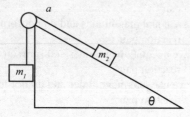

10. Consider the above configuration of masses attached via a massless rope and pulley over a frictionless inclined plane. What is the acceleration of the masses?

    (A) $(m_1 - m_2) g/(m_1 + m_2)$
    (B) $(m_1 - m_2 \sin \theta) g/(m_1 + m_2)$
    (C) $(m_1 - m_2 \cos \theta) g/(m_1 + m_2)$
    (D) g

11. A person is pulling a block of mass m with a force equal to its weight directed 30° above the horizontal plane across a rough surface, generating a friction f on the block. If the person is now pushing downward on the block with the same force 30° above the horizontal plane across the same rough surface, what is the friction on the block? (μ_k is the coefficient of kinetic friction across the surface.)

    (A) f
    (B) $1.5f$
    (C) $2f$
    (D) $3f$

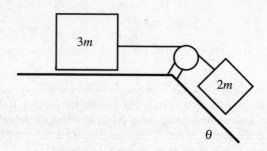

12. In the figure above, two blocks of mass $3m$ and $2m$ are attached together. The plane is frictionless and the pulley is frictionless and massless. The inclined portion of the plane creates an angle θ with the horizontal floor. What is the acceleration of the block $2m$ if both blocks are released from rest (gravity = g)?

    (A) $2mg$

    (B) $\left(\dfrac{2}{5}\right) g \sin \theta$

    (C) $\left(\dfrac{2}{3}\right) g \sin \theta$

    (D) $\left(\dfrac{3}{5}\right) g \sin \theta$

13. If a roller coaster cart of mass m was not attached to the track, it would still remain in contact with a track throughout a loop of radius r as long as

    (A) $v \le \sqrt{(rg)}$
    (B) $v \ge \sqrt{(rg)}$
    (C) $v \le \sqrt{(rg/m)}$
    (D) $v \ge \sqrt{(rg/m)}$

GO ON TO THE NEXT PAGE.

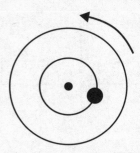

14. The diagram above shows a top view of an object of mass M on a circular platform of mass $2M$ that is rotating counterclockwise. Assume the platform rotates without friction. Which of the following best describes an action by the object that will increase the angular speed of the entire system?

(A) The object moves toward the center of the platform, increasing the total angular momentum of the system.
(B) The object moves toward the center of the platform, decreasing the rotational inertia of the system.
(C) The object moves away from the center of the platform, increasing the total angular momentum of the system.
(D) The object moves away from the center of the platform, decreasing the rotational inertia of the system.

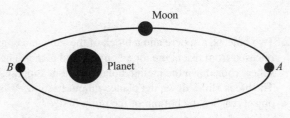

15. A moon has an elliptical orbit about the planet as shown above. At point A, the Moon has speed v_A and is at a distance r_A from the planet. At point B, the Moon has a speed of v_B. Which of the following correctly explains the method for determining the distance of the Moon from the planet at point B in the given quantities?

(A) Conservation of Angular Momentum, because the gravitational force exerted by the Moon on the planet is the same as that exerted by the planet on the moon
(B) Conservation of Angular Momentum, because the gravitational force exerted on the Moon is always directed toward the planet
(C) Conservation of Energy, because the gravitational force exerted on the Moon is always directed toward the planet
(D) Conservation of Energy, because the gravitational force exerted by the Moon on the planet is the same as that exerted by the planet on the Moon

16. A sphere starts from rest atop a hill with a constant angle of inclination and is allowed to roll without slipping down the hill. What force provides the torque that causes the sphere to rotate?

(A) Static friction
(B) Kinetic friction
(C) The normal force of the hill on the sphere
(D) Gravity

17. Which of the following correctly describes the motion of a real object in free fall? Assume that the object experiences drag force proportional to speed and that it strikes the ground before reaching terminal sped.

(A) It will fall with increasing speed and increasing acceleration.
(B) It will fall with increasing speed and decreasing acceleration.
(C) It will fall with decreasing speed and increasing acceleration.
(D) It will fall with decreasing speed and decreasing acceleration.

18. Which of the following concerning uniform circular motion is true?

(A) The centrifugal force is the action-reaction pair of the centripetal force.
(B) The centripetal acceleration and velocity point in the same direction.
(C) The velocity of the object in motion changes, whereas the acceleration of the object is constant.
(D) A satellite undergoing uniform circular motion is falling toward the center in a circular path.

GO ON TO THE NEXT PAGE.

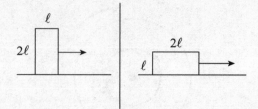

19. A girl of mass *m* and a boy of mass 2*m* are sitting on opposite sides of a see-saw with its fulcrum in the center. Right now, the boy and girl are equally far from the fulcrum, and it tilts in favor of the boy. Which of the following would NOT be a possible method to balance the see-saw?

    (A) Move the boy to half his original distance from the fulcrum.
    (B) Move the girl to double her original distance from the fulcrum.
    (C) Allow a second girl of mass *m* to join the first.
    (D) Move the fulcrum to half its original distance from the boy.

21. A wooden block experiences a frictional force, *f*, as it slides across a table. If a block of the same material with half the height and twice the length were to slide across the table, what would be the frictional force it experienced?

    (A) (1/2)*f*
    (B) *f*
    (C) 2*f*
    (D) 4*f*

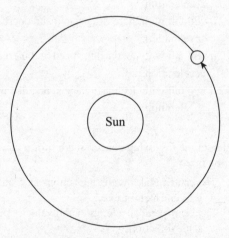

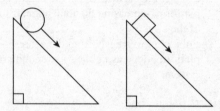

20. Given that the Earth's mass is *m*, its tangential speed as it revolves around the Sun is *v*, and the distance from the Sun to the Earth is *r*, which of the following correctly describes the work done by the centripetal force, W_c, in one year's time?

    (A) $W_c > 2r(mv^2/r)$
    (B) $W_c = 2r(mv^2/r)$
    (C) $W_c < 2r(mv^2/r)$
    (D) Cannot be determined

22. Two objects, a sphere and a block of the same mass, are released from rest at the top of an inclined plane. The sphere rolls down the inclined plane without slipping. The block slides down the plane without friction. Which object reaches the bottom of the ramp first?

    (A) The sphere, because it gains rotational kinetic energy, but the block does not
    (B) The sphere, because it gains mechanical energy due to the torque exerted on it, but the block does not
    (C) The block, because it does not lose mechanical energy due to friction, but the sphere does
    (D) The block, because it does not gain rotational kinetic energy, but the sphere does

GO ON TO THE NEXT PAGE.

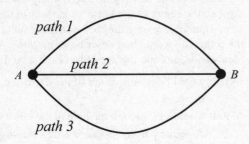

path 1

path 2

A   B

path 3

23. In the diagram above, a mass m starting at point A is projected with the same initial horizontal velocity v_0 along each of the three tracks shown here (with negligible friction) sufficient in each case to allow the mass to reach the end of the track at point B. (Path 1 is directed up, path 2 is directed horizontal, and path 3 is directed down.) The masses remain in contact with the tracks throughout their motions. The displacement A to B is the same in each case, and the total path length of path 1 and 3 are equal. If t_1, t_2, and t_3 are the total travel times between A and B for paths 1, 2, and 3, respectively, what is the relation among these times?

(A) $t_3 < t_2 < t_1$
(B) $t_2 < t_3 < t_1$
(C) $t_2 < t_1 = t_3$
(D) $t_2 = t_3 < t_1$

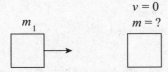

m_1

$v = 0$
$m = ?$

24. An object of mass m_1 experiences a linear, elastic collision with a stationary object of unknown mass. In addition to m_1, what is the minimum necessary information that would allow you to determine the mass of the second object?

(A) The final speed of object 1
(B) The initial speed of object 1
(C) The final speed of object 2
(D) Any 2 of the above values

25. A block is dragged along a table and experiences a frictional force, f, that opposes its movement. The force exerted on the block by the table is

(A) zero
(B) parallel to the table
(C) perpendicular to the table
(D) neither parallel nor perpendicular to the table

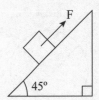

F

45°

26. A box of mass m is sitting on an incline of 45° and it requires an applied force F up the incline to get the box to begin to move. What is the maximum coefficient of static friction?

(A) $\left(\dfrac{\sqrt{2}F}{mg}\right) - 1$

(B) $\left(\dfrac{\sqrt{2}F}{mg}\right)$

(C) $\left(\dfrac{\sqrt{2}F}{mg}\right) + 1$

(D) $\left(\dfrac{2F}{mg}\right) - 1$

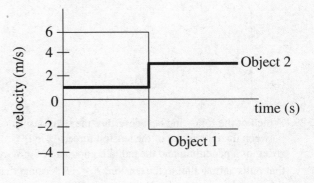

27. The graph above shows the velocities of two objects undergoing a head-on collision. Given that Object 1 has 4 times the mass of Object 2, which type of collision is it?

(A) Perfectly elastic
(B) Perfectly inelastic
(C) Inelastic
(D) Cannot be determined

GO ON TO THE NEXT PAGE.

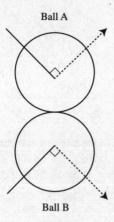

Ball A

Ball B

28. The picture above depicts the collision of two balls of equal mass. Which arrow best indicates the direction of the impulse on Ball A from Ball B during the collision?

(A)

(B)

(C)

(D)

29. Which of the following best describes the relationship between the magnitude of the tension force, F_T, in the string of a pendulum and the radial component of gravity that pulls antiparallel to the tension, $F_{g, \text{radial}}$? Assume that the pendulum is only displaced by a small amount.

(A) $F_T > F_{g, \text{radial}}$
(B) $F_T \geq F_{g, \text{radial}}$
(C) $F_T = F_{g, \text{radial}}$
(D) $F_T \leq F_{g, \text{radial}}$

30. Pretend someone actually managed to dig a hole straight through the center of the Earth all the way to the other side. If an object were dropped down that hole, which of the following would best describe its motion? Assume ideal conditions and that the object cannot be destroyed.

(A) It would fall to the center of the Earth and stop there.
(B) It would fall through the hole to the other side, continue past the opposite side's opening, and fly into space.
(C) It would oscillate back and forth from one opening to the other indefinitely.
(D) It would fall to the other side and stop there.

31. A sound wave with frequency f travels through air at speed v. With what speed will a sound wave with frequency $4f$ travel through the air?

(A) $v/4$
(B) v
(C) $2v$
(D) $4v$

32. If an object's kinetic energy is doubled what happens to its speed?

(A) It increases by a factor of $\sqrt{2}/2$.
(B) It increases by a factor of $\sqrt{2}$.
(C) It is doubled.
(D) It is quadrupled.

33. A toy car and a toy truck collide. If the toy truck's mass is double the toy car's mass, then, compared to the acceleration of the truck, the acceleration of the car during the collision will be

(A) double the magnitude and in the same direction
(B) double the magnitude and in the opposite direction
(C) half the magnitude and in the same direction
(D) half the magnitude and in the opposite direction

GO ON TO THE NEXT PAGE.

34. The Gravitron is a carnival ride that looks like a large cylinder. People stand inside the cylinder against the wall as it begins to spin. Eventually, it is rotating fast enough that the floor can be removed without anyone falling. Given then the coefficient of friction between a person's clothing and the wall is μ, the tangential speed is v, and the radius of the ride is r, what is greatest mass that a person can be to safely go on this ride?

    (A) $\mu v^2/(rg)$
    (B) $r^2v^2/(\mu g)$
    (C) $rg/(\mu v^2)$
    (D) None of the above.

35. In a spring-block oscillator, the maximum speed of the block is

    (A) proportional to amplitude
    (B) proportional to the square of amplitude
    (C) proportional to the square root of amplitude
    (D) inversely proportional to the square root of amplitude

36. A student is experimenting with a simple spring-block oscillator of spring constant k and amplitude A. The block attached to the spring has a mass of M. If the student places a small block of mass m on top of the original block, which of the following is true?

    (A) The small block is most likely to slide off when the original block is at maximum displacement from the equilibrium position, but will not slide off as long as the coefficient of static friction between the blocks is greater than $kA/[(M+m)g]$.
    (B) The small block is most likely to slide off when the original block is at the equilibrium position, but will not slide off as long as the coefficient of static friction between the blocks is greater than $kA/[(M+m)g]$.
    (C) The small block is most likely to slide off when the original block is at maximum displacement from the equilibrium position, but will not slide off as long as the coefficient of static friction between the blocks is greater than $(M+m)g/(kA)$.
    (D) The small block is most likely to slide off when the original block is at the equilibrium position, but will not slide off as long as the coefficient of static friction between the blocks is greater than $(M+m)g/(kA)$.

37. A flute supports standing waves with pressure nodes at each end. The lowest note a flute can play is 261.63 Hz. What is the approximate length of the flute? (speed of sound in air = 343 m/s)

    (A) 32.8 cm
    (B) 65.5 cm
    (C) 76.3 cm
    (D) 131 cm

38. You are standing on a railroad track as a train approaches at a constant velocity. Suddenly the engineer sees you, applies the brakes, and sounds the whistle. Which of the following describes the sound of the whistle as you hear it starting from that moment?

    (A) Loudness increasing, pitch increasing
    (B) Loudness increasing, pitch constant
    (C) Loudness decreasing, pitch increasing
    (D) Loudness increasing, pitch decreasing

GO ON TO THE NEXT PAGE.

Questions 39-41 refer to the following figure:

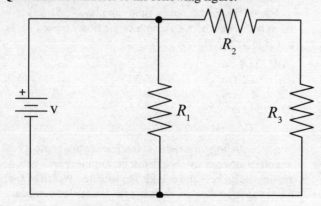

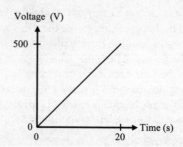

39. Determine the total power dissipated through the circuit shown above in terms of V, R_1, R_2, and R_3.

(A) $\dfrac{V^2}{R_1 + R_2 + R_3}$

(B) $\dfrac{R_1 + R_2 + R_3}{V^2}$

(C) $\dfrac{R_1(R_2 + R_3)}{V^2(R_1 + R_2 + R_3)}$

(D) $\dfrac{V^2(R_1 + R_2 + R_3)}{R_1(R_2 + R_3)}$

40. If $V = 100$ V, $R_1 = 50\ \Omega$, $R_2 = 80\ \Omega$ and $R_3 = 120\ \Omega$, determine the voltage across R_3.

(A) 100 V
(B) 60 V
(C) 40 V
(D) 20 V

41. If R_1 were to burn out, the current coming out from the battery would

(A) increase
(B) decrease
(C) stay the same
(D) There is no current, because the circuit is now incomplete.

42. A circuit consists of a 500 Ω resistor connected to a variable voltage source. The voltage is increased linearly from 0 V to 5 V over a period of 20 s, as shown in the graph above. Which of the following graphs corresponds to the power dissipated by the resistor as a function of time?

(A)

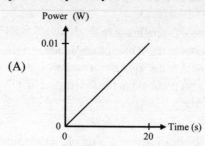

(B)

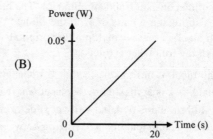

(C)

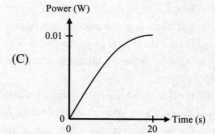

(D)

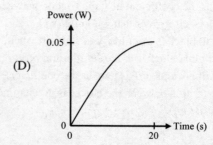

GO ON TO THE NEXT PAGE.

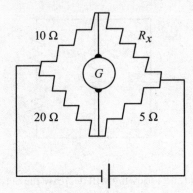

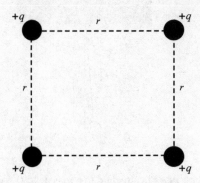

43. A Wheatstone bridge (diagram above) is a configuration of resistors and a sensitive current meter, called a *galvanometer*, that is used to determine the resistance of an unknown resistor. In the Wheatstone bridge shown here, find the value of R_x such that the current through galvanometer G is zero.

    (A) 25 Ω
    (B) 15 Ω
    (C) 10 Ω
    (D) 2.5 Ω

44. What happens to the electric force between two point charges if the magnitude of both charges are doubled, and the distance between them is halved?

    (A) The force is halved.
    (B) The force remains the same.
    (C) The force is quadrupled.
    (D) The force increases by a factor of 16.

45. In the figure above, four charges are arranged. If the magnitudes of all the charges q are all the same and the distance r between them is as shown above, what is the magnitude of the net force on the bottom right charge in terms of q, r, and k (where $k = \dfrac{1}{4\pi\varepsilon_0}$)?

    (A) $k\left(\dfrac{q^2}{2r^2}\right)(1 + \sqrt{2})$

    (B) $k\left(\dfrac{q^2}{r^2}\right)(1 + \sqrt{2})$

    (C) $k\left(\dfrac{q^2}{2r^2}\right)$

    (D) $k\left(\dfrac{q^2}{r^2}\right)$

Directions: For each of the questions 46-50, <u>two</u> of the suggested answers will be correct. Select the two answers that are best in each case, and then fill in both of the corresponding circles on the answer sheet.

46. An object traveling at x m/s can stop at a distance d m with a maximum negative acceleration. If the car is traveling at $2x$ m/s, which of the following statements are true? Select two answers.

    (A) The stopping time is doubled.
    (B) The stopping time is quadrupled.
    (C) The stopping distance is doubled.
    (D) The stopping distance is quadrupled.

GO ON TO THE NEXT PAGE.

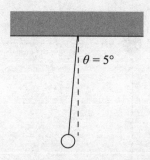

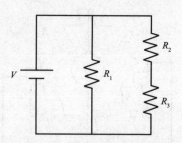

47. A 2 kg mass is attached to a massless, 0.5 m string and is used as a simple pendulum by extending it to an angle $\theta = 5°$ and allowing it to oscillate. Which of the following changes will change the period of the pendulum? Select two answers.

   (A) Replacing the mass with a 1 kg mass
   (B) Changing the initial extension of the pendulum to a 10° angle
   (C) Replacing the string with a 0.25 m string
   (D) Moving the pendulum to the surface of the Moon

48. N resistors ($N > 2$) are connected in parallel with a battery of voltage V_o. If one of the resistors is removed from the circuit, which of the following quantities will decrease? Select two answers.

   (A) The voltage across any of the remaining resistors
   (B) The current output by the battery
   (C) The total power dissipated in the circuit
   (D) The voltage supplied by the battery

49. Which of the following will decrease the current through R_3 in the circuit above? Select two answers.

   (A) A decrease in R_1
   (B) An increase in R_1
   (C) An increase in R_2
   (D) An increase in R_3

50. A sound wave travels through a metal rod with wavelength λ and frequency f. Which of the following is true? Select two answers.

   (A) When this sound wave passes into air, the frequency will change.
   (B) When this sound wave passes into air, the wavelength will change.
   (C) While in the metal rod, the λ and frequency f have a direct relationship.
   (D) While in the metal rod, the λ and frequency f have an inverse relationship.

END OF SECTION I

DO NOT CONTINUE UNTIL INSTRUCTED TO DO SO.

AP PHYSICS 1
SECTION II
Free-Response Questions

Time—90 minutes

Percent of total grade—50

<u>General Instructions</u>

Use a separate piece of paper to answer these questions. Show your work. Be sure to write CLEARLY and LEGIBLY. If you make an error, you may save time by crossing it out rather than trying to erase it.

GO ON TO THE NEXT PAGE.

AP PHYSICS 1

SECTION II

Directions: Questions 1, 2, and 3 are short free-response questions that require about 13 minutes to answer and are worth 8 points. Questions 4 and 5 are long free-response questions that require about 25 minutes each to answer and are worth 13 points each. Show your work for each part in the space provided after that part.

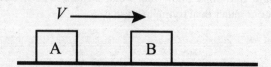

1. An experiment is conducted in which Block A with a mass of m_A is slid to the right across a frictionless table. Block A collides with Block B, which is initially at rest, of an unknown mass and sticks to it.

   (a) Describe an experimental procedure that determines the velocities of the blocks before and after a collision. Include all the additional equipment you need. You may include a labeled diagram of your setup to help in your description. Indicate what measurements you would take and how you would take them. Include enough detail so that the experiment could be repeated with the procedure you provide.

   (b) If Block A has a mass of 0.5 kg and starts off with a speed of 1.5 m/s and the experiment is repeated, the velocity of the blocks after the collide are recorded to be 0.25 m/s. What is the mass of Block B?

   (c) How much kinetic energy was lost in this collision from part (b)?

GO ON TO THE NEXT PAGE.

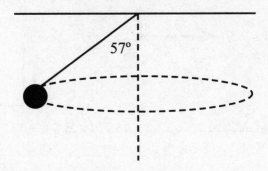

2. A conical pendulum is hanging from a string that is 2.2 meters long. It makes a horizontal circle. The mass of the ball at the end of the string is 0.5 kg.

    (a)   Below, make a free-body diagram for the ball at the point shown in the above illustration. Label each force with an appropriate letter.

    (b)   Write out Newton's Second Law in both the x- and y-direction in terms used in your free-body diagram.

    (c)   Calculate the centripetal acceleration from your free-body diagram.

    (d)   What is the radius of the circle that the ball is traveling in?

    (e)   What is the speed of the ball?

GO ON TO THE NEXT PAGE.

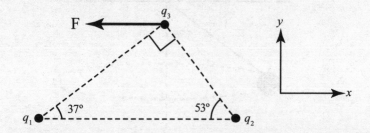

3. Three particles are fixed in place in a horizontal plane, as shown in the figure above. Particle 1 has a charge that has a magnitude of 4.0×10^{-6} C and the sign of the charge is unknown. Particle 2 has a charge that has a magnitude of 1.7×10^{-6} C and the sign of the charge is unknown. Particle 3 has a charge of $+1.0 \times 10^{-6}$ C. The distance between q_1 and q_2 is 5.0 m, the distance between q_2 and q_3 is 3.0 m, and the distance between q_1 and q_3 is 4.0 m. The electrostatic force $\mathbf{F}$ on particle 3 due to the other two charges is shown in the negative x-direction.

    (a)   Determine the signs of the charges of q_1 and q_2.

    (b)   On the diagram below, draw and label the force F_1 of the force exerted by Particle 1 on Particle 3 and the force F_2 of the force exerted by Particle 2 on Particle 3.

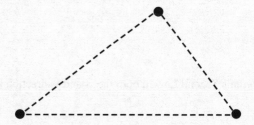

    (c)   Calculate the magnitude of the electrostatic force on Particle 3.

    (d)   Draw and label clearly where another positively charged particle could be placed so the net electrostatic force on Particle 3 is zero.

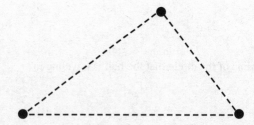

GO ON TO THE NEXT PAGE.

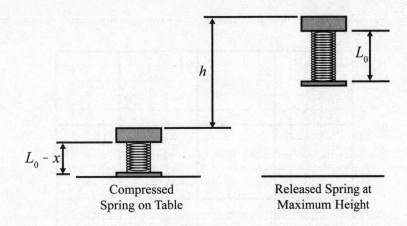

Compressed
Spring on Table

Released Spring at
Maximum Height

4. An experiment is designed to calculate the spring constant k of a vertical spring for a jumping toy. The toy is compressed a distance of x from its natural length of L_0, as shown on the left in the diagram, and then released. When the toy is released, the top of the toy reaches a height of h in comparison to its previous height and the spring reaches its maximum extension. The experiment is repeated multiple times and replaced with different masses m attached to the spring. The spring itself has negligible mass.

(a) Derive an expression for the height h in terms of m, x, k, and any other constants provided in the formula sheet.

(b) To standardize the experiment, the compressed distance x is set to 0.020 m. The following table shows the data for different values of m.

	m (kg)	h (m)
	0.020	0.49
	0.030	0.34
	0.040	0.28
	0.050	0.19
	0.060	0.18

(i) What quantities should be graphed so that the slope of a best-fit straight line through the data points can help us calculate the spring constant k?

(ii) Fill in the blank column in the table above with calculated values. Also include a header with units.

GO ON TO THE NEXT PAGE.

(c)   On the axes below, plot the data and draw the best-fit straight line. Label the axes and indicate scale.

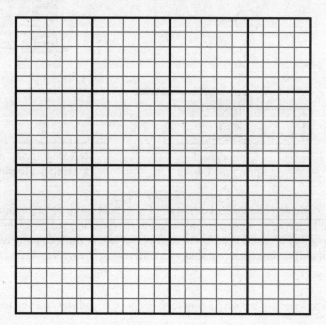

(d)   Using your best-fit line, calculate the numerical value of the spring constant.

(e)   Describe an experimental procedure that determines the height h in the experiment, given that the toy is only momentarily at that maximum height. You may include a labeled diagram of your setup to help in your description.

GO ON TO THE NEXT PAGE.

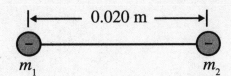

5. In the above diagram, two small objects, each with a charge of –4.0 nC, are held together by a 0.020 m length of insulating string. The objects are initially at rest on a horizontal, non-conducting, frictionless surface. The effects of gravity on each other can be considered negligible.

   (a)   Calculate the tension in the string.

   (b)   Illustrate the electric field by drawing electric field lines for the two objects on the following diagram.

         The masses of the objects are $m_1 = 0.030$ kg and $m_2 = 0.060$ kg. The string is now cut.

   (c)   Calculate the magnitude of the initial acceleration of each object.

   (d)   On the axes below, sketch a graph of the acceleration a of the object of mass m_2 versus the distance d between the objects after the string has been cut.

   (e)   In a brief paragraph, describe the speed of the objects as time increases, assuming that the objects remain on the horizontal, non-conducting frictionless surface.

STOP

Practice Test 2: Answers and Explanations

ANSWER KEY AP PHYSICS 1 TEST 2

1.	C	26.	A	
2.	D	27.	A	
3.	C	28.	D	
4.	C	29.	B	
5.	C	30.	C	
6.	B	31.	B	
7.	B	32.	B	
8.	C	33.	B	
9.	B	34.	D	
10.	B	35.	A	
11.	D	36.	A	
12.	B	37.	B	
13.	B	38.	D	
14.	B	39.	D	
15.	A	40.	B	
16.	A	41.	B	
17.	B	42.	D	
18.	D	43.	D	
19.	D	44.	D	
20.	C	45.	A	
21.	B	46.	A, D	
22.	D	47.	C, D	
23.	A	48.	B, C	
24.	D	49.	C, D	
25.	D	50.	B, D	

SECTION I

1. **C** Since acceleration due to gravity is 10 m/s^2, the vertical speed of the object after 2 seconds will be 20 m/s. Because the horizontal speed will not be affected, the direction will be mostly down, but slightly to the right.

2. **D** Mass is not changing, so you only need to consider the changes in velocity. For each segment in order, the change in velocity is +7 m/s, 0 m/s, −4 m/s, −3 m/s, and −2 m/s. The question does not specify magnitude only, so all negatives will come before the others.

3. **C** Acceleration is the change in velocity divided by the change in time. You found the changes in velocity for each segment in the previous question, and the times for each segment are, in order, 3 s, 2 s, 1 s, 3 s, and 2 s. Divide the results in the previous question by these values, and the highest result (ignoring sign because the problem specifies magnitude only) is for segment CD.

4. **C** In a velocity vs. time graph, displacement is the area beneath the curve. If the question had asked for displacement, then anything beneath the x-axis would be added as a negative displacement. However, distance cannot be negative, so there is no distinction between positive and negative velocity for this question.

   Split the area into triangles and rectangles like this:

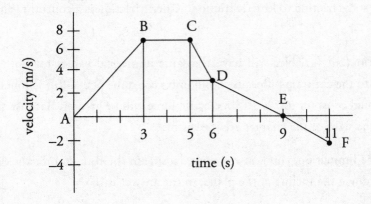

5. **C** At the highest point from the ground, the ball has a vertical velocity of zero. Therefore, applying the formula $v_y = v_{0y} + a_y t$ and rearranging it for t, it becomes $t = -v_{0y}/-g$. Substituting $v_{0y} = v \sin \theta$ into the equation, $t = v \sin \theta/g$.

6. **B** The distance travelled in the first second can be found using the Big Five,

$$d_{\text{first}} = v_0 t_{\text{first}} + \frac{1}{2} a t_{\text{first}}^2$$

Using $t_{\text{first}} = 1$ s and $v_0 = 0$ m/s since the bubble starts at rest, the equation becomes

$$d_{\text{first}} = \left(\frac{1}{2}\right) a$$

The distance traveled in the last second can be found using another Big Five equation.

$$d_{last} = v_{final}t_{last} - \frac{1}{2}at_{last}^2$$

Using $t_{last} = 1$ s and $v_{final} = v_0 + at_{total} = at_{total}$, the equation becomes

$$d_{last} = at_{total} - \left(\frac{1}{2}\right)a$$

The question is asking for the difference in the distance, so

$$d_{last} - d_{first} = at_{total} - \left(\frac{1}{2}\right)a - \left(\frac{1}{2}\right)a = at_{total} - a = (t-1)a$$

7. **B** Because the floor is horizontal the weight and normal force balance each other out, so these two forces do have equal magnitudes (eliminate (C) and (D)). These two forces, however, do not form an action-reaction pair. In order to form an action-reaction pair two forces must be acting on each other. The correct action-reaction pair in this situation is Earth's pull on the person, weight, and the person's pull on Earth.

8. **C** The force of friction on a stationary object will exactly oppose the force applied to it until it reaches a maximum. This eliminates (B) and (D). Once enough force has been applied, the friction will change from static friction to kinetic friction. Kinetic friction is a constant value, so this eliminates (A).

9. **B** In this situation, all variables will have the same numerical values for both forces. Therefore, it comes down to the constants. Because Coulomb's constant (9.0×10^9) is much, much greater than the gravitational constant (6.7×10^{-11}), electric force will be greater. Because the question specifies magnitude, the signs of the charges are irrelevant.

10. **B** By Process of Elimination, you know you need a sine in the equation for the downward pull of the second mass down the incline of the plane, so the answer is (B).

11. **D** The friction f on the block is represented by the formula $f = \mu_k N$, where N is the normal force on the block. When the force is applied 30° above the horizontal, $N = mg - mg \sin 30°$. Since $\sin 30°$ is 0.5, $N = mg - 0.5mg = 0.5mg$. Substituting N into the formula for friction, it becomes $f_1 = \mu_k(0.5mg)$. When the force is applied 30° below the horizontal, $N = mg + mg \sin 30° = mg + 0.5mg = 1.5mg$. Substituting N into the formula for friction, it becomes $f_2 = \mu_k(1.5mg) = 3f_1$.

12. **B** Because all the masses are attached and moving as a single unit, the acceleration of any block in the system is the same as that of any other. The net force on the blocks is $F_{net} = (2M)g \sin \theta$. Therefore,

$$F_{net} = ma \rightarrow (5M)a = (2M)g\sin\theta = \left(\frac{2}{5}\right)g\sin\theta$$

13. **B**   At the top of the loop, the cart will remain in contact as long as the centripetal force is at least as much as the force of gravity on the cart.

$$F_c \geq F_g$$
$$\frac{mv^2}{r} \geq mg$$
$$\frac{v^2}{r} \geq g$$
$$v \geq \sqrt{rg}$$

14. **B**   The only true statement in regards to rotational motion is (B). A mass that has greater inertia is harder to rotate. The further away the mass is from the axis of rotation, the greater the rotational inertia will be. So, when the object moves toward the center of rotation, rotational inertia is decreased.

15. **A**   By Newton's Third Law, the gravitational force exerted by the planet on the Moon will equal the gravitational force exerted by the Moon on the planet (eliminate (B) and (C)). Between Conservation of Angular Momentum and Conservation of Energy, the correct one to use to help you find distance is Conservation of Angular Momentum. Conservation of Energy will help you find the speed of the Moon. Conservation of Angular Momentum will help you find the distance of the Moon from the planet.

16. **A**   Draw out a free body diagram of all the forces acting on the sphere:

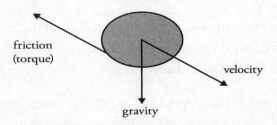

Gravity points straight down and does affect the object but it does not provide the torque (eliminate (D)). The normal force points perpendicular to the plane and does no work on the object nor does it provide the torque (eliminate (C)). The friction in this case does provide the torque. Since the ball is rolling down the hill and not sliding down it, it is not kinetic friction (eliminate (B)). This makes sense, because when you drive a car the tires grip the road and move you forward at one point. If there were no static friction on the road, you would never be able to go anywhere because the tires would not be able to grip onto anything.

17.   **B**   An object's terminal speed is the highest speed it can reach in free fall. If it has not yet reached that speed, its speed must still be increasing. This eliminates (C) and (D). The problem says that the object will experience a drag force that increases as speed increases. Therefore, the opposing force will increase over time, which means the net force will decrease (since the force from gravity will be constant). If net force decreases, so will acceleration.

18.   **D**   The centripetal force is a name given to the net force of an object undergoing uniform circular motion. Therefore, it is not a separate force and does not have an action-reaction pair. This eliminates (A) and (B). The speed of the object in uniform circular motion is constant, but its direction changes, therefore the velocity changes with time. However, the acceleration also changes because the direction of the centripetal acceleration always points to the center of the circle. This eliminates (C). A satellite undergoing uniform circular motion is in fact falling toward the center, but never accomplishes its goal due to its tangential velocity from its centripetal acceleration. Its velocity changes as a result, but it would always form a tangent to its circular path.

19.   **D**   To balance, the see-saw, you need to balance the torques. Since $T = Fr \sin \theta$, the boy currently provides double the torque. Choices (B) and (C) would double the torque on the girl's side, and (A) would cut the boy's torque in half. Choice (D) would cut the boy's torque in half, but it would also increase the girl's torque, creating a new imbalance.

20.   **C**   Centripetal force always acts perpendicular to the object's motion, which means it cannot ever do work.

21.   **B**   The same two materials would be in contact, so the coefficient of friction would remain the same. The object's mass would also be the same, so the normal force would remain constant. Since $F_f = \mu F_N$, and neither of those terms changed, the frictional force will also be unchanged.

22.   **D**   The energy is the same for both objects to go from the top of the incline to the bottom. With the sphere some of it will need to be rotational kinetic energy. The block has pure translational kinetic energy and will reach the bottom of the incline first.

23.   **A**   All the masses begin with the same speed v_0. For Path 1, the object must climb up which would decrease its speed, then come back down to reach its same initial speed. During this time, the speed is always below the initial speed. For Path 2, the object maintains the same initial speed the whole time. For Path 3, the object speeds up as it goes down the path and then slows down back to its initial speed when it climbs. During this time, the speed is always above the initial speed. If you compare the average speed, $v_3 > v_2 > v_1$, hence $t_3 < t_2 < t_1$.

24.   **D**   For an elastic collision, both kinetic energy and momentum will be conserved. Writing these statements as equations gives

$$\frac{1}{2}m_1 v_{1,0}^2 + \frac{1}{2}m_2 v_{2,0}^2 = \frac{1}{2}m_1 v_{1,f}^2 + \frac{1}{2}m_2 v_{2,f}^2$$

$$m_1 v_{1,0} + m_2 v_{2,0} = m_1 v_{1,f} + m_2 v_{2,f}$$

You're told m_1 is a given in the question, and $v_{2,0}$ is also known to be 0 m/s since the object started at rest. This leaves m_2, $v_{1,0}$, $v_{1,f}$ and $v_{2,f}$ as unknowns. With two equations, you can have two unknowns. Therefore, (D) is correct.

25. **D**   The table provides both the normal force and the frictional force. The normal will be perpendicular to the table's surface and the frictional force will be parallel to the surface. Therefore, the total force from the table (the sum of normal and friction) will be neither parallel nor perpendicular to the surface.

26. **A**   The force of static friction and the force of gravity are acting down the incline in this situation. When the box just begins to move upwards, the forces in both directions are equal and the force of static friction is at its maximum. Therefore, you have the equation

$$F = \mu_s mg \cos 45° + mg \sin 45° \quad \rightarrow \quad \mu_s = \left(\frac{\sqrt{2}\, F}{mg} \right) - 1$$

27. **A**   First, perfectly inelastic can be immediately eliminated because the objects have different velocities after the collision. Next, recall that a collision is perfectly elastic if kinetic energy is conserved.

$$K_0 = K_f$$

$$\frac{1}{2} m_1 v_{1,0}^2 + \frac{1}{2} m_2 v_{2,0}^2 = \frac{1}{2} m_1 v_{1,f}^2 + \frac{1}{2} m_2 v_{2,f}^2$$

$$\frac{1}{2} m_1 (6 \text{ m/s})^2 + \frac{1}{2} (4\,m_1)(1 \text{ m/s})^2 = \frac{1}{2} m_1 (2 \text{ m/s})^2 + \frac{1}{2} (4\,m_1)(3 \text{ m/s})^2$$

$$20\,m_1 \text{ J} = 20\,m_1 \text{ J}$$

Because kinetic energy was conserved, the collision must be perfectly elastic.

28. **D**   In the picture, neither ball experiences any change in its horizontal motion. Ball A is originally moving down, but moves up after the collision. Therefore, the impulse it received is entirely up.

29. **B**   Think of the pendulum as undergoing circular motion. The centripetal force would be equal to the difference between tension and the radial component of gravity.

$$F_c = F_t - F_{g,\,radial}$$

$$\frac{mv^2}{r} = F_t - F_{g,\,radial}$$

So long as the pendulum is in motion (as long as $v > 0$), tension must have a larger magnitude than the radial component of gravity. However, when $v = 0$ (at the maximum displacement), the two will be equal. Therefore, (B) is the correct answer.

30. **C**   This situation is equivalent to a spring-block system. It would start with some potential energy at one extreme edge, turn all of that energy into kinetic energy as it moved to the center, and continue to the other edge due to its momentum (turning the energy back into potential energy). This process would continue indefinitely in ideal conditions.

31. **B** Since the wave speed is independent of the frequency, the speed is still v. This is Big Wave Rule #1. The wave with frequency $4f$ will have 1/4 the wavelength, but the wave speed will stay the same.

32. **B** Kinetic energy is related to speed according to the equation $K = \frac{1}{2}mv^2$. Comparing the new kinetic energy, K', to the original kinetic energy yields:

$$\frac{K'}{K} = 2 = \frac{\frac{1}{2}mv'^2}{\frac{1}{2}mv^2}$$

$$2 = \left(\frac{v'}{v}\right)^2$$

$$\frac{v'}{v} = \sqrt{2}$$

33. **B** The forces during a collision are an example of an action-reaction pair. This means the force on the car will be equal in magnitude but opposite in direction compared to the force on the car. Because the forces are in opposite directions, the accelerations will be as well. Eliminate (A) and (C). Because the forces are equal in magnitude, Newton's Second Law states that the acceleration of the lighter object will be double the acceleration of the heavier object. Eliminate (D). Therefore, the answer is (B).

34. **D** First, draw a free-body diagram:

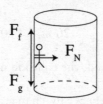

In order to stay in place, the force of friction needs to have a magnitude equal to the force of gravity's.

$$F_f = F_g$$
$$\mu F_N = mg$$

Next, solve for the normal force. This is possible because the person is undergoing uniform circular motion, so set the normal force equal to centripetal force.

$$F_c = F_N$$
$$\frac{mv^2}{r} = F_N$$

Finally, substitute this value into the previous result.

$$\mu\left(\frac{mv^2}{r}\right) = mg$$

Notice that the mass actually drops out of the equation. The only variables that matter are the coefficient of friction, speed, and the radius.

35.   A   Apply Conservation of Energy to compare when the block is at its equilibrium position to when the block is at greatest displacement:

$$E_1 = E_2$$

$$K_1 + V_1 = K_2 + V_2$$

$$K_1 = V_2$$

$$\frac{1}{2}mv^2_{max} = \frac{1}{2}kA^2$$

$$v = A\sqrt{\frac{k}{m}}$$

36.   A   First, in order for the small block to slide off, there would have to be a big force affecting the larger block. Hooke's Law ($F = -kx$) states that this will happen at maximum displacement from the equilibrium position. Eliminate (B) and (D).

Second, the equation given in (C) indicates that the heavier the top block is, the more likely it is to slide off. But this is obviously not true, so (A) must be correct.

37.   B   The lowest note that can be played is the fundamental frequency of the flute. For a standing wave with a node at both ends, the fundamental wavelength is twice the length of the flute, $2L$. Applying the wave speed equation,

$$v = \lambda f$$

$$343 \text{ m/s} = (2L)(261.63 \text{ Hz})$$

$$L = \frac{343 \text{ m/s}}{2(261.63 \text{ Hz})} = 0.655 \text{ m} = 65.5 \text{ cm}$$

38.   D   Intensity is inversely related to the radius squared, so the fact that the train is approaching you means the intensity (loudness) will increase. The pitch, however, is dictated by the Doppler effect. In this case, the detector is motionless, but the source is moving toward the detector. This motion results in a higher pitch at the detector. The brakes of the train will act to reduce the speed of the train though, which will continually diminish this effect, so the pitch will decrease.

39.   D   First, replace all the resistors with an equivalent resistor.

R_2 and R_3 are in series, so $R_{2 \text{ and } 3} = R_2 + R_3$.

R_1 and $R_{2\,and\,3}$ are in parallel, so $\dfrac{1}{R_{eq}} = \dfrac{1}{R_1} + \dfrac{1}{R_{2\,and\,3}} = \dfrac{1}{R_1} + \dfrac{1}{R_2 + R_3} \rightarrow R_{eq} = \dfrac{R_1(R_2 + R_3)}{R_1 + R_2 + R_3}$.

Second, calculate the current coming out of the battery,

$$I_{total} = \frac{V}{R_{eq}} = \frac{V(R_1 + R_2 + R_3)}{R_1(R_2 + R_3)}$$

Finally, calculate the power,

$$P = IV = \left[\frac{V(R_1 + R_2 + R_3)}{R_1(R_2 + R_3)}\right]V = \frac{V^2(R_1 + R_2 + R_3)}{R_1(R_2 + R_3)}$$

40.   **B**   Because R_1 and R_2 and R_3 are in parallel, the voltage across each matches the battery of 100 V. Using what you solved from problem 39 and your knowledge that current splits off in parallel, to calculate the current going into R_2 and R_3 use

$$I = V/R = V/R_{2\,and\,3} = V/(R_2 + R_3) = (100\text{ V})/(80\ \Omega + 120\ \Omega) = (100\text{ V})/(200\ \Omega) = 0.5\text{ A}$$

Now that you know the current going into this branch of the parallel circuit, you know that R_2 and R_3 are in series, and in series the current remains the same. So the current going across R_3 is 0.5 A. Using this you can calculate the voltage going across R_3,

$$V_3 = I_3 R_3 = (0.5\text{ A})(120\ \Omega) = 60\text{ V}$$

41.   **B**   If R_1 were to burn out, the total resistance in the circuit would increase. Because $I = V/R$ this increase in resistance would decrease the current.

42.   **D**   At 20 s, the voltage across the resistor is 5 V, so the current is

$$I = \frac{\Delta V}{R} = \frac{5\text{ V}}{500\ \Omega} = 0.01\text{ A}$$

The power at this time is given by

$$P = I\Delta V = (0.01\text{ A})(5\text{ V}) = 0.05\text{ W}$$

The shape of the graph can be obtained by combining these two equations:

$$P = I\Delta V = \left(\frac{\Delta V}{R}\right)\Delta V = \frac{(\Delta V)^2}{R}$$

which shows that the power dissipated by the resistor will increase quadratically with the voltage.

43.   **D**   The question specifies that the current through G must be 0. Recall that in parallel circuits current will distribute itself in a manner inversely proportional to the resistance in each path. Looking at the left-hand side of the configuration, you can see that the resistance on the top is half the

resistance of the bottom. That means the current will distribute itself with 2/3 along the top path and 1/3 along the bottom path. For no current to pass through G, this same distribution must also be true on the right-hand side, otherwise some current would pass through G to correct any changes to the relative resistances. The 5 Ω resistor must have the same proportional resistance to the unknown resistor that the 20 Ω resistor has to the 10 Ω resistor. Therefore, the unknown must be 2.5 Ω.

44. **D** Coulomb's Law is given by

$$\left|\vec{F}_E\right| = k\left|\frac{q_1 q_2}{r^2}\right|$$

Doubling each charge, and halving the distance between them yields:

$$\left|\vec{F}_E\right| = k\left|\frac{(2q_1)(2q_2)}{\left(\frac{r}{2}\right)^2}\right| = k\left|\frac{4q_1 q_2}{\frac{r^2}{4}}\right| = 16 \cdot k\left|\frac{q_1 q_2}{r^2}\right|$$

45. **A** Although you do not care about the direction of our net vector in the end, you do have to take orientation into account to find our magnitude. Number the charges from the top left and going clockwise as 1, 2, 3, and 4. Drawing out the different forces on the bottom right charge (charge 3), you get

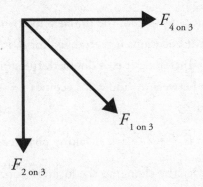

Now solve each of these force vectors

$$F_{4\ on\ 3} = k\frac{qq}{r^2} = k\frac{q^2}{r^2}$$

$$F_{2\ on\ 3} = k\frac{qq}{r^2} = k\frac{q^2}{r^2}$$

These two are the simpler force vectors to solve. In order to solve the charge 1's force on charge 3, the distance first needs to be solved. Because they are ordered in a square using a diagonal creates a 45°-45°-90° triangle. The distance in this case is $r\sqrt{2}$. So,

$$F_{1\ on\ 3} = k\frac{qq}{\left(r\sqrt{2}\right)^2} = \frac{1}{2}k\frac{q^2}{r^2}$$

First, combine the $F_{4 \text{ on } 3}$ with $F_{2 \text{ on } 3}$ to get an equivalent vector that will be in the same direction as $F_{1 \text{ on } 3}$.

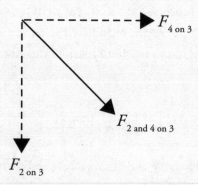

They form a 45°-45°-90° triangle when laid tail-to-end so,

$$F_{2 \text{ and } 4 \text{ on } 3} = k\frac{q^2}{r^2}\sqrt{2}$$

This vector points in the same direction as $F_{1 \text{ on } 3}$, so to get the magnitude of the net force, simply add $F_{2 \text{ and } 4 \text{ on } 3}$ with $F_{1 \text{ on } 3}$,

$$F_{\text{net}} = F_{2 \text{ and } 4 \text{ on } 3} + F_{1 \text{ on } 3} = k\frac{q^2}{r^2}\sqrt{2} + \frac{1}{2}k\frac{q^2}{r^2} = \frac{1}{2}k\frac{q^2}{r^2}(\sqrt{2} + 1) = k\left(\frac{q^2}{2r^2}\right)(1 + \sqrt{2})$$

46. **A, D** According to the formula $v^2 = v_0^2 + 2ad$, the initial velocity, v_0, can be related to the stopping distance, d. If v is zero, and the equations is rearranged for d, it becomes $d = v_0^2/2a$. Since the car is decelerating, a is negative. Therefore, if v_0 is doubled, the stopping distance, d, is quadrupled. To determine the relationship between v_0 and t, it becomes $t = -v_0/a$. Therefore, if v_0 is doubled, the stopping time, t, is doubled.

47. **C, D** The period for a pendulum is $T = 2\pi\sqrt{\dfrac{L}{g}}$. Making changes to the length of the string or changing the acceleration of gravity (by changing the location form the Earth to the Moon) will cause changes in the period.

48. **B, C** The resistors are all in parallel, meaning their voltage is the same across each resistor and matches the voltage of the battery. Eliminate (A). The voltage provided by the battery does not increase or decrease; the battery still maintains the same voltage. Eliminate (D). The resistors are in parallel, so eliminating one of the resistors will increase the equivalent resistance. Because $I = V/R$, this increase in resistance would decrease the current. The equation of power is $P = IV$. If the current decreases, the power decreases as well.

49. **C, D** The branch of the circuit containing R_1 is connected in parallel to the branch containing R_2 and R_3. Thus, the voltage across each branch is equal to V. R_2 and R_3 are connected in series, so their equivalent resistance is $R_2 + R_3$. The current through the both R_2 and R_3 is therefore given by:

$$I = \frac{\Delta V}{R} = \frac{V}{R_2 + R_3}$$

According to this equation, a decrease in V, or an increase in R_2 or R_3 would cause a decrease in the current through R_3. Any change to $R1$ would not have an impact on the current through R_3 because R_1 has no impact on the voltage across the branch of the circuit containing R_2 and R_3.

50. **B, D** Big Wave Rule #2 states that when a wave passes into another medium the frequency remains the same. Eliminate (A). The frequency does not change when passing into another medium; the speed and wavelength of the wave change. Choice (B) is correct.

Big Wave Rule #1 states the speed of a wave will be constant while in a single medium. So, while the sound wave is in the metal rod the speed will remain consistent. So at two different points in the rod,

$$v = \lambda f \rightarrow \lambda_1 f_1$$

This is an inverse relationship between our wavelength, λ, and frequency, f; (D) is also correct.

SECTION II

1. (a) One such example would be to mark off two distances on the table—one for Block A before the collision, and one for the combined blocks after the collision. Push Block A to give it an initial speed. Use a stopwatch to measure the time it takes for the blocks to cross the marked distances. The speeds are the distances divided by the time.

    (b)

    $$m_A v_i = (m_A + m_B) v_f$$

    $$m_B = \left(\frac{m_A v_i}{v_f} \right) - m_A = \left[\frac{(0.5\,\text{kg})(1.5\,\text{m/s})}{0.25\,\text{m/s}} \right] - 0.5\,\text{kg} = 2.5\,\text{kg}$$

    (c)

    $$K_{\text{initial}} = \frac{1}{2}(0.5\,\text{kg})(1.5\,\text{m/s})^2 = 0.56\,\text{J}$$

    $$K_{\text{final}} = \frac{1}{2}(3\,\text{kg})(0.25\,\text{m/s})^2 = 0.09\,\text{J}$$

    The total kinetic energy lost is equal to

    $$K_{\text{initial}} - K_{\text{final}} = 0.56\,\text{J} - 0.09\,\text{J} = 0.47\,\text{J}$$

2. (a) A free-body diagram would include only the tensions in the string and the force of gravity as shown below. Because the pendulum makes a horizontal circle, take some care to draw the direction of the force represented by the tension along the path of the string.

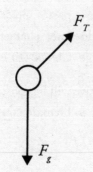

(b) Where θ is the angle between F_T and the normal,

$$\begin{array}{ll} \Sigma F_x = ma_x & \Sigma F_y = ma_y \\ F_T \sin\theta = ma_x \quad \text{And} & F_T \cos\theta - F_g = ma_y \end{array}$$

(c) The centripetal force is the net force in the x-direction. However, you need to use some information from the y-direction. Because the conical pendulum travels in a horizontal circle, there is no acceleration in the y-direction and so Newton's Second Law in the y-direction becomes

$$F_T \cos\theta - F_g = 0$$

or

$$F_T \cos\theta = F_g$$

$$F_T = \frac{F_g}{\cos\theta}$$

$$F_T = \frac{0.5 \text{ kg}(10 \text{ m/s}^2)}{\cos 57^\circ}$$

$$F_T = 9.2 \text{ N}$$

Knowing this you can solve for the centripetal acceleration. It is the same as the acceleration in the x-direction.

$$F_T \sin\theta = ma_c$$

$$\frac{F_T \sin\theta}{m} = a_c$$

$$\frac{9.2 \text{ N} \sin 57^\circ}{0.5 \text{ kg}} = a_c$$

$$15.4 \text{ m/s}^2 = a_c$$

(d) The radius the ball travels can be found using some geometry. The length of the string is 2.2 m at an angle of 57 degrees. This means

$$\sin \theta = \frac{opp}{hyp}$$

$$\sin 57° = \frac{r}{2.2 \text{ m}}$$

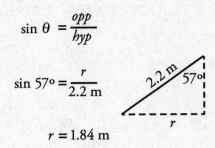

$$r = 1.84 \text{ m}$$

(e) The ball will travel at a speed of

$$a_c = \frac{v^2}{r}$$
$$v^2 = a_c r$$
$$v = \sqrt{a_c r}$$
$$v = \sqrt{15.4 \text{ m/s}^2 \times 1.84 \text{ m}}$$
$$v = 5.32 \text{ m/s}^2$$

3. (a) Because the force vector points to the left, the only logical solution is if q_1 is negative and q_2 is positive.

(b)

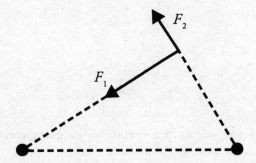

(c) Applying Coulomb's Law to determine F_1 and F_2,

$$F_1 = \frac{kq_1q_3}{r_{13}^2} = \frac{(9 \times 10^9 \text{ N} \cdot \text{m}^2/\text{C}^2)(4.0 \times 10^{-6} \text{ C})(1.0 \times 10^{-6} \text{ C})}{(4.0 \text{ m})^2} = 2.25 \times 10^{-3} \text{ N}$$

$$F_2 = \frac{kq_2q_3}{r_{23}^2} = \frac{(9 \times 10^9 \text{ N} \cdot \text{m}^2/\text{C}^2)(1.7 \times 10^{-6} \text{ C})(1.0 \times 10^{-6} \text{ C})}{(3.0 \text{ m})^2} = 1.70 \times 10^{-3} \text{ N}$$

F is the vector sum of the two forces, **F** = **F₁** + **F₂**. Since they are at right angles to each other, the magnitude can be found by applying the Pythagorean Theorem

$$F = \sqrt{F_1^2 + F_2^2} = \sqrt{(2.25 \times 10^{-3}\,\mathrm{N})^2 + (1.70 \times 10^{-3}\,\mathrm{N})^2}$$

$$F = 2.8 \times 10^{-3}\,\mathrm{N}$$

(d)

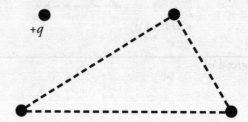

From the setup, you know that the current net force is entirely to the left. Therefore, you need to cancel this out with a force that is entirely to the right. The question specifies that you must use a positively charged particle to do this, and positive charges will create repulsive forces in this case because q_3 is itself positive. So if you're trying to repel an object and make it move right, you need to be to that object's left. Without additional numerical information in the question, it's impossible to determine exactly how far from q_3 this charge should be placed, but it must be somewhere along the line that extends directly to the left of q_3.

4. (a) The spring observes Conservation of Mechanical Energy,

$$mgh = \frac{1}{2}kx^2$$

$$h = \frac{kx^2}{2mg}$$

(b) (i) The value of $1/m$ would help us in determining the spring constant when we take the slope of the line.

(ii)

$1/m$ (kg⁻¹)	m (kg)	h (m)
50	0.020	0.49
30	0.030	0.34
25	0.040	0.28
20	0.050	0.19
17	0.060	0.18

(c)

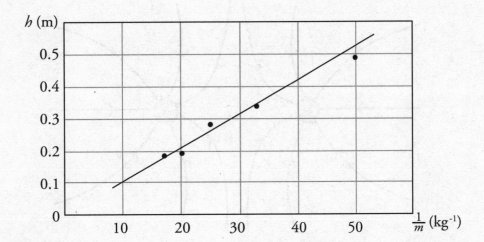

(d)   To calculate the slope, select two points on our best-fit line. Here is one example,

$$\text{slope} = \frac{(0.42 - 0.10)\text{ m}}{(40 - 10)\text{ kg}^{-1}} = \frac{0.32\text{ m}}{30\text{ kg}^{-1}} = 1.07 \times 10^{-2}\text{ m} \cdot \text{kg}$$

From part (a),

$$h = \frac{kx^2}{2mg} \rightarrow \text{slope} = \frac{kx^2}{2g}$$

Note: Values between 450 N/m and 550 N/m are acceptable.

$$k = \frac{2g(\text{slope})}{x^2} = \frac{2(9.8\text{ m/s}^2)(1.07 \times 10^{-2}\text{ m} \cdot \text{kg})}{(0.02\text{ m})^2} = 524\text{ N/m}$$

(e)   One example would be to use a meter stick and video camera. Hold the meter stick next to the toy and allow it to jump up. Record the toy with a video camera. With this you can watch the video in slow motion later to determine its height. You can also do an alternate method of finding the height by recording the time t it takes to fall from its highest point to the lowest. Then, using kinematics equations, you can determine the height.

5.   (a)   Applying Newton's Second Law to one of the objects

$$T = F_E$$

$$T = k\frac{q^2}{r^2} = \frac{(9.0 \times 10^9\text{ N} \cdot \text{m}^2/\text{C}^2)(-4.0 \times 10^9\text{ C})^2}{(0.020\text{ m})^2} = 3.6 \times 10^{-4}\text{ N}$$

(b)

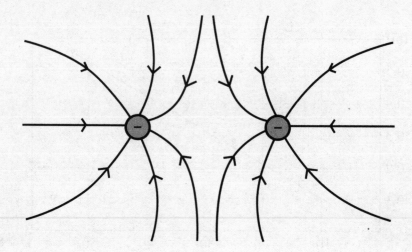

(c)   The acceleration is caused by the electrostatic force, and initially that force has the same magnitude as the one calculated in part (A).

$$F_{net} = F_E = T$$

$$ma = T$$

Now calculate the acceleration on each object,

$$a_1 = \frac{T}{m_1} = \frac{\left(3.6 \times 10^{-4}\ \text{N}\right)}{\left(0.030\ \text{kg}\right)} = 1.2 \times 10^{-2}\ \text{m/s}^2$$

$$a_2 = \frac{T}{m_2} = \frac{\left(3.6 \times 10^{-4}\ \text{N}\right)}{\left(0.060\ \text{kg}\right)} = 6.0 \times 10^{-3}\ \text{m/s}^2$$

(d)

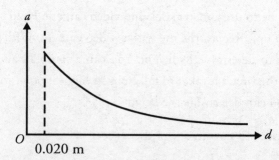

(e)   As time increases, the speed of the objects increase. The objects move further apart, but their speeds increase at a slower rate. This is because the electrostatic force decreases as the objects move further apart. Hence, the acceleration also decreases. The speed approaches a constant value as the distance between the objects, d, approaches infinity.

The Princeton Review®

Completely darken bubbles with a No. 2 pencil. If you make a mistake, be sure to erase mark completely. Erase all stray marks.

1.

YOUR NAME: _____
(Print)   Last   First   M.I.

SIGNATURE: _____   DATE: ___/___/___

HOME ADDRESS: _____
(Print)   Number and Street

_____
City   State   Zip Code

PHONE NO.: _____
(Print)

IMPORTANT: Please fill in these boxes exactly as shown on the back cover of your test book.

2. TEST FORM

6. DATE OF BIRTH

Month	Day		Year	
JAN				
FEB	⓪	⓪	⓪	⓪
MAR	①	①	①	①
APR	②	②	②	②
MAY	③	③	③	③
JUN		④	④	④
JUL		⑤	⑤	⑤
AUG		⑥	⑥	⑥
SEP		⑦	⑦	⑦
OCT		⑧	⑧	⑧
NOV		⑨	⑨	⑨
DEC				

3. TEST CODE

⓪	Ⓐ	Ⓙ	⓪	⓪		
①	Ⓑ	Ⓚ	①	①		
②	Ⓒ	Ⓛ	②	②		
③	Ⓓ	Ⓜ	③	③		
④	Ⓔ	Ⓝ	④	④		
⑤	Ⓕ	Ⓞ	⑤	⑤		
⑥	Ⓖ	Ⓟ	⑥	⑥		
⑦	Ⓗ	Ⓠ	⑦	⑦		
⑧	Ⓘ	Ⓡ	⑧	⑧		
⑨			⑨	⑨		

4. REGISTRATION NUMBER

⓪	⓪	⓪	⓪	⓪	⓪	⓪	⓪
①	①	①	①	①	①	①	①
②	②	②	②	②	②	②	②
③	③	③	③	③	③	③	③
④	④	④	④	④	④	④	④
⑤	⑤	⑤	⑤	⑤	⑤	⑤	⑤
⑥	⑥	⑥	⑥	⑥	⑥	⑥	⑥
⑦	⑦	⑦	⑦	⑦	⑦	⑦	⑦
⑧	⑧	⑧	⑧	⑧	⑧	⑧	⑧
⑨	⑨	⑨	⑨	⑨	⑨	⑨	⑨

7. SEX

◯ MALE
◯ FEMALE

The Princeton Review®

5. YOUR NAME

First 4 letters of last name				FIRST INIT	MID INIT
Ⓐ	Ⓐ	Ⓐ	Ⓐ	Ⓐ	Ⓐ
Ⓑ	Ⓑ	Ⓑ	Ⓑ	Ⓑ	Ⓑ
Ⓒ	Ⓒ	Ⓒ	Ⓒ	Ⓒ	Ⓒ
Ⓓ	Ⓓ	Ⓓ	Ⓓ	Ⓓ	Ⓓ
Ⓔ	Ⓔ	Ⓔ	Ⓔ	Ⓔ	Ⓔ
Ⓕ	Ⓕ	Ⓕ	Ⓕ	Ⓕ	Ⓕ
Ⓖ	Ⓖ	Ⓖ	Ⓖ	Ⓖ	Ⓖ
Ⓗ	Ⓗ	Ⓗ	Ⓗ	Ⓗ	Ⓗ
Ⓘ	Ⓘ	Ⓘ	Ⓘ	Ⓘ	Ⓘ
Ⓙ	Ⓙ	Ⓙ	Ⓙ	Ⓙ	Ⓙ
Ⓚ	Ⓚ	Ⓚ	Ⓚ	Ⓚ	Ⓚ
Ⓛ	Ⓛ	Ⓛ	Ⓛ	Ⓛ	Ⓛ
Ⓜ	Ⓜ	Ⓜ	Ⓜ	Ⓜ	Ⓜ
Ⓝ	Ⓝ	Ⓝ	Ⓝ	Ⓝ	Ⓝ
Ⓞ	Ⓞ	Ⓞ	Ⓞ	Ⓞ	Ⓞ
Ⓟ	Ⓟ	Ⓟ	Ⓟ	Ⓟ	Ⓟ
Ⓠ	Ⓠ	Ⓠ	Ⓠ	Ⓠ	Ⓠ
Ⓡ	Ⓡ	Ⓡ	Ⓡ	Ⓡ	Ⓡ
Ⓢ	Ⓢ	Ⓢ	Ⓢ	Ⓢ	Ⓢ
Ⓣ	Ⓣ	Ⓣ	Ⓣ	Ⓣ	Ⓣ
Ⓤ	Ⓤ	Ⓤ	Ⓤ	Ⓤ	Ⓤ
Ⓥ	Ⓥ	Ⓥ	Ⓥ	Ⓥ	Ⓥ
Ⓦ	Ⓦ	Ⓦ	Ⓦ	Ⓦ	Ⓦ
Ⓧ	Ⓧ	Ⓧ	Ⓧ	Ⓧ	Ⓧ
Ⓨ	Ⓨ	Ⓨ	Ⓨ	Ⓨ	Ⓨ
Ⓩ	Ⓩ	Ⓩ	Ⓩ	Ⓩ	Ⓩ

1. Ⓐ Ⓑ Ⓒ Ⓓ
2. Ⓐ Ⓑ Ⓒ Ⓓ
3. Ⓐ Ⓑ Ⓒ Ⓓ
4. Ⓐ Ⓑ Ⓒ Ⓓ
5. Ⓐ Ⓑ Ⓒ Ⓓ
6. Ⓐ Ⓑ Ⓒ Ⓓ
7. Ⓐ Ⓑ Ⓒ Ⓓ
8. Ⓐ Ⓑ Ⓒ Ⓓ
9. Ⓐ Ⓑ Ⓒ Ⓓ
10. Ⓐ Ⓑ Ⓒ Ⓓ
11. Ⓐ Ⓑ Ⓒ Ⓓ
12. Ⓐ Ⓑ Ⓒ Ⓓ
13. Ⓐ Ⓑ Ⓒ Ⓓ

14. Ⓐ Ⓑ Ⓒ Ⓓ
15. Ⓐ Ⓑ Ⓒ Ⓓ
16. Ⓐ Ⓑ Ⓒ Ⓓ
17. Ⓐ Ⓑ Ⓒ Ⓓ
18. Ⓐ Ⓑ Ⓒ Ⓓ
19. Ⓐ Ⓑ Ⓒ Ⓓ
20. Ⓐ Ⓑ Ⓒ Ⓓ
21. Ⓐ Ⓑ Ⓒ Ⓓ
22. Ⓐ Ⓑ Ⓒ Ⓓ
23. Ⓐ Ⓑ Ⓒ Ⓓ
24. Ⓐ Ⓑ Ⓒ Ⓓ
25. Ⓐ Ⓑ Ⓒ Ⓓ
26. Ⓐ Ⓑ Ⓒ Ⓓ

27. Ⓐ Ⓑ Ⓒ Ⓓ
28. Ⓐ Ⓑ Ⓒ Ⓓ
29. Ⓐ Ⓑ Ⓒ Ⓓ
30. Ⓐ Ⓑ Ⓒ Ⓓ
31. Ⓐ Ⓑ Ⓒ Ⓓ
32. Ⓐ Ⓑ Ⓒ Ⓓ
33. Ⓐ Ⓑ Ⓒ Ⓓ
34. Ⓐ Ⓑ Ⓒ Ⓓ
35. Ⓐ Ⓑ Ⓒ Ⓓ
36. Ⓐ Ⓑ Ⓒ Ⓓ
37. Ⓐ Ⓑ Ⓒ Ⓓ
38. Ⓐ Ⓑ Ⓒ Ⓓ
39. Ⓐ Ⓑ Ⓒ Ⓓ

40. Ⓐ Ⓑ Ⓒ Ⓓ
41. Ⓐ Ⓑ Ⓒ Ⓓ
42. Ⓐ Ⓑ Ⓒ Ⓓ
43. Ⓐ Ⓑ Ⓒ Ⓓ
44. Ⓐ Ⓑ Ⓒ Ⓓ
45. Ⓐ Ⓑ Ⓒ Ⓓ
46. Ⓐ Ⓑ Ⓒ Ⓓ
47. Ⓐ Ⓑ Ⓒ Ⓓ
48. Ⓐ Ⓑ Ⓒ Ⓓ
49. Ⓐ Ⓑ Ⓒ Ⓓ
50. Ⓐ Ⓑ Ⓒ Ⓓ

1.

YOUR NAME: _____
(Print)   Last   First   M.I.

SIGNATURE: _____   DATE: ___ / ___ / ___

HOME ADDRESS: _____
(Print)   Number and Street

_____
City   State   Zip Code

PHONE NO.: _____
(Print)

IMPORTANT: Please fill in these boxes exactly as shown on the back cover of your test book.

2. TEST FORM

6. DATE OF BIRTH

Month	Day		Year	
○ JAN				
○ FEB	⓪	⓪	⓪	⓪
○ MAR	①	①	①	①
○ APR	②	②	②	②
○ MAY	③	③	③	③
○ JUN		④	④	④
○ JUL		⑤	⑤	⑤
○ AUG		⑥	⑥	⑥
○ SEP		⑦	⑦	⑦
○ OCT		⑧	⑧	⑧
○ NOV		⑨	⑨	⑨
○ DEC				

3. TEST CODE   **4. REGISTRATION NUMBER**

⓪	Ⓐ	Ⓙ	⓪	⓪	⓪	⓪	⓪	⓪	⓪	⓪
①	Ⓑ	Ⓚ	①	①	①	①	①	①	①	①
②	Ⓒ	Ⓛ	②	②	②	②	②	②	②	②
③	Ⓓ	Ⓜ	③	③	③	③	③	③	③	③
④	Ⓔ	Ⓝ	④	④	④	④	④	④	④	④
⑤	Ⓕ	Ⓞ	⑤	⑤	⑤	⑤	⑤	⑤	⑤	⑤
⑥	Ⓖ	Ⓟ	⑥	⑥	⑥	⑥	⑥	⑥	⑥	⑥
⑦	Ⓗ	Ⓠ	⑦	⑦	⑦	⑦	⑦	⑦	⑦	⑦
⑧	Ⓘ	Ⓡ	⑧	⑧	⑧	⑧	⑧	⑧	⑧	⑧
⑨			⑨	⑨	⑨	⑨	⑨	⑨	⑨	⑨

7. SEX

○ MALE
○ FEMALE

The **Princeton** Review®

5. YOUR NAME

First 4 letters of last name				FIRST INIT	MID INIT
Ⓐ	Ⓐ	Ⓐ	Ⓐ	Ⓐ	Ⓐ
Ⓑ	Ⓑ	Ⓑ	Ⓑ	Ⓑ	Ⓑ
Ⓒ	Ⓒ	Ⓒ	Ⓒ	Ⓒ	Ⓒ
Ⓓ	Ⓓ	Ⓓ	Ⓓ	Ⓓ	Ⓓ
Ⓔ	Ⓔ	Ⓔ	Ⓔ	Ⓔ	Ⓔ
Ⓕ	Ⓕ	Ⓕ	Ⓕ	Ⓕ	Ⓕ
Ⓖ	Ⓖ	Ⓖ	Ⓖ	Ⓖ	Ⓖ
Ⓗ	Ⓗ	Ⓗ	Ⓗ	Ⓗ	Ⓗ
Ⓘ	Ⓘ	Ⓘ	Ⓘ	Ⓘ	Ⓘ
Ⓙ	Ⓙ	Ⓙ	Ⓙ	Ⓙ	Ⓙ
Ⓚ	Ⓚ	Ⓚ	Ⓚ	Ⓚ	Ⓚ
Ⓛ	Ⓛ	Ⓛ	Ⓛ	Ⓛ	Ⓛ
Ⓜ	Ⓜ	Ⓜ	Ⓜ	Ⓜ	Ⓜ
Ⓝ	Ⓝ	Ⓝ	Ⓝ	Ⓝ	Ⓝ
Ⓞ	Ⓞ	Ⓞ	Ⓞ	Ⓞ	Ⓞ
Ⓟ	Ⓟ	Ⓟ	Ⓟ	Ⓟ	Ⓟ
Ⓠ	Ⓠ	Ⓠ	Ⓠ	Ⓠ	Ⓠ
Ⓡ	Ⓡ	Ⓡ	Ⓡ	Ⓡ	Ⓡ
Ⓢ	Ⓢ	Ⓢ	Ⓢ	Ⓢ	Ⓢ
Ⓣ	Ⓣ	Ⓣ	Ⓣ	Ⓣ	Ⓣ
Ⓤ	Ⓤ	Ⓤ	Ⓤ	Ⓤ	Ⓤ
Ⓥ	Ⓥ	Ⓥ	Ⓥ	Ⓥ	Ⓥ
Ⓦ	Ⓦ	Ⓦ	Ⓦ	Ⓦ	Ⓦ
Ⓧ	Ⓧ	Ⓧ	Ⓧ	Ⓧ	Ⓧ
Ⓨ	Ⓨ	Ⓨ	Ⓨ	Ⓨ	Ⓨ
Ⓩ	Ⓩ	Ⓩ	Ⓩ	Ⓩ	Ⓩ

1. Ⓐ Ⓑ Ⓒ Ⓓ
2. Ⓐ Ⓑ Ⓒ Ⓓ
3. Ⓐ Ⓑ Ⓒ Ⓓ
4. Ⓐ Ⓑ Ⓒ Ⓓ
5. Ⓐ Ⓑ Ⓒ Ⓓ
6. Ⓐ Ⓑ Ⓒ Ⓓ
7. Ⓐ Ⓑ Ⓒ Ⓓ
8. Ⓐ Ⓑ Ⓒ Ⓓ
9. Ⓐ Ⓑ Ⓒ Ⓓ
10. Ⓐ Ⓑ Ⓒ Ⓓ
11. Ⓐ Ⓑ Ⓒ Ⓓ
12. Ⓐ Ⓑ Ⓒ Ⓓ
13. Ⓐ Ⓑ Ⓒ Ⓓ

14. Ⓐ Ⓑ Ⓒ Ⓓ
15. Ⓐ Ⓑ Ⓒ Ⓓ
16. Ⓐ Ⓑ Ⓒ Ⓓ
17. Ⓐ Ⓑ Ⓒ Ⓓ
18. Ⓐ Ⓑ Ⓒ Ⓓ
19. Ⓐ Ⓑ Ⓒ Ⓓ
20. Ⓐ Ⓑ Ⓒ Ⓓ
21. Ⓐ Ⓑ Ⓒ Ⓓ
22. Ⓐ Ⓑ Ⓒ Ⓓ
23. Ⓐ Ⓑ Ⓒ Ⓓ
24. Ⓐ Ⓑ Ⓒ Ⓓ
25. Ⓐ Ⓑ Ⓒ Ⓓ
26. Ⓐ Ⓑ Ⓒ Ⓓ

27. Ⓐ Ⓑ Ⓒ Ⓓ
28. Ⓐ Ⓑ Ⓒ Ⓓ
29. Ⓐ Ⓑ Ⓒ Ⓓ
30. Ⓐ Ⓑ Ⓒ Ⓓ
31. Ⓐ Ⓑ Ⓒ Ⓓ
32. Ⓐ Ⓑ Ⓒ Ⓓ
33. Ⓐ Ⓑ Ⓒ Ⓓ
34. Ⓐ Ⓑ Ⓒ Ⓓ
35. Ⓐ Ⓑ Ⓒ Ⓓ
36. Ⓐ Ⓑ Ⓒ Ⓓ
37. Ⓐ Ⓑ Ⓒ Ⓓ
38. Ⓐ Ⓑ Ⓒ Ⓓ
39. Ⓐ Ⓑ Ⓒ Ⓓ

40. Ⓐ Ⓑ Ⓒ Ⓓ
41. Ⓐ Ⓑ Ⓒ Ⓓ
42. Ⓐ Ⓑ Ⓒ Ⓓ
43. Ⓐ Ⓑ Ⓒ Ⓓ
44. Ⓐ Ⓑ Ⓒ Ⓓ
45. Ⓐ Ⓑ Ⓒ Ⓓ
46. Ⓐ Ⓑ Ⓒ Ⓓ
47. Ⓐ Ⓑ Ⓒ Ⓓ
48. Ⓐ Ⓑ Ⓒ Ⓓ
49. Ⓐ Ⓑ Ⓒ Ⓓ
50. Ⓐ Ⓑ Ⓒ Ⓓ

NOTES

International Offices Listing

China (Beijing)
1501 Building A,
Disanji Creative Zone,
No.66 West Section of North 4th Ring Road Beijing
Tel: +86-10-62684481/2/3
Email: tprkor01@chol.com
Website: www.tprbeijing.com

China (Shanghai)
1010 Kaixuan Road
Building B, 5/F
Changning District, Shanghai, China 200052
Sara Beattie, Owner: Email: sbeattie@sarabeattie.com
Tel: +86-21-5108-2798
Fax: +86-21-6386-1039
Website: www.princetonreviewshanghai.com

Hong Kong
5th Floor, Yardley Commercial Building
1-6 Connaught Road West, Sheung Wan, Hong Kong
(MTR Exit C)
Sara Beattie, Owner: Email: sbeattie@sarabeattie.com
Tel: +852-2507-9380
Fax: +852-2827-4630
Website: www.princetonreviewhk.com

India (Mumbai)
Score Plus Academy
Office No.15, Fifth Floor
Manek Mahal 90
Veer Nariman Road
Next to Hotel Ambassador
Churchgate, Mumbai 400020
Maharashtra, India
Ritu Kalwani: Email: director@score-plus.com
Tel: + 91 22 22846801 / 39 / 41
Website: www.score-plus.com

India (New Delhi)
South Extension
K-16, Upper Ground Floor
South Extension Part–1,
New Delhi-110049
Aradhana Mahna: aradhana@manyagroup.com
Monisha Banerjee: monisha@manyagroup.com
Ruchi Tomar: ruchi.tomar@manyagroup.com
Rishi Josan: Rishi.josan@manyagroup.com
Vishal Goswamy: vishal.goswamy@manyagroup.com
Tel: +91-11-64501603/ 4, +91-11-65028379
Website: www.manyagroup.com

Lebanon
463 Bliss Street
AlFarra Building - 2nd floor
Ras Beirut
Beirut, Lebanon
Hassan Coudsi: Email: hassan.coudsi@review.com
Tel: +961-1-367-688
Website: www.princetonreviewlebanon.com

Korea
945-25 Young Shin Building
25 Daechi-Dong, Kangnam-gu
Seoul, Korea 135-280
Yong-Hoon Lee: Email: TPRKor01@chollian.net
In-Woo Kim: Email: iwkim@tpr.co.kr
Tel: + 82-2-554-7762
Fax: +82-2-453-9466
Website: www.tpr.co.kr

Kuwait
ScorePlus Learning Center
Salmiyah Block 3, Street 2 Building 14
Post Box: 559, Zip 1306, Safat, Kuwait
Email: infokuwait@score-plus.com
Tel: +965-25-75-48-02 / 8
Fax: +965-25-75-46-02
Website: www.scorepluseducation.com

Malaysia
Sara Beattie MDC Sdn Bhd
Suites 18E & 18F
18th Floor
Gurney Tower, Persiaran Gurney
Penang, Malaysia
Email: tprkl.my@sarabeattie.com
Sara Beattie, Owner: Email: sbeattie@sarabeattie.com
Tel: +604-2104 333
Fax: +604-2104 330
Website: www.princetonreviewKL.com

Mexico
TPR México
Guanajuato No. 242 Piso 1 Interior 1
Col. Roma Norte
México D.F., C.P.06700
registro@princetonreviewmexico.com
Tel: +52-55-5255-4495
+52-55-5255-4440
+52-55-5255-4442
Website: www.princetonreviewmexico.com

Qatar
Score Plus
Office No: 1A, Al Kuwari (Damas)
Building near Merweb Hotel, Al Saad
Post Box: 2408, Doha, Qatar
Email: infoqatar@score-plus.com
Tel: +974 44 36 8580, +974 526 5032
Fax: +974 44 13 1995
Website: www.scorepluseducation.com

Taiwan
The Princeton Review Taiwan
2F, 169 Zhong Xiao East Road, Section 4
Taipei, Taiwan 10690
Lisa Bartle (Owner): lbartle@princetonreview.com.tw
Tel: +886-2-2751-1293
Fax: +886-2-2776-3201
Website: www.PrincetonReview.com.tw

Thailand
The Princeton Review Thailand
Sathorn Nakorn Tower, 28th floor
100 North Sathorn Road
Bangkok, Thailand 10500
Thavida Bijayendrayodhin (Chairman)
Email: thavida@princetonreviewthailand.com
Mitsara Bijayendrayodhin (Managing Director)
Email: mitsara@princetonreviewthailand.com
Tel: +662-636-6770
Fax: +662-636-6776
Website: www.princetonreviewthailand.com

Turkey
Yeni Sülün Sokak No. 28
Levent, Istanbul, 34330, Turkey
Nuri Ozgur: nuri@tprturkey.com
Rona Ozgur: rona@tprturkey.com
Iren Ozgur: iren@tprturkey.com
Tel: +90-212-324-4747
Fax: +90-212-324-3347
Website: www.tprturkey.com

UAE
Emirates Score Plus
Office No: 506, Fifth Floor
Sultan Business Center
Near Lamcy Plaza, 21 Oud Metha Road
Post Box: 44098, Dubai
United Arab Emirates
Hukumat Kalwani: skoreplus@gmail.com
Ritu Kalwani: director@score-plus.com
Email: info@score-plus.com
Tel: +971-4-334-0004
Fax: +971-4-334-0222
Website: www.princetonreviewuae.com

Our International Partners

The Princeton Review also runs courses with a variety of partners in Africa, Asia, Europe, and South America.

Georgia
LEAF American-Georgian Education Center
www.leaf.ge

Mongolia
English Academy of Mongolia
www.nyescm.org

Nigeria
The Know Place
www.knowplace.com.ng

Panama
Academia Interamericana de Panama
http://aip.edu.pa/

Switzerland
Institut Le Rosey
http://www.rosey.ch/

All other inquiries, please email us at
internationalsupport@review.com